The Best American Science
and Nature Writing 2011

The Best American Science and Nature Writing™ 2011

Edited and with an Introduction
by Mary Roach

Tim Folger, Series Editor

A Mariner Original

HOUGHTON MIFFLIN HARCOURT
BOSTON · NEW YORK 2011

www.hmhbooks.com

ISSN 1530-1508
ISBN 978-0-547-35063-9

Printed in the United States of America

DOC 10 9 8 7 6 5 4 3 2 1

Contents

Contents

Foreword

ONE DAY LAST February a team of NASA scientists made an astonishing announcement. A space telescope named Kepler, launched in March 2009, had found 1,235 planets orbiting other stars. Astronomers cautioned that they would need years to confirm the discovery completely, but there's little doubt that most of the new planets will survive scrutiny. In one stroke the number of planets in our galaxy more than doubled; as recently as twenty years ago the only known planets were those in our own solar system. For the first time in history we now know that planetary systems are common in the universe, which raises considerably the odds that life — perhaps even intelligent life — may exist elsewhere in the cosmos. Geoff Marcy, an astronomer on the Kepler team, could barely contain his excitement.

"It really is a historic moment," he told me during a chat in his office on the University of California campus in Berkeley. "I really think February 3, 2011" — the date NASA released the news — "will be remembered for a long time. It was a moment when all the interested members of our species, no matter what continent they lived on, realized that the Milky Way galaxy is just teeming with Earth-size planets. What Kepler is doing is literally finding new worlds — not metaphorical worlds, but actual worlds. This really is an extraordinary new chapter in human history."

The new chapter — one of the great discoveries of the ages — made headlines for a day, and then . . . our collective attention moved on. The Super Bowl was only a few days away; protesters were massing in Cairo's Tahrir Square. A thousand new worlds, some of which might be habitable? Yesterday's news. It made me wonder what cultural life might have been like, say, in Galileo's time, if Renaissance Italy had been plagued by a twenty-four-hour news cycle. Would endlessly recycled rumors of a papal scandal have eclipsed Galileo's discovery of the moons of Jupiter in 1610? Or maybe a hit reality stage show— *Growing Up Medici*—would have so distracted officials of the Inquisition that they wouldn't have cared when Galileo claimed that Earth orbited the sun, and the old man would have been spared his sentence of lifetime house arrest.

So give the Inquisitors some credit. They tortured; they burned; they broke wills—but they knew a turning point in history when they saw one, even if they did their best to suppress it. They must have thought they were doing the right thing. Maybe they even imagined that future generations would be grateful for their efforts. But who now remembers the name of the pope who reigned when Galileo was imprisoned? (It was Paul V, but I had to look it up.) Four hundred years from now, if our species survives, it's probably safe to assume that the wars of our time—not to mention our leaders and celebrities—will be forgotten. The year 2011 may well be remembered as the time when we learned that our world was but one of many in the galaxy. And who knows? In another four centuries we may have found that we're not alone in the universe.

Although astronomers don't yet know whether any of the planets discovered by the Kepler space telescope harbor life, there's good evidence that the raw ingredients for life are as common as space dust. One of the articles in this year's anthology, "Cosmic Blueprint of Life," by Andrew Grant, informs us that some of the complex organic molecules that are essential for life apparently form spontaneously in interstellar space. These compounds may have been the seeds from which life on Earth began—and if it happened here . . .

After pondering the question of life's beginnings, you might want to consider a provocative meditation on the nature of reality presented by Stephen Hawking and Leonard Mlodinow in "The (Elusive) Theory of Everything." In just two mind-bending pages they make a compelling case that physicists will never develop an overarching "theory of everything." The universe, they argue, is too complex ever to be tamed by any single theory, no matter how elegant. I don't know what the Inquisitors would have thought about this, but I think house arrest would be the best that Hawking and Mlodinow could hope for.

As exciting as it would be to find evidence of life on other worlds, our own planet abounds with strange—and disturbing—examples of the often savage forms life can take. Jill Sisson Quinn's transcendent "Sign Here If You Exist" challenges us to consider what we can learn about the nature of reality, the existence of God, and our notions of immortality from the fierce ichneumon wasp. It's an unforgettable story.

And there are many others in this collection, from Tim Zimmermann's gripping account of a fatal encounter with a killer whale at a marine park to Christopher Ketcham's surprising "New Dog in Town," in which we learn that the resurgence of a certain wily predator may be a sign that something is seriously wrong with wildlife habitat in North America. Mary Roach will tell you more about these and other stories in the pages ahead, and you would be hardpressed to find a wittier, more knowledgeable guide to the world of science and nature writing in this corner of the Milky Way.

I hope that readers, writers, and editors will nominate their favorite articles for next year's anthology at http://timfolger.net/forums. The criteria for submissions and deadlines, and the address to which entries should be sent, can be found in the "news and announcements" forum on my website. Once again this year I'm offering an incentive to enlist readers to scour the nation in search of good science and nature writing: send me an article that I haven't found, and if the article makes it into the anthology, I'll mail you a free copy of next year's edition. I'll sign it, and so will Mary. Right, Mary? I also encourage readers to use the forums to leave feedback about the collection and to discuss all things scientific. The best way for publications to guarantee that their articles are considered for inclusion in the anthology is to place me on their subscription list, using the address posted in the news and announcements section of the forums.

It has been a delight working with Mary Roach; I'm looking forward to reading her latest book, *Packing for Mars*. Once again this year I'm grateful to Amanda Cook and Meagan Stacey at Houghton Mifflin Harcourt for making this book possible, and for their forgiveness of missed deadlines. And, as always, I would be lost without my beauteous wife, Anne Nolan.

TIM FOLGER

Introduction

AT A WRITING conference in 2007, the *New Yorker* writer Burkhard Bilger told the story of an article he'd recently written, about cheese makers who flout government regulations by using unpasteurized milk. Bilger described a moment that, I think, goes a long way toward explaining the consistent excellence of his work. He had just spent three days with a cheese maker in Indiana. As he got into the car to drive home, a creeping gloom descended. The material was flat—he didn't have it. But at some point in the next round of phone calls, he heard tell of a cloistered convent in Connecticut making raw cow's-milk cheese. Nuns! Bilger got back into his car.

For another writer, the sexy lawlessness of the "raw milk underground," as Bilger cleverly phrased it in his subtitle, might have been enough. As cheese stories go, that is practically a potboiler. For Bilger, it wasn't. *His* raw milk underground would be populated by Benedictine sisters. (To my delight, Bilger has returned to the topic of edible rot in 2010, with dumpster-diving, road kill–eating "opportunivores" standing in for nuns.)

Very few people, I think, would tell you that they'd like to read a 5,000-word article about food safety and fermentation, which is, in essence, what Bilger's two pieces are about. People are not drawn to his writing for the science as much as for the things that bring it to life: characters, settings, stories, wit. These are the sugar, to be all cliché about it, that makes the medicine go down. Make no mistake, good science writing is medicine. It is a cure for ignorance and fallacy. Good science writing peels away the blinders, generates wonder, brings the open palm to the forehead: *Oh! Now I get it!* And sometimes it does much more than that.

In "The Love That Dare Not Squawk Its Name," Jon Mooallem finds his nuns in an albatross colony on Kaena Point, Oahu. The birds and the woman who studies them become the narrative framework for a look at the science of same-sex animal pairings. Mooallem's wry eye for detail immediately wins the reader over. "At any given moment . . . ," he writes of the scene at Kaena Point, "some birds may be just turning up while others sit there killing time. It feels like an airport baggage-claim area."

The science going on at Kaena Point is a launch pad for something loftier: a meditation on politics and human nature. We have a strong inclination to view animals as mute, hairy (or feathered) versions of ourselves and to use — or distort — zoologists' findings to validate our beliefs and even to promote our political agendas. When the albatross researcher's paper came out (har! unintended!), it was lauded by gay rights activists as a call for equality, and at the same time it was condemned by social conservatives as propaganda. (Even Stephen Colbert got in on the act, denouncing "albatresbians" and their "Sappho-avian agenda.") The researcher was blindsided. "The study is about albatross," she said to Mooallem. "The study is not about humans." In Mooallem's hands, it is about both. As clearheadedly as he takes on the social coils and kinks, he delves into the conundrum of a behavior that seems, on the surface, to be at cross-purposes with natural selection. The result is one of the most balanced, far-reaching but never meandering, thoughtful, deep but witty pieces of writing I've seen in a while.

Like Mooallem, Jaron Lanier, in "The First Church of Robotics," examines our eagerness to see humanness where it is does not really exist. As Lanier smoothly argues, the personification — and occasional deification — of computer intelligence not only overinflates the machine, it debases the human being. "Lulled by the concept of ever-more intelligent AI's, [we] are expected to trust algorithms to assess our aesthetic choices, the progress of a student, the credit risk of a homeowner . . . In doing so we only end up misreading the capability of our machines and distorting our own capabilities as human beings." Lanier's calm, effective prose lowers the flame beneath the AI hype, all the dire prognostications of superintelligent robotic takeovers. *Settle down,* it seems to say. *Let's look at the facts.*

The mixture of science and religion—so spicy, so impossible to emulsify—has also, this year, drawn in the immense talents of Jill Sisson Quinn. "Sign Here If You Exist" begins with a description of the surreally complex egg-laying behavior of the ichneumon wasp. Had Quinn written about nothing more than that, the beauty of her phrasings would still have left me speechless. But her musings on the ichneumon led her to reframe her views on God and afterlife. She comes to see body, not soul, as immortal. "I got my elements from stars: mass from water, muscles from beans, thoughts from fish and olives. When Edward Abbey died, his body was buried in nothing more than an old sleeping bag in the southern Arizona desert. He said, 'If my decomposing carcass helps nourish the roots of a juniper tree or the wings of a vulture—that is immortality enough for me.'"

First-rate essays like Quinn's tempt the nonbeliever to adopt science as a kind of God stand-in—infallible, omniscient, eternal, good. "Lies, Damned Lies, and Medical Science" is a bucket of ice water to the face. David H. Freedman profiles the Greek gadfly and meta-researcher John Ioannidis. In a 2005 *Journal of the American Medical Association* article, Ioannidis looked at forty-five of the most often cited and highly regarded medical research papers of the past thirteen years: among them, studies recommending hormone replacement therapy for menopausal women, vitamin E to reduce the risk of heart disease, and daily low-dose aspirin to prevent heart attacks and strokes. Forty-one percent of these, Ioannidis found, "had been convincingly shown to be wrong or significantly exaggerated."

Science writing regularly honors the accomplishments of science, but it is equally important to expose its shortcomings. Atul Gawande's "Letting Go" profiles the American physician's overcommitment to preserving life. There is a tendency among surgeons and oncologists to equate death with failure. Death is someone else's job. Too often, Gawande argues, doctors do not merely shy away from death. They get in the way of it, and by their obfuscation or omissions, their rosy-hued prognoses, they deprive patients of a dignified death.

Gawande's and Freedman's articles happened to be back to back in the stack by my bed. Happily, Jon Cohen's piece on how to collect semen from a chimpanzee was next on the pile.

In simpler times, good, careful writing about the natural world fostered awe and wonder, the necessary starting points for good stewardship. Today's nature writing, more and more, fosters disquiet and concern. Much of this year's best nature writing is a call to action, the writers' work artfully spotlighting environmental atrocities both within and far outside our normal gaze. If not for their work, I would not have realized the extent to which broken space hardware clutters our heavens and jellyfish clog our oceans. The plight of songbirds on Cyprus netted by the thousands for diners would have remained distant and abstract.

Two of my favorite pieces manage to impart urgency without sowing despair. The authors use fine narrative writing, humor even, to draw readers in and quietly open their eyes to the direness of the situation. In "New Dog in Town," Christopher Ketcham charms the reader with his opening description of coyotes on the Van Cortlandt Park Golf Course in the Bronx. "They apparently like to watch the players tee off among the Canada geese. They hunt squirrels and rabbits and wild turkeys along the edge of the forest surrounding the course, where there are big old hardwoods and ivy that looks like it could strangle a man—good habitat in which to den, skulk, plan. Sometimes in summer the coyotes emerge from the steam of the woods to chew golf balls and spit them onto the grass in disgust." Ketcham meets a golfer who describes a coyote trotting alongside his golf cart, stopping and starting when he does. The man mimes the coyote watching the golf balls fly—"following with his head the coyote-tracked ball's trajectory up and up, along the fairway, then its long arc down." Ketcham goes on to explain how, over millennia, the coyote's "elasticity" has allowed it to expand its range, while the less adaptable wolf has dwindled. The coyote has no need for a pack or for cover of woods. It sleeps anywhere, eats anything ("garbage, darkness, rats, air"). Coyotes have been around since North America teemed with saber-toothed cats, mastodons, and the glyptodon, "a turtle as big as a Volkswagen." All but the coyote perished in a mysterious mass extinction.

By the end of Ketcham's essay, we realize that the coyote on the golf course, the coyote who howls along with the ambulance sirens, is a wily portent of doom. Another massive die-off is underway. Nothing loud or catastrophic, no asteroids, just the slow, expanding vacuum of habitat loss. The "weed species" thrive as thousands of others blink out, all but unnoticed, year upon year. One day, millennia from now, scientists will ponder the massive die-off of the late Holocene. *Whatever could have happened?*

In "Fish Out of Water"—so titled for the Asian carp's alarm response of leaping fifteen feet into the air, "knocking boaters' glasses off and breaking their noses and chipping their teeth and leaving body bruises in the shape of fish"—Ian Frazier charts the well-meaning, largely ineffectual, occasionally comical efforts to brake the spread of this swiftly invasive, ecologically disastrous species. These efforts include exporting them to China ("Rick Smith belongs to the very small number of motorcycle-rally food vendors who also ink multimillion-dollar deals with the Chinese"), erecting electric fish barriers, and attempting to legislate them out of existence. (The Obama administration has a "zero tolerance policy" for Asian carp.)

The hardest science in Frazier's piece is the DNA analysis being used to track carpal creep through Illinois's waterways. "I did not follow all the science of it," Frazier admits. He tends to spend his word count on the places that make for better reading. Had this piece run in a science magazine rather than *The New Yorker,* I suspect the explanation of the DNA work would have been expanded and we would probably not have read about the Redneck Fishing Tournament (prizes awarded for most carp taken in a two-day period, and costumes). That would be a shame. Readers would have missed "the smell of ketchup and mud," the sound of "crushed blue and white Busch beer cans disappearing into the mud, crinkling underfoot," and the winning costumes—"devils from Greenview, Illinois, and cavemen from Michigan."

There is nothing wrong with an article on the intricacies of DNA analysis for monitoring invasive species. But you risk missing the waterways for the fish. Environmental science happens in a context. Frazier handily captures the complexities of the challenge—

logistical and political—and the flavor of the grass-roots creativity being applied to overcome it. Explaining DNA science clearly to readers is a laudable skill, but equally so is reporting the issue in a way that roots the science in a place—in people and their situations. That can be, in its way, even more complex than sequencing DNA. "Through waving weedbeds of bureaucracy and human cross-purposes," writes Frazier, "the fish swims."

When your characters are galaxies or subatomic particles, different standards apply. If your topic is "time" or "the elusive theory of everything," simply helping the reader to understand will more than suffice. Stephen Hawking and Leonard Mlodinow got me all the way to the top of the second column before they lost me. ("According to quantum physics, the past, like the future, is indefinite and exists only as a spectrum of possibilities.") This was long before they dropped the bomb that there are no less than five different string theories, and that all five are about to be scrapped for a network of theories collectively known as M-theory. It may make you feel better to know that although physicists are on the verge of assembling a unifying network of theories to explain the workings of the universe, none of them "seems to know what the M stands for."

Hawking and Mlodinow's central metaphor is the goldfish in the glass bowl, believing its view of reality to be an accurate representation of the external world. He has the reader imagine the goldfish as they "formulate scientific laws from their distorted frame of reference that would always hold true and that would enable them to make predictions about the future motion of objects outside the bowl." This was a good year for fish humor in science writing.

MARY ROACH

The Best American Science and Nature Writing 2011

YUDHIJIT BHATTACHARJEE

The Organ Dealer

FROM *Discover*

ELENI DAGIASI FLEW from Athens to Delhi in January 2008 on a mission to save her life. With her husband, Leonidas, she took a taxi from the airport past sparkling multiplexes and office buildings to a guesthouse in the booming exurb of Gurgaon. A kitchen staff was on hand, the rooms had cable, and there was a recreation area with billiards, providing patients with creature comforts while kidney transplants were arranged. Over the next week, as her operation was scheduled, Dagiasi went to a makeshift hospital for dialysis. Then one night, while she was watching TV with her husband, a chef turned off the lights and urged everyone to leave. Shortly afterward, ten policemen stormed in. "We were too stunned to react," says Leonidas Dagiasi, a former fisherman who borrowed money from his employer to finance the trip. The couple and other guests were hauled off for questioning. The Gurgaon hospital, it turned out, was the hub of a thriving black market in kidneys. The organs were harvested from poor Indian workers, many of whom had been tricked or forced into selling the organ for as little as $300.

The mastermind, India's Central Bureau of Investigation (CBI) charged, was Amit Kumar—a man who performed the surgeries with no more formal training than a degree in Ayurveda, the ancient Indian system of medicine. In a career spanning two decades, Kumar had established one of the world's largest kidney-trafficking rings, with a supply chain that extended deep into the Indian countryside. Some of his clients were from India. Many came from Greece, Turkey, the Middle East, Canada, and the United States.

At parties in India and abroad, Kumar introduced himself as one of India's foremost kidney surgeons, said Rajiv Dwivedi, a CBI investigator based in Delhi. The claim wasn't entirely illegitimate: investigators estimate that Kumar has performed hundreds of successful transplants, a practice so lucrative that he was able to finance Bollywood movies and had to fend off extortion threats from the Mumbai mafia. Two weeks after the police crackdown in Gurgaon, Kumar was arrested at a wildlife resort in Nepal and brought back to India, where he now awaits trial.

Kumar's operation was a microcosm of the vast, shadowy underworld of transplant trafficking that extends from the favelas of São Paulo to the slums of Manila. The tentacles of the trade crisscross the globe, leaving no country untouched, not even the United States, as evidenced by the July 2009 arrest of a New York rabbi who has been charged with arranging illegal transplants in this country by bringing in poor Israelis to supply kidneys.

In June 2008 I traveled to India to get an inside view of Kumar's ring and examine the perverse enterprise that fueled its rise. How did Kumar build his organ empire, and how was he able to run it for so long? The answers, I learned, lay in the grinding poverty and entrenched corruption of India, the desperation of patients on dialysis, and the transnational nature of the black-market transplant business—which, though dominated by the kidney exchange, includes livers and hearts as well. The factors at play in India allow the kidney trade to thrive around the world, despite efforts by various governments to stamp it out.

Life without a working kidney is harsh. We are born with two of these internal filters, located below the rib cage, to remove waste and excess water from our blood. Patients with kidney failure—often the result of diabetes and high blood pressure—can die within days from the buildup of toxins in the bloodstream and the bloating of organs. To avoid this outcome, modern medicine offers dialysis, a process in which blood is cleansed at least three times a week by pumping it through an external or internal filter. This grueling routine comes with dietary restrictions and side effects like itching, fatigue, and risk of infection. Theoretically you can live on dialysis for decades; in reality, though, the risks are so great that without a new kidney, premature death is the frequent result.

No wonder that those needing a kidney vastly exceed the number of kidneys available from deceased donors. In the United States, some 88,000 individuals were on the waiting list as of early 2010, with 34,000 names typically added every year. The wait averages five years. The situation in Greece is similarly dire: Eleni Dagiasi put her name on a list around 2006 and expected a waiting period of five years or more. In the meantime, she needed dialysis three days a week—a treatment requiring that she live in Athens, more than seventy-five miles and three hours' travel from her husband, who works on Andros Island as a caretaker of yachts. After Eleni learned of the India option through one of Kumar's brokers, the couple saw it as a way out.

They could have gone elsewhere: to Pakistan, where entire villages are populated by men who have been stripped of a kidney; to China, where kidney harvesting from executed prisoners has supported a booming transplant industry; or to the Philippines, where transplant tourism flourished until May 2008, when the government banned the trade. Transplant tourism today accounts for as much as 10 percent of all donor kidneys transplanted, says Luc Noël, coordinator for the Department of Essential Health Technologies at the World Health Organization (WHO). Often lured by middlemen (or drugged, beaten, and otherwise coerced), donors end up with a few hundred to a few thousand dollars and a scar at the waist that has become an emblem of exploitation and human indignity.

The kidney trade has its origins not in the underworld but in the bright light of medical advancement and the globalization of health care. It began in a hospital in Boston in 1954, when a medical team led by the plastic surgeon Joseph Murray conducted the first successful kidney transplant from one identical twin to the other. There was no immune rejection to contend with because the donor's and recipient's organs had coexisted happily in their mother's womb. Through the 1950s and '60s, researchers attempted to make transplants work in patients who were unrelated to their donors. To help the new organ withstand the assault from the recipient's natural defenses, doctors developed tissue-type matching, a technique to determine if the chemistry of the donor's immune system, defined by antigens on the surface of cells, was similar to that of the recipient's. Doctors also bombarded the recipient with X-rays and used a variety of drugs to beat the immune system into submission.

In the early 1980s, transplants became feasible on a wider scale. What changed the scene was an immunosuppressant molecule called cyclosporin, developed by researchers at the Swiss pharmaceutical company Sandoz. It became the foundation for new drugs that could counter organ rejection with unprecedented effectiveness.

The possibilities quickly became evident. "Doctors realized that with cyclosporin, you did not need a related donor," says Lawrence Cohen, an anthropologist at the University of California at Berkeley. Clinics were able to cast a wider net for donors, and kidney transplants became an established surgery around the world. Soon kidneys were a commodity. The first reports of kidney selling began to surface in India around 1985. With its large base of doctors and an expanding health care industry, India had already been attracting medical tourists from the rest of South Asia and the Middle East. Now there was a growing stream of patients from these countries checking into hospitals in Chennai, Mumbai, and Bangalore for kidney transplants. "There were lots and lots of sales," Cohen says.

In Southeast Asia another kidney-trading corridor had opened up, with the Philippines as the hub. Patients from Japan and elsewhere traveled to Manila to buy kidneys. The organs often came from jailed felons, according to Nancy Scheper-Hughes, an anthropologist at the University of California, Berkeley, who has documented the trade in various countries. "Guards would pick out the healthiest-looking prisoners," she says. Some reports allege that buyers negotiated with the prisoners' families, not the prisoners themselves. Meanwhile, China became the destination for patients from Singapore, Taiwan, and Korea. Under a rule approved by the Chinese government in 1984, kidneys and other organs were harvested from executed prisoners. Human rights activists became concerned that China might have been ramping up its executions through the 1980s and '90s in order to boost its organ supply.

The practice was gaining notoriety, but interventions urged by WHO and others often failed. India legislated its Transplantation of Human Organs Act, banning the buying and selling of organs, in 1994. But the law did not eliminate the practice; it simply drove it underground. By the end of the decade, the kidney trade was thriving, largely due to the Internet. Websites touting "transplant packages" priced from about $20,000 to $70,000 sent patients, many of them Americans, flocking to hospitals in the Philippines, Pakistan, and China. In 2001, when authorities apprehended a criminal syndicate trafficking kidneys from slum dwellers in Brazil and poor villagers in Moldova to Israelis, it became evident that the trade had spread far and wide.

When Kumar was arrested on February 7, 2008, he struck a defiant pose for the cameras as Nepalese officials prepared to escort him to Delhi. In the weeks that followed, he became a media celebrity, with investigators leaking stories about his flamboyant lifestyle. News reports alluded to Kumar's owning properties in India, Hong Kong, Australia, and Canada. Sher Bahadur Basnet, a Nepalese police official who apprehended Kumar, told me that he made frequent trips to nightclubs and casinos in Kathmandu. In an ironic twist, an Indian news channel discovered a clip from an obscure 1991 Hindi flick titled *Khooni Raat* ("bloody night") in which Kumar—who harbored ambitions of becoming a movie star—plays a bit role as an upstanding police officer.

The first time I saw Kumar was in June 2008 at a court in Ambala, some 125 miles northwest of Delhi. The court had yet to open its doors when I arrived, and I sat outside on a bench. Two of the ring's employees—a driver named Harpal and a cook who worked at the guesthouse, Suresh—showed up, wiping their faces with handkerchiefs. They told me they had no idea that Kumar had been conducting illegal transplants.

A police van drove up carrying Kumar and his accomplices, including his youngest brother, Jeevan Raut (Kumar's original family name), who has a degree in homeopathy; a middleman named Gyasuddin; and a physician named Upender Dublish. Dressed in a beige shirt, Kumar waved at his lawyers from behind the vehicle's rusty iron-mesh window. He had shaved off his mustache, and his eyes looked bulbous. He glanced about furtively while talking with his lawyers.

I inched closer to the van under the gaze of policemen standing nearby, one of whom told me sternly that reporters were not allowed. Nonetheless, I introduced myself to Kumar, who responded with a nervous smile. "I never forced anybody to donate a kidney," he told me before a potbellied guard shooed me away. Inside the courtroom, Kumar kept primping his hair and smoothing out his shirt while the prosecutor, Ashok Singh, presented the charges, citing complaints by seven men alleging that they were tricked or forced into selling their kidneys. Kumar looked crestfallen as he was led back to the van, but his brother walked with a swagger, yelling out to me, "They have no case against us!"

At the start of their investigation, CBI officials found it difficult to believe Kumar had been conducting transplants on his own. "We thought his role was to provide donors and clients," Dwivedi told me. The agency assembled a panel comprising a surgeon, a nephrologist, and a forensic-medicine expert to probe Kumar's self-proclaimed expertise. In a two-hour interview not unlike a qualifying exam, the panelists asked Kumar to walk them through the steps involved in a transplant, from removing a kidney to hooking it up inside a recipient. By the end they were convinced. "Kumar had adequate theoretical knowledge about the surgical process" for kidney transplants, the panel said in a report filed in court.

Yet when Kumar entered the transplant business in the 1980s, it was not as a surgeon but as an entrepreneur. To understand how a leading organ dealer got his start, I visited Kumar's second-youngest brother, Ganesh Raut, a real estate developer who now lives in the same Mumbai apartment where Kumar set up a hospital in 1984. Although Ganesh has not always been on good terms with his brother, he has stood by him since the arrest, accompanying Kumar's lawyers to court hearings and meeting with Kumar in jail.

Nobody answered when I rang the bell, but the door was unlocked and I ventured in to find a friendly basset hound wagging its tail in the hallway. Ganesh, a portly, clean-shaven man, was in the middle of morning prayers in front of a miniature temple. I took in the smell of burning incense as he lit a diya and completed the ritual. Then he told me of Kumar's childhood ambitions.

"He would often tell us that he wanted to do something extraordinary in life," Ganesh said. After college in 1977, Kumar went to work at the M. A. Podar Ayurvedic Hospital in Mumbai. He began assisting with simple outpatient surgical procedures that Ayurvedic practitioners are licensed to handle, nothing more complicated than removing hemorrhoids, and later began freelancing as a surgical attendant at mainstream hospitals. Within a few years, he had made enough money and contacts to start his own hospital. Ganesh showed me the room that had once been the hospital's oper-ation theater, where surgeons hired by Kumar performed head-and-neck and abdominal surgeries in the early years, before he turned it into more of a transplant center. Now it was a sitting area furnished with wicker chairs.

Ganesh was less forthcoming about Kumar's kidney venture, so I went to see Rakesh Maria, a top official of the Mumbai police, who shut down Kumar's first foray into transplants in 1995. By the late 1980s, Kumar (then using his birth name, Santosh Raut) had be-come well aware of medical tourists streaming in from the Middle East. "He knew at once that if he could tap this market, he would hit the jackpot," Maria said. To do so, he reached out to cab drivers at the city's international airport, paying them a commission to find clients.

There was no shortage of potential donors to Kumar's makeshift hospital in Mumbai. He targeted homeless beggars, handcart pull-ers, sweepers. He hired two men to scout the slums and offer any-where from a paltry $300 to just over $1,000 for a kidney. Later he began paying the agents a fixed amount of a few thousand dollars per donor. To boost their cut, the agents started paying donors less and less.

Through the early 1990s, the business flourished, and the doz-ens of transplants performed at the hospital (many of them by a Mumbai Hospital surgeon named Yogesh Kothari) provided Ku-mar with the equivalent of a surgical residency. He attended con-ferences, read up on nephrology, and learned about different kinds of sutures and immunosuppressant drugs. In 1994 he per-formed his first transplant under Kothari's watch. But after a short-changed donor filed a complaint, police raided Kumar's hospital that August, arresting eleven people on charges of cheating and criminal conspiracy. Kumar himself was charged with conducting illegal foreign-exchange transactions through two frontmen—a grocer and a silversmith—who had helped him receive payments from overseas clients to the tune of $60,000.

After four months in a Mumbai prison, Kumar was freed on bail. In 1995, after Maharashtra (the state that contains the city of Mum-bai) banned organ selling, he was busted again and charged with conducting illegal transplants. Finally he fled to Jaipur, 800 miles to the north. There he assumed his new name and conducted thir-teen illegal transplants before being arrested yet again, in 1996.

Released on bail, he disappeared from view and did not resurface until 1999, in Gurgaon. There Kumar built an empire bigger and more organized than ever as the Indian legal system churned sluggishly on.

While Kumar was setting up business in Gurgaon, Scheper-Hughes, Cohen, and two other academics were launching Organs Watch, a program to document exploitative organ transplantation worldwide. As Kumar's business boomed, the researchers visited Brazil, Turkey, the Philippines, and elsewhere to interview hundreds of kidney sellers (including victims of coercion and fraud) along with buyers, doctors, and police.

As a result of new regulations, the climate for kidney dealers has gotten tougher. China implemented regulations in 2007 to stop the organ trade. Pakistan's last president, Pervez Musharraf, issued an ordinance that same year to outlaw organ trafficking, which in April 2009 survived a legal challenge by a petitioner who called it un-Islamic. And in May 2008, the Philippines, which for years had promoted transplant tourism, followed suit. Yet, like a multiheaded Hydra, the industry seems to be surviving legal onslaughts in countries like India and branching into newer, more vulnerable places such as Iraq and sub-Saharan Africa. "It gets snuffed out in one place and crops up in another," Scheper-Hughes says.

The trade is adapting to the changing legal climate. One shift has been venue: transplant surgeries seem to have moved out of large hospitals to small clinics away from the public eye. "We even have anecdotal reports of surgery being transferred to private homes," WHO's Noël says.

Another change is in marketing. Gone are the blatant online advertisements for transplant tours. Instead, dealers posing as private sellers post personal ads on Facebook. When the social networking site Orkut outlawed transplant-related communities, it saw a proliferation of user names containing the word *kidney,* such as "kidney4u." The Orkut pages of these users are typically blank and presumably exist only as a place for potential buyers to leave a message.

Americans are not just buyers; some may be sellers as well. Scheper-Hughes discovered "an older brother in the Los Angeles area offering to sell his younger brother's kidney online on Craigslist." The phenomenon has opened up a new form of trade, diffuse and fragmented, that has already increased the number of commercial transactions manifold and ushered in new kinds of abuse. And money is not the only inducement, says Monir Moniruzzaman, an anthropologist at Michigan State University in East Lansing. In a study of the organ trade, he has found numerous ads that promise jobs, visas, and even citizenship in the developed world.

Driving from Delhi to Gurgaon in June 2008, I glimpsed the soaring skyscrapers of Gurgaon's economic boom, but shanties closer to the ground spoiled the glossy view. Finally I entered the gated residential community where Kumar's hospital, sealed by the CBI, sat in a row of villas. It was enclosed by a wall and draped in bougainvillea vines whose pink flowers stirred in the muggy afternoon breeze. Peeking through the locked iron gate, I saw a stretcher left behind on the grounds. Having escaped from Mumbai and Jaipur, Kumar set up a clandestine facility here, putting up a sign advertising a lawyer's services. "The idea was to keep nosy neighbors away," CBI's Dwivedi said.

Kumar did not work alone. One of his main accomplices, the physician Upender Dublish, ran a hospital in Ballabgarh, an industrial town about two hours away. Bald and rotund, Dublish projects the resigned air of a man terribly wronged but too beaten to protest. At the Ambala courtroom, he told me with a pained look that he needed medical attention for a heart ailment. Dublish ran Kumar's supply chains, in which the more enterprising donors became middlemen themselves. The result was a pyramid scheme, which Scheper-Hughes says is common to kidney markets around the world. She has seen the pattern in Brazil and Moldova. "Middlemen will say to prospective donors, 'I lived through it and made money; you can too.'" An early recruit was Gyasuddin, whom I saw with Kumar in Ambala. He told me he sold his kidney for $1,000 and became a node in Kumar's network, earning about $110 "for every donor I could find." He brought dozens of donors to Kumar from his hometown, Meerut, an old city about forty miles from Delhi.

Even as old charges from past arrests slogged through the legal system, Kumar's layers of protection grew. In 2004 he set up an in-house tissue-matching facility at the hospital to avoid dealing with external diagnostic labs, which he felt would draw the attention of law enforcement officials. But what really prevented exposure was corruption and lack of government oversight. In October 2003, a fifty-five-year-old Turkish patient named Mehmet Bayzit died at the hospital. Bayzit's relatives wanted the body flown back home, which meant that it had to be embalmed. There was a problem, though: embalming centers need a death certificate, which in turn requires a doctor's report.

The situation threatened to attract scrutiny of Kumar's operations. Dublish came to the rescue. He wrote a report on his Ballabgarh hospital letterhead, stating that Bayzit came to his hospital after experiencing chest pains on his way back from a trip to see the Taj Mahal. The report, citing cardiac arrest as the cause of death, enabled Dublish to obtain a death certificate from local municipal authorities. When two more Turkish patients died, in 2004 and 2005, Dublish used the same story as a cover-up. Indian authorities started an investigation only in 2006, when the Turkish embassy in Delhi, prodded by the patients' relatives, requested an inquiry. Still Kumar remained unscathed.

In fact, Kumar did not just evade the law; he turned to it for help. In his testimony to the CBI, he claimed that he received extortion threats from men working for Chota Shakeel, a Mumbai mobster. In 2007, after Kumar lodged a complaint, the Anti-Extortion Cell of the Mumbai police arrested two men suspected of issuing threats on Shakeel's behalf.

The level of official incompetence and corruption in Kumar's case is too outrageous to believe. Since his wealth made him a target for income tax authorities, the government dispatched tax investigators to study his accounts in August 2006. They ignored the fact that the hospital was unlicensed and that it was a venue for illegal transplants. Instead, flabbergasted that Kumar had been paying no taxes for years, the Indian tax department sent him a notice that legal action would follow. Kumar promptly sent the tax department four checks totaling $1.4 million, but when the department tried cashing the checks, they bounced. In a further blow, Delhi police apprehended one of Kumar's accomplices in January 2008. Kumar reportedly paid a bribe of about $40,000 to a group of police officers in exchange for the man's release.

Kumar's luck turned on January 24, 2008, when a constable in

Moradabad, a town near Meerut, witnessed a fight between Gya-suddin, the middleman, and a donor named Vidya Prakash; Prakash charged that Gyasuddin had stolen his kidney. The two were taken to the nearby police station, where Gyasuddin directed police to the hospital and guesthouse, resulting in raids and, ultimately, Kumar's arrest.

I went to Meerut to see Prakash, the man who brought down the ring. To get to his house, I walked through a quiet alley where barefoot children played cricket with a muddy tennis ball and a woman in a burqa emerged to dump vegetable peels and eggshells into the street. I zigzagged through lanes that got progressively narrower, passing a dimly lit lathe shop, to finally arrive at a cluster of houses attached to one another like calcifications on a bone.

There, in a two-story structure, I found the Prakash home. Vidya once made a living repairing analog fare meters for auto-rickshaws. When digital meters came on line, he fell on hard times, he said, and had to sell his blood just to eat. One day he came down with a fever, and a local blood broker turned kidney hunter offered to take him to Gurgaon for treatment. After some lab tests at the Gurgaon hospital, he claims, he was taken for "stomach surgery." When he regained consciousness, there was a scar on his waist. Prakash says that Kumar's men, including Gyasuddin, threatened him physically when he complained.

But other kidney sellers in Meerut told me that Prakash had been a part of the pyramid. Among those he had allegedly lured was Mohan Lal (not his real name), a sixteen-year-old who had run away from his village after his father discovered his affair with a girl from a different caste. Lal was wandering through Meerut's streets in search of work when, he claims, Prakash offered him a job as a dishwasher at the Gurgaon guesthouse. Once Lal reported for work, he says, he was taken for "free" medical tests and falsely diagnosed with a gallstone that Kumar's assistants told him needed immediate surgery. (Piyush Anand, a senior CBI official who supervised the investigation, said a number of complainants were similarly misled.) After he was relieved of a kidney, Lal reluctantly accepted cash and continued to work at the guesthouse for several months.

According to Lal and two other kidney sellers, Prakash complained to the police in Moradabad only after a falling-out with

Gyasuddin over commission. Prakash, a key witness in the case against Kumar, denies these allegations, but the irony is rich: Kumar's operation may have been undone not by proactive law enforcement but by squabbling within the ranks.

A week after I saw Kumar in Ambala, officials took him to Mumbai for questioning on the charges filed back in 1995. They tried to duck the media by switching trains, but I managed to follow along. On a June evening, I sat outside the police station where Kumar had been taken for interrogation and jotted down questions for a written interview. Through the windows of the jailhouse, I watched Kumar sitting in front of a desk, across from a police officer who had taken my queries to him. He was dressed in prison clothes: a striped vest and pajamas. He looked haggard. Between sips of tea, he scribbled answers.

"Being a good surgeon, that was always my dream," he wrote. He explained his Ayurveda degree: his high school exam scores had been just shy of what he needed to get into medical school. He bragged about his self-taught skills. "I helped recipients get well, and in turn they helped donors lead somewhat better lives."

Kumar's defense echoes arguments made by those proposing legalization of the kidney trade, such as the Oxford ethicist and philosopher Janet Radcliffe Richards. In a 1996 paper in the *Journal of Medicine and Philosophy,* Richards wrote that it was a fallacy to view the trade in terms of the "greedy rich" and the "exploited poor." Such misplaced moral indignation, she wrote, "leave[s] behind one trail of people dying who might have been saved, and another of people desperate enough to offer their organs thrust back into the wretchedness they were hoping to alleviate."

The debate is not an easy one to settle, as I learned when I met with Govind Verma (not his real name), a father from Delhi whose eight-year-old son, Rahul, received a transplant from Kumar in January 2008. After I repeatedly rang the doorbell of his house one morning, Verma appeared on a second-floor balcony and nervously looked down at me, then reluctantly ushered me into his living room. In a tremulous voice tinged with guilt, he told me that his son's kidneys had failed at age five from infections caused by urinary reflux. The boy hardly went to school. In the fall of 2007, Verma took him to Chennai in hopes of getting a transplant, but the surgeon was arrested, and kidney sales in that city were stopped. A family acquaintance sent him to Kumar, who, out of sympathy, charged a discount price of $12,000 for the job. Rahul no longer needed dialysis, Verma told me, and was seeing friends again. "I had no choice," Verma said. As he led me out to the gate, his eyes shone with tears. "There were times when I thought of killing myself," he said, motioning with his hand to suggest jumping from the balcony upstairs.

Rahul's life may have improved, but none of the donors I met in Meerut or Mumbai seemed to have climbed out of poverty. Rakesh, a homeless man in Meerut, said the $1,000 he got after being tricked into giving up a kidney allowed him to buy drugs and alcohol for two months without having to work. Along with two other homeless donors, he was living in the compound of an abandoned government building. He told me that he was too ashamed to go back to his village.

What I found is borne out by Madhav Goyal, a Johns Hopkins doctor and bioethicist who interviewed more than three hundred Indian men compelled to sell their kidneys in the mid-1990s. The bulk of their fees—an average of $1,070—was used to settle the debts that had driven them to make the sale in the first place. The rest was gobbled up by basic needs like food and clothing. Using the cash to start a small business, such as a tea stall or an auto-rickshaw service, remained a pipe dream.

Free-trade proponents argue that donors would fare much better under a government-regulated program, which could provide incentives such as lifelong health insurance or college tuition for the donor's children. Scheper-Hughes counters that the concept amounts to a "kidney tax" on the poor, exacting an unfair price for basic services and opportunities. She points to Iran, which has allowed kidney donations for cash since 1988, virtually eliminating the waiting list for the organ. A survey of five hundred Iranian donors who received $1,200 and a year of medical insurance from the government found that their quality of life, as measured by factors like financial condition and psychological health, remained poor three to six months after the donation. "Nobody denies that most donors live in extreme poverty; many are drug addicts," says WHO's Noël. "There is candid recognition in Iran now that the scheme is working well only for the recipients."

Indian officials find it hard to imagine a financially compensated donor program that would be free of exploitation. Singh, the CBI lawyer, says that initially recipients might pay the price set by the government, but soon they would start looking for bargains. It would amount to giving the likes of Kumar a hunting license, he argues.

Sale of an organ will always be "an unequal transaction," the CBI director Vijay Shankar, who has since retired, told me. Legalizing the trade would further institutionalize India's glaring social inequalities while providing an unfair advantage to rich nations like the United States. His words echoed in my head the next morning when I stepped outside my hotel in Gurgaon, where guests were eating a breakfast of aloo parathas, omelets, and juice. At the hotel gate, a man from the shantytown across the street was sharpening knives for the hotel's kitchen, a weekly assignment that earned him three dollars. As sparks flew from his grindstone, he told me that even with Gurgaon's booming economy, he was struggling to feed his family. When he rode off, I imagined how life would change for him if he sold a kidney.

BURKHARD BILGER

Nature's Spoils

FROM *The New Yorker*

THE HOUSE AT 40 Congress Street wouldn't have been my first choice for lunch. It sat on a weedy lot in a disheveled section of Asheville, North Carolina. Abandoned by its previous owners, condemned by the city, and minimally rehabilitated, it was occupied — perhaps infested is a better word — by a loose affiliation of opportunivores. The walls and ceilings, chicken coop, and solar oven were held together with scrap lumber and drywall. The sinks, disconnected from the sewer, spilled their effluent into plastic buckets, providing water for root crops in the gardens. The whole compound was painted a sickly greenish gray — the unhappy marriage of twenty-three cans of surplus paint from Home Depot. "We didn't put in the pinks," Clover told me.

Clover's pseudonym both signaled his emancipation from a wasteful society and offered a thin buffer against its authorities. "It came out of the security culture of the old Earth First! days," another opportunivore told me. "If the Man comes around, you can't give him any incriminating information." Mostly, though, the names fit the faces: Clover was pale, slender, and sweet-natured, with fine blond hair gathered in a bun. His neighbor Catfish had droopy whiskers and fleshy cheeks. There were four men and three women in all, aged twenty to thirty-five, crammed into seven small bedrooms. Only one had a full-time job, and more than half received food stamps. They relied mostly on secondhand bicycles for transportation, and each paid two hundred dollars or less in rent. "We're just living way simple," Clover said. "Super low-impact, deep green."

Along one wall of the kitchen, rows of pine and wire shelves were crowded with dumpster discoveries, most of them pristine: boxes of organic tea and artisanal pasta, garlic from Food Lion, baby spring mix from Earth Fare, tomatoes from the farmers' market. About half the household's food had been left somewhere to rot, Clover said, and there was often enough to share with Asheville's other opportunivores. (A couple of months earlier, they'd unearthed a few dozen cartons of organic ice cream; before that, enough Odwalla juices to fill the bathtub.) Leftovers were pickled or composted, brewed into mead, or, if they looked too dicey, fed to the chickens. "We have our standards," a young punk with a buzz-cut scalp and a skinny ponytail told me. "We won't dumpster McDonald's." But he had eaten a good deal of scavenged sushi, he said—it was all right, as long as it didn't sit in the dumpster overnight—and his housemate had once scored a haggis. "Oh no, no," she said, when I asked if she'd eaten it. "It was canned."

Lunch that day was lentil soup, a bowl of which was slowly congealing on the table in front of me. The carrots and onions in it had come from a dumpster behind Amazing Savings, as had the lentils, potatoes, and most of the spices. Their color reminded me a little of the paint on the house. Next to me, Sandor Katz scooped a spoonful into his mouth and declared it excellent. A self-avowed "fermentation fetishist," Katz travels around the country giving lectures and demonstrations, spreading the gospel of sauerkraut, dill pickles, and all foods transformed and ennobled by bacteria. His two books— *Wild Fermentation* and *The Revolution Will Not Be Microwaved*—have become manifestos and how-to manuals for a generation of underground food activists, and he's at work on a third, definitive volume. Lunch with the opportunivores was his idea.

Katz and I were on our way to the Green Path, a gathering of herbalists, foragers, raw-milk drinkers, and roadkill eaters in the foothills of the Smoky Mountains. The groups in Katz's network have no single agenda or ideology. Some identify themselves as punks, others as hippies, others as evangelical Christians; some live as rustically as homesteaders—the "techno-peasantry," they call themselves; others are thoroughly plugged in. If they have a connecting thread, it's their distrust of "dead, anonymous, industrialized, genetically engineered, and chemicalized corporate food," as Katz has written. Americans are killing themselves with cleanliness, he believes. Every year we waste 40 percent of the food we produce, and we process, pasteurize, or irradiate much of the rest, sterilizing the live cultures that keep us healthy. Lunch from a dumpster isn't just a form of conservation; it's a kind of inoculation.

"This is a modern version of the ancient tradition of gleaning," Katz said. "When the harvest is over, the community has a common-law right to pick over what's left." I poked at the soup with my spoon. The carrots seemed a little soft—whether overcooked or overripe, I couldn't tell—but they tasted all right. I asked the kid with the ponytail if he'd ever brought home food that was spoiled. "Oh, hell, yes!" he said, choking back a laugh. "Jesus Christ, yes!" Then he shrugged, suddenly serious. "It happens: diarrhea, food poisoning. But I think we've developed pretty good immune systems by now."

To most cooks, a kitchen is a kind of battle zone—a stainless steel arena devoted to the systematic destruction of bacteria. We fry them in oil and roast them in ovens, steam them, boil them, and sluice them with detergents. Our bodies are delicate things, easily infected, our mothers taught us, and the agents of microscopic villainy are everywhere. They lurk in raw meat, raw vegetables, and the yolks of raw eggs, on the unwashed hand and in the unmuffled sneeze, on the grimy countertop and in the undercooked pork chop.

Or maybe not. Modern hygiene has prevented countless colds, fevers, and other ailments, but its central premise is hopelessly outdated. The human body isn't besieged; it's saturated, infused, with microbial life at every level. "There is no such thing as an individual," Lynn Margulis, a biologist at the University of Massachusetts at Amherst, told me recently. "What we see as animals are partly just integrated sets of bacteria." Nearly all the DNA in our bodies belongs to microorganisms: they outnumber our own cells nine to one. They process the nutrients in our guts, produce the chemicals that trigger sleep, ferment the sweat on our skin and the glucose in our muscles. ("Humans didn't invent fermentation," Katz likes to say. "Fermentation created us.") They work with the immune system to mediate chemical reactions and drive out the most common infections. Even our own cells are kept alive by mitochondria—the tiny microbial engines in their cytoplasm. Bacteria are us.

"Microbes are the minimal units, the basic building blocks of life on Earth," Margulis said. About half a billion years ago, land verte-brates began to encase themselves in skin, and their embryos in protective membranes, sealing off the microbes inside them and fostering ever more intimate relations with them. Humans are the acme of that evolution—walking, talking microbial vats. By now the communities we host are so varied and interdependent that it's hard to tell friend from enemy—the bacteria we can't live with from those we can't live without. *E. coli, Staphylococcus aureus,* and the bacteria responsible for meningitis and stomach ulcers all live peaceably inside us most of the time, turning dangerous only on rare occasions and for reasons that are poorly understood. "This cliché nonsense about good and bad bacteria, it's so insidious," Margulis said. "It's this Western, dichotomized, Cartesian thing . . . Like Jesus rising."

In the past decade, biologists have embarked on what they call the second human-genome project, aimed at identifying every bac-terium associated with people. More than a thousand species have been found so far in our skin, stomach, mouth, guts, and other body parts. Of those, only fifty or so are known to harm us, and they have been studied obsessively for more than a century. The rest are mostly new to science. "At this juncture, biologists cannot be blamed for finding themselves in a kind of 'future shock,'" Mar-garet McFall-Ngai, an expert in symbiosis at the University of Wis-consin at Madison, wrote in *Nature Reviews Microbiology* two years ago. Or, as she put it in an earlier essay, "We have been looking at bacteria through the wrong end of the telescope."

Given how little we know about our inner ecology, carpet-bomb-ing it might not always be the best idea. "I would put it very bluntly," Margulis told me. "When you advocate your soaps that say they kill all harmful bacteria, you are committing suicide." The bacteria in the gut can take up to four years to recover from a round of antibi-otics, recent studies have found, and the steady assault of deter-gents, preservatives, chlorine, and other chemicals also takes its toll. The immune system builds up fewer antibodies in a sterile en-vironment; the deadliest pathogens can grow more resistant to an-tibiotics; and innocent bystanders such as peanuts or gluten are more likely to provoke allergic reactions. All of which may explain why a number of studies have found that children raised on farms are less susceptible to allergies, asthma, and autoimmune diseases. The cleaner we are, it sometimes seems, the sicker we get.

"We are living in this cultural project that's rarely talked about," Katz says. "We hear about the war on terror. We hear about the war on drugs. But the war on bacteria is much older, and we've all been indoctrinated into it. We have to let go of the idea that they're our enemies." Eating bacteria is one of life's great pleasures, Katz says. Beer, wine, cheese, bread, cured meats, coffee, chocolate: our best-loved foods are almost all fermented. They start out bitter, bland, cloying, or indigestible and are remade by microbes into something magnificent.

Fermentation is a biochemical magic trick—a benign form of rot. It's best known as the process by which yeast turns sugar into alcohol, but an array of other microorganisms and foods can ferment as well. In some fish dishes, for instance, the resident bacteria digest amino acids and spit out ammonia, which acts as a preservative. Strictly speaking, all fermentation is anaerobic (it doesn't consume oxygen); most rot is aerobic. But the two are separated less by process than by product. One makes food healthy and delicious; the other not so much.

Making peace with microorganisms has its risks, of course. *E. coli* can kill you. Listeria can kill you. Basic hygiene and antibiotic overkill aren't hard to tell apart at home, but the margin of error shrinks dramatically in a factory. Less than a gram of the bacterial toxin that causes botulism, released into the American milk supply, could poison a hundred thousand people, the National Academy of Sciences estimated in 2005. And recent deaths and illnesses from contaminated beef, spinach, and eggs have persuaded food regulators to clamp down even harder. While Katz's followers embrace their bacterial selves, the Obama administration has urged Congress to pass a comprehensive new set of food-safety laws, setting the stage for a culture war of an unusually literal sort. "This is a revolution of the everyday," Katz says, "and it's already happening."

When Katz picked me up in Knoxville at the beginning of our road trip, the back seat of his rented Kia was stacked with swing-top bottles and oversized Mason jars. They were filled with foamy, semiopaque fluids and shredded vegetables that had been fermenting in his kitchen for weeks. A sour, pleasantly funky aroma pervaded the cabin, masking the new-car smell of industrial cleaners and off-gassing plastics. It was like driving around in a pickle barrel.

Physically speaking, food activists tend to present a self-negating argument. The more they insist on healthy eating, the unhealthier they look. The pickier they are about food, the more they look like they could use a double cheeseburger. Katz was an exception. At forty-eight, he had clear blue eyes, a tightly wound frame, and ropy forearms. His hands were callused and his skin was ruddy from hours spent weeding his commune's vegetable patches and herding its goats. He wore his hair in a stubby Mohawk, his beard in bushy muttonchops. If not for his multiple earrings and up-to-the-minute scientific arguments, he might have seemed like a figure out of the nineteenth century, selling tonics and bromides from a painted wagon.

Katz was a political activist long before he was a fermentation fetishist. Growing up on New York's Upper West Side, the eldest son of progressive Polish and Lithuanian Jews, he was always involved in one campaign or another. At the age of ten, in 1972, he spent his afternoons on street corners handing out buttons for George McGovern. At eleven he was a campaign volunteer for the mayoral candidate Al Blumenthal. When he reached sixth grade and found that one of the city's premier programs for gifted students, Hunter College High School, was only for girls, he helped bring an antidiscrimination suit that forced it to turn coed. He later served on the student council with Elena Kagan, the future Supreme Court justice. "The staggered lunch hour was our big issue," he says.

At Brown as an undergraduate, Katz became a well-known figure: a bearish hippie in the Abbie Hoffman mold, with a huge head of curly hair. His causes were standard issue for the time: gay rights, divestment from South Africa, U.S. out of Central America (as a senior, he and a group of fellow-activists placed a CIA recruiter under citizen's arrest). Yet Katz lacked the usual stridency of the campus radical. "I remember a particular conversation in 1982 or '83," his classmate Alicia Svigals, who went on to found the band the Klezmatics, told me. "We were standing on a street corner in Providence, and I said, 'Sandy, I think I might be a lesbian.' And he said, 'Oh, I think I might be gay.' At the time, that was a huge piece of news. It wasn't something you said lightly. But his reaction was

'How wonderful and exciting! How fantastic! This is going to be so much fun!' The world was about to be made new—and so easily."

After graduation, Katz moved back to New York. He took a job as the executive director of Westpride, a lobbying group opposed to a massive development project on the Upper West Side. (The developer, Donald Trump, was eventually forced to scale down his plans.) As the AIDS epidemic escalated, in the late eighties, Katz became an organizer for ACT UP and a columnist for the magazine *OutWeek*. His efforts on both fronts caught the eye of Ruth Messinger, the Manhattan borough president, who hired him in 1989 as a land-use planner and as a de facto liaison to the gay community. "He was a spectacular person," Messinger told me recently. "Creative and flamboyant and fun to be around. He just had a natural instinct and talent as an organizer of people." Messinger was thinking of running for mayor (she won the Democratic nomination in 1997, only to get trounced by Rudolph Giuliani), and Katz's ambitions rose with hers. "I would fantasize about what city agency I wanted to administer," he recalls. Then, in 1991, he found out he was HIV-positive.

Katz had never been particularly promiscuous. He'd had his first gay sexual experience at the age of twenty-one, crossing the country on a Green Tortoise bus, and had returned to New York just as its bathhouse days were waning. He'd never taken intravenous drugs and had avoided the riskiest sexual activities. The previous HIV tests he'd taken had come back inconclusive—perhaps, he reasoned, because of a malarial infection that he'd picked up in West Africa. "I have no idea how it happened," he told me. "I remember walking out of the doctor's office in such a daze. I was just utterly shell-shocked."

The virus wasn't necessarily a death sentence, though an effective treatment was years away. But it did transform Katz's political ambitions. "They just dematerialized," he told me. For all his iconoclasm, he had always dreamed of being a United States senator. Now he focused on curing himself. He cut back his hours and moved from his parents' apartment to the East Village. So many of his friends had died while on AZT and other experimental drugs that he decided to search for alternatives. He had already taken up yoga and switched to a macrobiotic diet. Now he began to consult with herbalists, drink nettle tea, and wander around Central Park

gathering medicinal plants. "I got skinny, skinny, skinny," he says.
"My friends thought I was wasting away."

New York's relentless energy had always helped drive his ambitions, but now he found that it wore him out. About a year after his
diagnosis, Katz went to visit some friends in New Orleans who had
rented a crash pad for Mardi Gras. Among the characters there,
he met a man from a place near Nashville called Hickory Knoll
(I've changed the name at Katz's request). Founded in the early
seventies by a group of back-to-the-landers, Hickory Knoll was
something of a legend in the gay community: a queer sanctuary in
the heart of the Bible Belt. "I was a typical New Yorker," Katz says. "I
considered the idea of living in Tennessee absurd." Still, he was in-
trigued. Hickory Knoll had no television or hot running water —
just goats, vegetable gardens, and gay men. Maybe it was just what
he needed.

Hickory Knoll lies just up the road from a Bible camp, in an airy
forest of tulip poplar and dogwood, maple, mountain laurel, paw-
paw, and persimmon. The camp and the commune share a hilltop,
a telephone cable, and, if nothing else, a belief in spiritual renewal:
"Want a new life?" a sign in front of a local church asked as I drove
past. "God accepts trade-ins." When Katz first arrived, in the spring
of 1992, the paulownia trees were in bloom, scattering the ground
with lavender petals. As he walked down the gravel path, the forest
canopy opened up and a cabin of hand-hewn chestnut logs, built
in the 1830s, appeared in the sunlight below. "It was a beautiful arrival," he says.

The commune had a shifting cast of about fifteen members,
some of whom had lived there for decades. It billed itself as a radical faerie sanctuary, though the term was notoriously slippery —
the faerie movement, begun in the late seventies by gay rights activists, embraced everyone from transvestites to pagans and anarchists,
their common interest being a focus on nature and spirituality.
Street kids from San Francisco, nudists from Nashville, a Mexican
minister coming out of the closet: all found their way, somehow, to
central Tennessee. Most were gay men, though anyone was welcome, and the great majority had never lived on the land before.
"Sissies in the wood," one writer called it, after tussling over camping arrangements with a drag queen in four-inch heels.

New arrivals stayed in the cabin "downtown," which had been fitfully expanded to encompass a library, living room, dining room, and kitchen, with four bedrooms upstairs. Farther down the path were a swaybacked red barn, a communal shower, a pair of enormous onion-domed cisterns, and a four-seater outhouse. The charge for room and board was on a sliding scale starting at seventy-five dollars a month, with a tacit agreement, laxly enforced, to pitch in — milking goats, mending fences, or just greeting new arrivals. Those who stayed eventually built houses along the ridges or bought adjacent land and started homesteads and communes of their own. In the spring, at the annual May Day celebration, their numbers grew to several hundred. "The gayborhood just keeps on growing," Weeder, Hickory Knoll's oldest member, told me one evening as we were sitting on the front porch of the cabin. "We're a pretty good voting bloc."

Inside, half a dozen men were preparing dinner. Food is the great marker of the day at Hickory Knoll — the singular goal toward which most labor and creativity tend. On my visit the kitchen seemed to be staffed by at least three cooks at all times, cutting biscuits, baking vegan meat loaf, washing kale; one of them, a gangly Oklahoman named Lady Now, worked in the nude. "Real estate determines culture," Katz likes to say, and the maxim is doubly true among underground food movements. Urban squatters gravitate toward freeganism and dumpster diving, homesteaders toward raw milk and roadkill. At Hickory Knoll the slow pace, lush gardens, and communal isolation are natural incubators for fermented food, though Katz didn't realize it right away. "It took a while for the New York City to wear off," Weeder told me. "Overanalyzing everything. Where am I going to go tonight? There really is nowhere to go."

That first year, a visitor named Crazy Owl brought some miso as barter for his stay, inspiring Katz to make some of his own. Miso, like many Asian staples, is usually made of fermented soybeans. The beans are hard to stomach alone, no matter how long they're cooked. But once inoculated with koji — the spores of the *Aspergillus oryzae* mold — they become silken and delicious. The enzymes in the mold predigest the beans, turning starches into sugars, breaking proteins into amino acids, unlocking nutrients from leaden compounds. A lowly bean becomes one of the world's great foods.

Katz experimented with more and more fermented dishes after

that. He made tempeh, natto, kombucha, and kefir. He recruited friends to chew corn for chicha—an Andean beer brewed with the enzymes in human saliva. At the Vanderbilt library in Nashville, he worked his way through the *Hand-book of Indigenous Fermented Foods* (1983), by the Cornell microbiologist Keith Steinkraus. When he'd gathered a few dozen recipes, Katz printed a pamphlet and sold some copies to a bookstore in Maine and a permaculture magazine in North Carolina. The pamphlet led to a contract from a publisher, Chelsea Green, and the release of *Wild Fermentation*, in 2003. The book was only a modest success at first, but it sold more copies each year—some 70,000 in all. Soon Katz was crisscrossing the country in his car, shredding cabbage in the aisles of Whole Foods or Trader Joe's, preaching the glories of sauerkraut.

"Fermented foods aren't culinary novelties," he told me one morning. "They aren't cupcakes. They're a major survival food." We were standing in his test kitchen in the basement of a farmhouse a few miles down the road from Hickory Knoll. Katz had rented the space two years earlier, when his classes and cooking projects outgrew the commune's kitchen, and outfitted it with secondhand equipment: a triple sink, a six-burner stove, a freezer, and two refrigerators, one of them retrofitted as a tempeh incubator. Along one wall a friend had painted a psychedelic mural showing a man conversing with a bacterium. Along another, Katz had pinned a canticle to wild fermentation, written by a Benedictine nun in New York. A haunch of venison hung in back, curing for prosciutto, surrounded by mismatched jars of sourdough, goat kefir, sweet potato fly, and other ferments, all bubbling and straining at their lids. "It's like having pets," Katz said.

The kitchen had the same aroma as Katz's car, only a few orders of magnitude funkier: the smell of life before cold storage. "We are living in the historical bubble of refrigeration," Katz said, pulling a jar of bright pink and orange sauerkraut off the shelf. "Most of these food movements aren't revolutionary so much as conservative. They want to bring back the way food has been."

Fermentation, like cooking with fire, is one of the initial conditions of civilization. The alcohol and acids it produces can preserve fruits and grains for months and even years, making sedentary society possible. The first ferments happened by accident—honey water turned to mead, grapes to vinegar—but people soon learned to re-create them. By 5400 B.C. the ancient Iranians were making wine. By 1800 B.C. the Sumerians were worshiping Ninkasi, the goddess of beer. By the first century B.C., the Chinese were making a precursor to soy sauce.

Katz calls fermentation the path of least resistance. "It's what happens when you do nothing," he says. Or, rather, if you do one or two simple things. A head of cabbage left on a counter will never turn to sauerkraut, no matter how long it sits there. Yeasts, molds, and a host of bacteria will attack it, digesting the leaves till all that's left is a puddle of black slime. To ferment, most food has to be protected from the air. It can be sealed in a barrel, stuffed in a casing, soaked in brine, or submerged in its own juices—anything, as long as oxygen doesn't touch it. The sauerkraut Katz was holding had been made ten days earlier. I'd watched him shred the cabbage—one head of red and one of green—sprinkle it with two tablespoons of salt to draw out the water, and throw in a few grated carrots. He'd scrunched everything together with his hands, to help release the juice, and packed it in a jar until the liquid rose to the top. "I would suggest not sealing it too tightly," he said, as he clamped down the lid. "Some jars will explode."

Three waves of bacteria had colonized the kraut since then, each one changing the chemical environment just enough to attract and fall victim to the next—like yuppie remodelers priced out of their own neighborhood. Sugars had been converted to acids, carbon dioxide, and alcohol. Some new nutrients had been created: B vitamins, for instance, and isothiocyanates, which laboratory studies have found to inhibit lung, liver, breast, and other cancers. Other nutrients were preserved, notably vitamin C. When Captain Cook circled the globe between 1772 and 1775, he took along 30,000 pounds of sauerkraut, and none of his crew died of scurvy.

I tried a forkful from Katz's jar, along with a slab of his black-rice tempeh. The kraut was crunchy and tart—milder than any I'd had from a store and much fresher tasting. "You could eat it after two weeks, you could eat it after two months, and if you lived in a cold environment and had a root cellar you could eat it after two years," Katz said. The longer it fermented the stronger it would get. His six-month-old kraut, made with radishes and Asian greens, was meaty, pungent, and as tender as pasta—the enzymes in it had broken down the pectin in its cell walls. Some people like it that way, he said. "When this Austrian woman tasted my six-week-old sauerkraut, she said, 'That's okay—for coleslaw.'"

While we were eating, the front door banged open and a young man walked in carrying some baskets of fresh-picked strawberries. He had long blond hair and hands stained red with juice. His name was Jimmy, he said. He lived at Hickory Knoll but was doing some farming up the road. "We originally grew herbs and flowers and planted them in patterns," he said. "But people were like, 'What are those patterns you're makin'? They don't look Christian to me.'" The locals were usually pretty tolerant, Katz said. In eighteen years, the worst incidents that he could recall were a few slashed tires and some teenagers yelling "Faggots!" from the road and shooting shotguns in the air. Rural Tennessee is a "don't ask, don't tell" sort of place, where privacy is the one inalienable right. But Jimmy's fancy crop might have counted as a public display. He laughed and handed me a berry, still warm from the sun. "They're not only organic," he said. "They're grown with gay love."

The fruit was sugar-sweet and extravagantly fragrant—a distillation of spring. But the sauerkraut was the more trustworthy food. An unwashed fruit or vegetable may host as many as a million bacteria per gram, Fred Breidt, a microbiologist with the United States Department of Agriculture and a professor of food science at North Carolina State University, told me. "We've all seen the cases," Katz said. "The runoff from agriculture gets onto a vegetable, or there's fecal matter from someone who handled it. Healthy people will get diarrhea; an elderly person or a baby might get killed. That's a possibility with raw food." If the same produce were fermented, its native bacteria would drive off the pathogens, and the acids and alcohol they produce would prevent any further infection. Breidt has yet to find a single documented case of someone getting sick from contaminated sauerkraut. "It's the safest food there is," Katz said.

Sauerkraut is Katz's gateway drug. He lures in novices with its simplicity and safety, then encourages them to experiment with livelier cultures, more offbeat practices. *The Revolution Will Not Be Microwaved* moves from anodyne topics such as seed saving and urban gardening to dirt eating, feral foraging, cannabis cookery, and the raw-milk underground. Unlike many food activists, Katz has a clear respect for peer-reviewed science, and he prefaces each discussion with the appropriate caveats. Yet his message is clear: "Our food system desperately demands subversion," he writes. "The more we sterilize our food to eliminate all theoretical risk, the more we diminish its nutritional quality."

On the first day of our road trip, not long after our lunch with the opportunivores, Katz and I paid a visit to a man he called one of the kingpins of underground food in North Carolina. Garth, as I'll call him, was a pale, reedy figure in his fifties with wide, spectral eyes. His linen shirt and suspenders hung slackly on his frame, and his sunken cheeks gave him the look of a hardscrabble farmer from a century ago. "I was sick for seventeen years," he told us. "Black circles under my eyes, weighed less than a hundred pounds. It didn't seem like I'd get very far." Doctors said that he had severe chemical sensitivities and a host of ailments—osteoporosis, emphysema, edema, poor circulation—but they seemed incapable of curing him. He tried veganism for a while, but only got weaker. "It's just not a good diet for skinny people," he said. So he went to the opposite extreme.

Inside his bright country kitchen, Garth carefully poured us each a glass of unpasteurized goat milk, as if proffering a magic elixir. The milk was pure white and as thick as cream. It had a long, flowery bloom and a faint tanginess. Raw milk doesn't spoil like pasteurized milk. Its native bacteria, left to multiply at room temperature, sour it into something like yogurt or buttermilk, only much richer in cultures. It was the mainstay of Garth's diet, along with raw butter, cream, and daily portions of raw liver, fish, chicken, or beef. He was still anything but robust, but he had enough energy to work long hours in the garden for the first time in years. "It enabled me to function," he said.

Raw milk brings the bacterial debate down to brass tacks. Drinking it could be good for you. Then again it could kill you. Just where the line between risk and benefit lies is a matter of fierce dispute—not to mention arrests, lawsuits, property seizures, and protest marches. In May, for instance, raw-milk activists, hoping to draw attention to a recent crackdown by Massachusetts agricultural authorities, milked a Jersey cow on Boston Common and staged a drink-in.

Retail sales of raw milk are illegal in most states, including North Carolina, but people drink it anyway. Some dairy owners label the milk for pet consumption only (though at two to five times the cost of pasteurized, it's too rich for most cats). Others sell it at farm stands or through herd-share programs. In my neighborhood in Brooklyn, the raw-milk cooperative meets every month in the aisles of a gourmet deli. The milk is trucked in from Pennsylvania—a violation of federal law, which prohibits the interstate transport of raw milk—but no one seems to mind. Garth buys milk from a local farmer and sells it out of his house. "It's illegal," he told me. "But it gets to the point where living is illegal."

The nutritional evidence both for and against raw milk is somewhat sketchy; much of it dates from before World War II, when raw milk was still legal. The Food and Drug Administration, in a fact sheet titled "The Dangers of Raw Milk," insists that pasteurization "DOES NOT reduce milk's nutritional value." The temperature of the process, well below the boiling point, is meant to kill pathogens and leave nutrients intact. Yet raw-milk advocacy groups, such as the Weston A. Price Foundation, in Washington, DC, point to a number of studies that suggest the opposite. An array of vitamins, enzymes, and other nutrients are destroyed, diminished, or denatured by heat, they say. Lactase, for instance, is an enzyme that breaks down lactose into simpler sugars that the body can better digest. Raw milk often contains *lactobacilli* and *bifidobacteria* that produce lactase, but neither the bacteria nor the enzyme can survive pasteurization. In one survey of raw-milk drinkers in Michigan and Illinois, 82 percent of those who had been diagnosed as lactose intolerant could drink raw milk without digestive problems. (A more extreme view, held by yet another dietary faction, is that people shouldn't be drinking milk at all—that it's a food specifically designed for newborns of other species, and as such is inimical to humans.)

To the FDA, the real problem with milk isn't indigestion but contamination. Poor hygiene and industrial production are a toxic combination. One sick cow, one slovenly worker, can contaminate the milk of a dozen dairies. In 1938 a quarter of all disease outbreaks from contaminated food came from milk, which had been known to carry typhoid, tuberculosis, diphtheria, and a host of other diseases. More recently, between 1998 and 2008, raw milk was responsible for eighty-five disease outbreaks in more than twenty states, including more than sixteen hundred illnesses, nearly two hundred hospitalizations, and two deaths. "Raw milk is inherently dangerous," the FDA concludes. "It should not be consumed by anyone at any time for any purpose."

Thanks in large part to pasteurization, dairy products now account for less than 5 percent of the food-borne disease outbreaks in America every year. Smoked seafood is six times more likely than pasteurized milk to contain listeria; hot dogs are sixty-five times more likely, and deli meats seventy-seven times more likely. "Every now and then, I meet people in the raw-milk movement who say, 'We have to end pasteurization now!'" Katz told me. "We can't end pasteurization. It would be the biggest disaster in the world. There would be a lot of dead children around."

Still, he says, eating food will always entail a modicum of risk. In an average year, there are 76 million cases of food poisoning in America, according to the Centers for Disease Control. Raw milk may be more susceptible to contamination than most foods (though it's still ten times less likely to contain listeria than deli meat is). But just because it can't be produced industrially doesn't mean it can't be produced safely, in smaller quantities. Wisconsin has some 13,000 dairies, about half of which, local experts estimate, are owned by farmers who drink their own raw milk. Yet relatively few people have been known to get sick from it. "If this were such a terrible cause and effect, we would be in the newspaper constantly," Scott Rankin, the chairman of the food-science department at the University of Wisconsin at Madison and a member of the state's raw-milk working group, told me. "Clearly there is an argument to be made in the realm of, yeah, this is a tiny risk."

The country's largest raw-milk dairy is Organic Pastures, in Fresno, California. Its products are sold in 375 stores and serve 50,000 people a week. "Nobody's dying," the founder and CEO, Mark McAfee, told me. In ten years, only two of McAfee's customers have reported serious food poisoning, he says, and none of the bacteria in those cases could be traced to his dairy. Raw milk is rigorously tested in California and has to meet strict limits for bacterial count. The state's standards hark back to the early days of pasteurization, when many doctors considered raw milk far more nutritious than pasteurized, and separate regulations insured its cleanliness. Dealing with live cultures, Katz and McAfee argue, forces dairies to do what all of agriculture should be doing anyway: downsize, localize, clean up production. "We need to go back a hundred and fifty years," McAfee told me. "Going back is what's going to help us go forward."

A century and a half is an eternity in public-health terms, but to followers of the so-called primal diet it's not nearly long enough. Humans have grown suicidally dainty, many of them say, and even a diet enriched by fermented foods and raw milk is too cultivated by half. Our ancestors were rough beasts: hunters, gatherers, scavengers, and carrion eaters, built to digest any rude meal they could find. Fruits and vegetables were a rarity, grains nonexistent. The human gut was a wild kingdom in those days, continually colonized and purged by parasites, viruses, and other microorganisms picked up from raw meat and from foraging. What didn't kill us, as they say, made us stronger.

A few miles north of downtown Asheville, in a small white farmhouse surrounded by trees, two of Katz's acquaintances were doing their best to emulate early man. Steve Torma ate mostly raw meat and raw dairy. His partner, Alan Muskat, liked to supplement his diet with whatever he could find in the woods: acorns, puffballs, cicadas and carpenter ants, sumac leaves, and gypsy-moth caterpillars. Muskat was an experienced mushroom hunter who had provisioned a number of restaurants in Asheville, and much of what he served us was surprisingly good. The ants, collected from his woodpile in the winter when they were too sluggish to get away, had a snappy texture and bright, tart flavor—like organic Pop Rocks. (They were full of formic acid, which gets its name from the Latin word for ant.) He brought us a little dish of toasted acorns, cups of honey-sweetened sumac tea, and goblets of a musky black broth made from decomposed inky-cap mushrooms. I felt, for a moment, as if I'd stumbled upon a child's tea party in the woods.

The primal diet has found a sizable following in recent years, particularly in southern California and, for some reason, Chicago. Its founder, Aajonus Vonderplanitz, a sixty-three-year-old former soap-opera actor and self-styled nutritionist, claims that it cured him of autism, angina, dyslexia, juvenile diabetes, multiple myeloma, and stomach cancer, as well as psoriasis, bursitis, osteoporo-

sis, tooth decay, and "mania created by excessive fruit." Vonderplanitz recommends eating roughly 85 percent animal products by
volume, supplemented by no more than one fruit a day and a pint
or so of "green drink"—a purée of fruits and vegetable juices.
(Whole-vegetable fibers, he believes, are largely indigestible.) The
diet's most potent component, though, is an occasional serving of
what Vonderplanitz calls "high meat."

Torma ducked into the back of the house and returned with a
swing-top jar in his hands. Inside lay a piece of organic beef, badly
spoiled. It was afloat in an ocher-colored puddle of its own decay,
the muscle and slime indistinguishable, like a slug. High meat is
the flesh of any animal that has been allowed to decompose. Torma
keeps his portions sealed for up to several weeks before ingesting
them, airing them out every few days. (Like the bacteria in sauerkraut, those that cause botulism are anaerobic; fermentation de-
stroys them, but they sometimes survive in sealed meats—*botulus*,
in Latin, means sausage.) Vonderplanitz says that he got high meat
and its name from the Eskimos, who savor rotten caribou and seal.
A regular serving of decayed heart or liver can have a "tremendous
Viagra effect" on the elderly, Vonderplanitz told me recently. The
first few bites, though, can be rough going. "I still have some resistance to it," Torma admitted. "But the health benefits! I'm fifty-two
now. I started this when I was forty-two, and I feel like I'm in my
twenties."

Primal eating has its detractors: the *Times* of London recently
dubbed it "the silliest diet ever." Most of us find whole vegetables
perfectly digestible. The notion that parasites and viruses are good
for us would be news to most doctors. And even Vonderplanitz and
his followers admit that high meat sometimes leaves them ill and
explosively incontinent. They call it detoxification.

Still, radical measures like these have had some surprising suc-
cesses. In a case published last year in the *Journal of Clinical Gastro-
enterology*, a sixty-one-year-old woman was given an entirely new set
of intestinal bacteria. The patient was suffering from severe diarrhea, Janet Jansson, a microbial ecologist at Lawrence Berkeley National Laboratory, told me. "She lost twenty-seven kilograms and
was confined to a diaper and a wheelchair." To repopulate her colon with healthy microbes, Jansson and her collaborators arranged
for a fecal transplant from the patient's husband. "They just put it

in a Waring blender and turned it into a suppository," Jansson says.
"It sounds disgusting, but it cured her. When we got another sample from her, two days later, she had adopted his microbial community." By then her diarrhea had disappeared.

Other experiments have been even more dramatic. At Washington University in St. Louis, the biologist Jeff Gordon has found that
bacteria can help determine body weight. In a study published in
Nature in 2006, Gordon and his lab mates, led by Peter Turnbaugh,
took a group of germ-free mice, raised in perfect sterility, and divided it in two. One group was inoculated with bacteria from normal mice; the other with bacteria from mice that had been bred
to be obese. Both groups gained weight after the inoculation, but
those with bacteria from obese mice had nearly twice the percentage of body fat by the end of the experiment. Later the same lab
took normal mice, fed them until they were fat, and transplanted
their bacteria into other normal mice. Those mice grew fat, too,
and the same pattern held true when the mice were given bacteria
from obese people. "It's a positive-feedback loop," Ruth Ley, a biologist from Gordon's lab who now teaches at Cornell, told me.
"Whether you're genetically obese or obese from a high-fat diet,
you end up with a microbial community that is particularly good at
extracting calories. It could mean that an obese person can extract
an extra five or ten calories out of a bowl of Cheerios."

Biologists no longer doubt the depth of our dependence on bacteria. Jansson avoids antibiotics unless they're the only option and
eats probiotic foods like yogurt and prebiotic foods like yacón, a
South American root that nourishes bacteria in the gut. Until we
understand more about this symbiosis, she and others say, it's best
to ingest the cultures we know and trust. "What's beautiful about
fermenting vegetables is that they're naturally populated by lactic-
acid bacteria," Katz said. "Raw flesh is sterile. You're just culturing
whatever was on the knife."

When Torma unclamped his jar, a sickly sweet miasma filled the
air—an odor as natural as it was repellent. Decaying meat produces its own peculiar scent molecules, I later learned, with names
like putrescine and cadaverine. I could still smell them on my
clothes hours later. Torma stuck two fingers down the jar and
fished out a long, wet sliver. "Want a taste?" he said.

It was the end of a long day. I'd spent most of it consuming everything set before me: ants, acorns, raw milk, dumpster stew, and seven kinds of mead, among other delicacies. But even Katz took a pass on high meat. While Torma threw back his head and dropped in his portion, like a seal swallowing a mackerel, we quietly took our leave. "You have to trust your senses," Katz said, as we were driving away. "To me, that smelled like death."

Katz has lived with HIV for almost two decades. For many years he medicated himself with his own ferments and local herbs — chickweed, yellow dock, violet leaf, burdock root. But periodic tests at the AIDS clinic in Nashville showed that his T-cell count was still low. Then, in the late nineties, he began to lose weight. He often felt listless and mildly nauseated. At first, he assumed that he was just depressed, but the symptoms got worse. "I started feeling lightheaded a lot and I had a couple of fainting episodes," he told me. "It dawned on me very slowly that I was suffering from classic AIDS wasting syndrome."

By then an effective cocktail of AIDS drugs had been available for almost three years. Katz had seen it save the life of one of his neighbors in Tennessee. "It was a really dramatic turnaround," he told me. "But I didn't want my life to be medically managed. I had a real reluctance to get on that treadmill." In the summer of 1999, he took a road trip to Maine to visit friends, hoping to snap out of his funk. By the time he got there, he was so exhausted that he couldn't get up for days. "I remember what really freaked me out was trying to balance my checkbook," he says. "I couldn't even do simple subtraction. It was like my brain wasn't functioning anymore." He had reached the end of his alternatives.

Katz doesn't doubt that the cocktail saved his life. In pictures from that trip, his eyes are hollowed out, his neck so thin that it juts from his woolen sweater like a broomstick. He got worse before he got better, he says — "It was like I had an anvil in my stomach." But one morning, about a month after his first dose, he woke up and felt like going for a walk. A few days after that, he had a strong urge to chop wood. He now takes three antiretroviral and protease-inhibitor drugs every day and hasn't had a major medical problem in ten years. He still doesn't have the stamina he'd like, and his forehead is often beaded with sweat, even on cool evenings around the commune's dinner table. "I wish this weren't my reality," he told me. "I don't feel great that my life is medically managed. But if that's what's keeping me alive, hallelujah."

It's this part that incenses some of his readers: having sung the praises of sauerkraut, revealed the secrets of kombucha, and gestured toward the green pastures of raw milk, Katz has surrendered to the false promise of Western medicine. His drug dependence is a sellout, they say—an act of bad faith. "Every two months or so, I get a letter from some well-meaning person who's decided that they have to tell me that I'm believing a lie," he told me. "That the HIV is meaningless and doesn't make people sick. That if I follow this link and read the truth, I will be freed from that lie and will stop having to take toxic pills and live happily ever after." Live cultures have been part of his healing, he said. They may even help prevent diseases like cancer. "But that doesn't mean that kombucha will cure your diabetes. It doesn't mean that sauerkraut cured my AIDS."

The trouble with being a diet guru, it seems, is that the more reasonable you try to be, the more likely you are to offend your most fervent followers. *The Revolution Will Not Be Microwaved* includes a chapter called "Vegetarian Ethics and Humane Meat." It begins, "I love meat. The smell of it cooking can fill me with desire, and I find its juicy, rich flavor uniquely satisfying." Katz goes on to describe his dismay at commercial meat production, his respect for vegetarianism, and his halfhearted attempts to embrace it. "When I tried being vegan, I found myself dreaming about eggs," he writes. "I could find no virtue in denying my desires. I now understand that many nutrients are soluble only in fats, and animal fats can be vehicles of rich nourishment."

Needless to say, this argument didn't fly with much of his audience. Last year the Canadian vegan punk band Propagandhi released a song called "Human(e) Meat (The Flensing of Sandor Katz)." Flensing is an archaic locution of the sort beloved by metal bands: it means to strip the blubber from a whale. "I swear I did my best to insure that his final moments were swift and free from fear," the singer yelps. "But consideration should be made for the fact that Sandor Katz was my first kill." He goes on to describe searing every hair on Katz's body, boiling his head in a stockpot, and turning it into a spreadable headcheese. "It's a horrible song," Katz told me. "When it came out, I was not amused. I had a little fear that

some lost vegan youth would try to find meaning by carrying out
this fantasy. But it's grown on me."

The moon was in Sagittarius on the last night of April, the stars out
in their legions. Katz and I had arrived in the Smoky Mountains to
join the gathering of the Green Path. About sixty people were
camped on a sparsely wooded slope half an hour west of Asheville.
Tents, lean-tos, and sleeping bags were scattered among the trees
below an open shed where meals were served: dandelion greens,
nettle pesto, kava brownies—the usual. In a clearing nearby, an
oak branch had been stripped and erected as a maypole, and a fire
pit dug for the night's ceremony: the ancient festival of Beltane, or
Walpurgisnacht.

We'd spent the day going on plant walks, taking wildcrafting lessons, and listening to a succession of seekers and sages—Turtle,
7Song, Learning Deer. Every few hours a cry would go up, and the
tribe would gather for an adult version of what kindergartners
call Circle Time: everyone holding hands and exchanging expressions of self-conscious wonder. The women wore their hair long
and loose or bobbed like pixies'; their noses were pierced and their
bodies wrapped in rag scarves and patterned skirts. The men, in
dreadlocks and piratical buns, talked of Babylon and polyamory.
The children ran heedlessly through the woods, needing no instruction in the art of absolute freedom. "Is your son home-
schooled?" I asked one mother, who crisscrossed the country with
her two children and a tepee and was known as the Queen of Roadkill. She laughed. "He's unschooled," she said. "He just learns as
he goes."

The Green Path was part ecological retreat and part pagan re-
vival meeting, but mostly it was a memorial for its founder, Frank
Cook, who had died a year earlier. Cook was a botanist and teacher
who traveled around the world collecting herbal lore, then writing
and lecturing about it back in the U.S. He lived by barter and do-
nation, refusing to be tied down by full-time work or a single residence, and was, by all accounts, an uncommonly gifted teacher.
(He and Katz often taught seminars together.) As a patron saint,
though, Cook had left his flock with an uneasy legacy. When he
died at forty-six, it was owing to a tapeworm infection acquired
on his travels. Antibiotics might have cured him, but he mostly

avoided them. By the time his mother and friends forced him to go to a hospital last spring, his brain was riddled with tapeworm larvae and the cysts that formed around them. "Frank was pretty dogmatic about Western medicine," Katz said. "And I really think that's why he's dead."

Around the bonfire that night, I could see Katz on the other side of the circle, holding hands with his neighbors. After eighteen years in the wilderness, he couldn't imagine moving back to New York City—a weekend there could still wear him out. Yet his mind had never entirely left the Upper West Side, and his voice, clipped and skeptical, was a welcome astringent here. After a while a woman stepped into the firelight, dressed in a long white gown with a crown of vines and spring flowers in her hair. Beltane was a time of ancient ferment, she said, when the powers of the sky come down and the powers of the earth rise up to meet them. She took two goblets and carried them to opposite points in the circle. They were full of May wine steeped with sweet woodruff, once considered an aphrodisiac. On this night, by our own acts of love and procreation, we would remind the fields and crops to grow.

There was more along those lines, though I confess that I didn't hear it. I was watching the kid to my left—a scruffy techno-peasant dressed in what looked like sackcloth and bark—take a swig of the wine. The goblets were moving clockwise around the circle, I'd noticed. By the time the other goblet reached me, thirty or forty people would have drunk from it.

I heard a throat being cleared somewhere in the crowd, and a cough quickly stifled. Embracing live cultures shouldn't mean sacrificing basic hygiene, Katz had told me that afternoon, after his sauerkraut seminar. "Part of respecting bacteria is recognizing where they can cause us problems." And so, when the kid had drunk his fill, I tapped his shoulder and asked for the goblet out of turn. I took a quick sip, sweet and bitter in equal measure. Then I watched as the wine made its way around the circle—teeming, as all things must, with an abundance of invisible life.

DEBORAH BLUM

The Chemist's War

FROM *Slate*

IT WAS CHRISTMAS EVE 1926, the streets aglitter with snow and lights, when the man afraid of Santa Claus stumbled into the emergency room at New York City's Bellevue Hospital. He was flushed, gasping with fear: Santa Claus, he kept telling the nurses, was just behind him, wielding a baseball bat.

Before hospital staff realized how sick he was — the alcohol-induced hallucination was just a symptom — the man died. So did another holiday partygoer. And another. As dusk fell on Christmas, the hospital staff tallied up more than sixty people made desperately ill by alcohol and eight dead from it. Within the next two days, yet another twenty-three people died in the city from celebrating the season.

Doctors were accustomed to alcohol poisoning by then, a routine of life in the Prohibition era. The bootlegged whiskeys and so-called gins often made people sick. The liquor produced in hidden stills frequently came tainted with metals and other impurities. But this outbreak was bizarrely different. The deaths, as investigators would shortly realize, came courtesy of the U.S. government.

Frustrated that people continued to consume so much alcohol even after it was banned, federal officials had decided to try a different kind of enforcement. They ordered the poisoning of industrial alcohols manufactured in the United States, products regularly stolen by bootleggers and resold as drinkable spirits. The idea was to scare people into giving up illicit drinking. Instead, by the time Prohibition ended in 1933, the federal poisoning program, by some estimates, had killed at least 10,000 people.

Although mostly forgotten today, the "chemist's war of Prohibition" remains one of the strangest and most deadly decisions in American law-enforcement history. As one of its most outspoken opponents, Charles Norris, the chief medical examiner of New York City during the 1920s, liked to say, it was "our national experiment in extermination." Poisonous alcohol still kills—sixteen people died just this month after drinking lethal booze in Indonesia, where bootleggers make their own brews to avoid steep taxes—but that's due to unscrupulous businessmen rather than government order.

I learned of the federal poisoning program while researching my new book, *The Poisoner's Handbook,* which is set in jazz-age New York. My first reaction was that I must have gotten it wrong. "I never heard that the government poisoned people during Prohibition, did you?" I kept saying to friends, family members, colleagues.

I did, however, remember the U.S. government's controversial decision in the 1970s to spray Mexican marijuana fields with Paraquat, a herbicide. It was primarily intended to destroy crops, but government officials also insisted that awareness of the toxin would deter marijuana smokers. They echoed the official position of the 1920s—if some citizens ended up poisoned, well, they'd brought it upon themselves. Although Paraquat wasn't really all that toxic, the outcry forced the government to drop the plan. Still, the incident created an unsurprising lack of trust in government motives, which reveals itself in the occasional rumors circulating today that federal agencies, such as the CIA, mix poison into the illegal drug supply.

During Prohibition, however, an official sense of higher purpose kept the poisoning program in place. As the *Chicago Tribune* editorialized in 1927: "Normally, no American government would engage in such business . . . It is only in the curious fanaticism of Prohibition that any means, however barbarous, are considered justified." Others, however, accused lawmakers opposed to the poisoning plan of being in cahoots with criminals and argued that bootleggers and their law-breaking alcoholic customers deserved no sympathy. "Must Uncle Sam guarantee safety first for souses?" asked Nebraska's *Omaha Bee.*

The saga began with ratification of the Eighteenth Amendment, which banned the manufacture, sale, or transportation of alcoholic beverages in the United States. High-minded crusaders and antialcohol organizations had helped push the amendment through in 1919, playing on fears of moral decay in a country just emerging from war. The Volstead Act, spelling out the rules for enforcement, passed shortly afterward, and Prohibition itself went into effect on January 1, 1920.

But people continued to drink—and in large quantities. Alcoholism rates soared during the 1920s; insurance companies charted the increase at more than 300 percent. Speakeasies promptly opened for business. By the decade's end, some 30,000 existed in New York City alone. Street gangs grew into bootlegging empires built on smuggling, stealing, and manufacturing illegal alcohol. The country's defiant response to the new laws shocked those who sincerely (and naively) believed that the amendment would usher in a new era of upright behavior.

Rigorous enforcement had managed to slow the smuggling of alcohol from Canada and other countries. But crime syndicates responded by stealing massive quantities of industrial alcohol—used in paints and solvents, fuels and medical supplies—and redistilling it to make it potable.

Well, sort of. Industrial alcohol is basically grain alcohol with some unpleasant chemicals mixed in to render it undrinkable. The U.S. government started requiring this "denaturing" process in 1906 for manufacturers who wanted to avoid the taxes levied on potable spirits. The U.S. Treasury Department, charged with overseeing alcohol enforcement, estimated that by the mid-1920s, some 60 million gallons of industrial alcohol were stolen annually to supply the country's drinkers. In response, in 1926, President Calvin Coolidge's government decided to turn to chemistry as an enforcement tool. Some seventy denaturing formulas existed by the 1920s. Most simply added poisonous methyl alcohol into the mix. Others used bitter-tasting compounds that were less lethal, designed to make the alcohol taste so awful that it became undrinkable.

To sell the stolen industrial alcohol, the liquor syndicates employed chemists to "renature" the products, returning them to a drinkable state. The bootleggers paid their chemists a lot more than the government did, and they excelled at their job. Stolen and redistilled alcohol became the primary source of liquor in the country. So federal officials ordered manufacturers to make their products far more deadly.

By mid-1927 the new denaturing formulas included some nota-

ble poisons—kerosene and brucine (a plant alkaloid closely related to strychnine), gasoline, benzene, cadmium, iodine, zinc,
mercury salts, nicotine, ether, formaldehyde, chloroform, cam-
phor, carbolic acid, quinine, and acetone. The Treasury Department also demanded that more methyl alcohol be added—up to
10 percent of the total product. It was the last that proved most
deadly.

The results were immediate, starting with that horrific holiday
body count in the closing days of 1926. Public health officials responded with shock. "The government knows it is not stopping
drinking by putting poison in alcohol," the New York City medical
examiner Charles Norris said at a hastily organized press conference. "Yet it continues its poisoning processes, heedless of the fact
that people determined to drink are daily absorbing that poison.
Knowing this to be true, the United States government must be
charged with the moral responsibility for the deaths that poisoned
liquor causes, although it cannot be held legally responsible."

His department issued warnings to citizens, detailing the dangers in whiskey circulating in the city: "Practically all the liquor that
is sold in New York today is toxic," read one 1928 alert. He publicized every death by alcohol poisoning. He assigned his toxicolo-
gist, Alexander Gettler, to analyze confiscated whiskey for poisons
—that long list of toxic materials I cited came in part from studies
done by the New York City medical examiner's office.

Norris also condemned the federal program for its dispropor-
tionate effect on the country's poorest residents. Wealthy people,
he pointed out, could afford the best whiskey available. Most of
those sickened and dying were those "who cannot afford expensive
protection and deal in low grade stuff."

And the numbers were not trivial. In 1926 in New York City,
1,200 were sickened by poisonous alcohol; 400 died. The following
year, deaths climbed to 700. These numbers were repeated in cities
around the country as public-health officials nationwide joined in
the angry clamor. Furious anti-Prohibition legislators pushed for a
halt in the use of lethal chemistry. "Only one possessing the instincts of a wild beast would desire to kill or make blind the man
who takes a drink of liquor, even if he purchased it from one violat-
ing the Prohibition statutes," proclaimed Senator James Reed of
Missouri.

Officially, the special denaturing program ended only when the Eighteenth Amendment was repealed in December 1933. But the chemist's war itself faded away before then. Slowly, government officials quit talking about it. And when Prohibition ended and good grain whiskey reappeared, it was almost as if the craziness of Prohibition—and the poisonous measures taken to enforce it—had never quite happened.

JON COHEN

Fertility Rites

FROM *The Atlantic*

PASCAL GAGNEUX, one of the few laboratory scientists who has studied wild chimpanzees, is a walking encyclopedia of chimpanzee/human differences. Ever since scientists began studying chimpanzees, they have emphasized our similarities, which *are* striking. But today neither Darwinists nor conservationists need such similarities to further their respective causes: abundant genomic evidence supports Darwinian evolution, and laws and regulations are in place to protect all endangered species, regardless of whether they are cute enough to excite human sympathy. This has paved the way for chimp researchers like Gagneux to focus on what separates us from chimps. Their goal is to sharpen our understanding of what makes a human human.

In Gagneux's case, he and his colleagues are hoping to use their analysis of the differences between human and chimp sperm—especially the sugars that adorn the sperms' surfaces and let them bind to cells in the walls of the uterus or a fallopian tube—to unlock one of the riddles of human infertility: does sperm sometimes have components that undermine its ability to fertilize an egg? Perhaps the differences between chimp and human sperm can help explain why humans miscarry nearly 50 percent of all conceptions, while chimps seem rarely to lose an embryo or fetus.

To get at such questions, Gagneux has spent many hours fashioning devices to coax sperm from chimpanzees. He began by sculpting a silicone version of a female chimp's rear end. But the male chimpanzees at the Primate Foundation of Arizona that were recruited to help with the project did not see it that way, and the model sat unmolested on a counter. "It's a nice chimp butt, but I thought it was a bonobo butt when I first saw it," Jim Murphy, the foundation's colony manager at the time, admitted to me when I visited a few years ago. "Maybe that's why they don't like it."

Gagneux's next attempt relied more on medical science than on art. He modified a piece of PVC pipe to create a variation on what's known as a Penrose drain, which is used to remove pus and other liquid discharge from wounds. For the chimps the pipe was rigged with a compartment that holds warm water; latex coated with K-Y Jelly lined the interior.

On this day Rachel Borman, who had worked at the foundation for ten years as an animal handler, was given the job of selecting a sperm donor and encouraging him to produce a sample. Borman first "gowned up" to protect her clothes. The target donor today was a sixteen-year-old named Shahee. Borman asked me not to follow her into the space that held the caged chimps, as the presence of a stranger might break the mood. So I peered through the glass portal in a door. "I'm just going to go in there with these other guys to make him jealous," Borman told me as she entered the chimp space. She did a quick pass by Shahee's rivals and returned to the supply room for the modified Penrose drain. With it in one hand and a training clicker in the other, Borman walked toward Shahee. (Trainers use clickers in tandem with positive reinforcement, usually food, to condition animals to perform a specific behavior—in this case, masturbation.) After a few clicks, Shahee stuck his erect penis through the bars. Borman held up the PVC pipe and said, "Good boy! Good boy!" She then gave him an M&M and walked back to the lab. "He did it," Borman said proudly.

Borman cracked open the tube. Lying on the tan latex was a chunk of chimp sperm about the size of a small wine cork. I say "chunk" because most of it had coagulated into what is known as a plug, about one-quarter of which usually melts in the warm vaginal vault. Using a Popsicle stick, Borman transferred the ejaculate into three vials. "It's fun for the chimps to do this," Borman explained as she capped the vials. "They love it."

My job was to shuttle the vials to San Diego when I flew home that night and then drive them to Gagneux's lab at the University of California at San Diego so he could study them while the sperm were still alive. As we exited the enclosure, we passed Shahee. He spat on me.

On the way to the airport, I realized that the chimp sperm created something of a dilemma. I had the vials in my day pack, the only bag I had brought for my short trip to Arizona. If I wanted to carry the bag with me onto the plane, I would have to pass it through security, and surely the screeners would question the liquid in my vials. What would I say? It was hair conditioner? Packed in laboratory vials? If I told the truth, would they think I was a modern Ilya Ivanovich Ivanov, the Russian scientist who tried to breed a "humanzee"? But if I checked my small day pack as luggage, would they suspect that I was a drug smuggler or some such and escalate to a search and a humiliating outing?

I gambled that the security checkpoint was a higher risk, and I checked my day pack at the ticket counter. My bet paid off. Before I knew it, I was back in San Diego, sperm in hand, at a late-night rendezvous with Gagneux.

People tend to think of sperm as cylindrical, but they are actually paddle-shaped, Gagneux told me. "When they move around, they resemble a surfboard tumbling around in the waves," he said. He prepared some of the sperm I'd flown in, placing it on a microscope slide. The microscope was connected to a computer screen, so I could watch in real time. The sperm did not resemble surfboards rumbling in the waves so much as bugs flittering about on the top of a pond. "Wow, look at that," said Gagneux. "It's pretty sweet, huh? There's nowhere near that many in humans."

Gagneux's lab space was adjacent to that of his collaborator Ajit Varki, who had helped uncover the functioning of the sugars, known as sialic acids, on cell surfaces. The sialic acids on the surfaces of human and chimp sperm have become the focus of Gagneux's work, too. Humans, as Varki discovered, have lost the ability to make one sialic acid, Neu5Gc, and Gagneux suspected that Neu5Gc played a role in fertilization. He hypothesized that Neu5Gc helped female chimpanzees, in a process called "cryptic female choice," get the benefit of the most compatible, highest-quality sperm. The sugar acted like the fuzzy part of Velcro and attached to barbs formed by sugar-binding proteins on the surface of the cells in the uterus or fallopian tubes. Neu5Gc, as Gagneux imagined it, might "sweet-talk" the female reproductive system.

Gagneux's Neu5Gc ideas had a critical implication for human fertility. Although we have lost the ability to synthesize Neu5Gc, we ingest the sugar when we eat meat and dairy products, and it, in turn, can then be incorporated into our cells. Does Neu5Gc coat the surface of human sperm? Is it found more readily on the sperm of men who eat lots of animal products? Does the extremely foreign Neu5Gc then trigger in women an immune response that selects against the survival of the sperm? "It could be that men who eat loads of meat pass a threshold and become infertile," suggested Gagneux.

I left Gagneux shortly after midnight, and he was cranking away on the fresh chimp- and human-sperm samples he had received during the day. Science has few "Eureka!" moments, and Gagneux did not solve any great mysteries that night. But profound insights time and again come from asking simple questions that, once raised, seem abundantly obvious. *Do the different sugars on the surfaces of chimp and human sperm impact fertility?* is one of those obvious, beautiful questions. And it may just lead to an unobvious explanation for one of the more vexing problems that modern humans face.

LUKE DITTRICH

The Brain That Changed Everything

FROM *Esquire*

THE LABORATORY AT NIGHT, the lights down low. An iMac streams a Pat Metheny version of an Ennio Morricone tune while Dr. Jacopo Annese, sitting in front of his ventilated biosafety cabinet, a small paintbrush in his hand, teases apart a crumpled slice of brain. The slice floats in saline solution in a shallow black plastic tray, and at first it looks exactly like a piece of ginger at a good sushi restaurant, one where they don't dye the ginger but leave it pale. Then Annese's brush, with its practiced dabs and tugs, gently unfurls it. The slice becomes a curlicued silhouette, recognizable for what it is, the organ it comes from, even if you are not, as Annese is, a neuroanatomist.

He loves quiet nights like these, when his lab assistants set him up with everything he needs—the numbered twist-off specimen containers, the paintbrushes, the empty glass slides—and then leave him alone with his music and his work.

Annese coaxes the slice into position above the glass slide that lies half submerged in the tray, cocking his head, peering at it from different angles, checking to see that he has the orientation right. The left hemisphere must, when you're looking directly at the slide, be on the right side of your field of view, just as it would be were you staring into the eyes of the brain's owner. Although brains are roughly symmetrical, they are not entirely so, and Annese has become familiar with the individual topography of this one, all its subtly asymmetric sulci.

For additional reference, Annese occasionally glances at a digital photograph on the screen of the computer. The photograph, two months old, was taken during another late night at the lab. It shows a bladed machine, a cryomicrotome, similar to a meat slicer in a deli. The machine holds a block of frozen gelatin, and the block of frozen gelatin holds a brain. Annese had sheared the whole brain into 2,401 70-micron-thin slices, the camera snapping once before each pass of the blade, and this particular image captures the moment before he sheared off the slice that now floats before him. The picture provides a useful comparison for Annese now, showing the slice as it looked in situ.

That night of slicing had not been as solitary as this one. Not only had all his lab assistants been here, but thousands of other people had been present in another sense, since Annese had streamed the event live online. A colleague of Annese's later confessed that he had watched the Web stream with anxiety, hoping that Annese wouldn't become famous as the second doctor to screw up this particular brain.

A few more gentle prods, and then Annese begins to slowly pull the glass out of the tray. Before he trained as a scientist, he worked as a cook, and he often uses cooking analogies to explain his work. The art of histology is a lot like baking, he says, since everything from the samples to the temperature to the tools must be finely calibrated, with little room for improvisation. And he is proud, justifiably so, of his skills. Soon the slide, with its burden perfectly positioned, is resting safely on the tepid surface of a warmer, where it will be left to dry overnight.

Annese reaches for another cryogenic vial, number 451, and screws off the lid. Just before he tips the next slice into the tray, he smiles and turns to me.

"See," he says, "how much work I have to do to clean up the mess your grandfather made?"

There were things Henry loved to do.

He loved to pet the animals. The Bickford Health Care Center was one of the first Eden Alternative elder-care facilities in Connecticut, which means that along with its forty-eight or so patients, the center housed three cats, four or five birds, a bunch of fish, a rabbit, and a dog named Sadie. Henry would spend hours sitting in his wheelchair in the courtyard with the rabbit on his lap and Sadie by his side.

Another thing he loved was to watch the trains go by. His old room, 133, is on the far side of the center, and its window looks out on the tracks, where several times a day the Amtrak roars past. The canal for which the town of Windsor Locks was named runs parallel to and just beyond the tracks, and beyond the canal looms the abandoned red-brick husk of an old paper mill.

He loved word games. Many of the scientific papers that have been written about Henry describe his avidity for crossword puzzles, though in his later years, according to a nurse at the center, Henry found crossword puzzles too great a challenge and started doing simple find-a-word puzzles instead.

He loved old movies. Bogart and Bacall, that era. The classics, we would call them, though of course they were not really classics to him. He'd ask to see one of these old movies, and one of the nurses or attendants would pop in a videocassette and press play on the remote control. Television sets were no shock to him, TV being a technology that developed during his time. But he never figured out how to operate a remote control.

There was one TV series he loved to watch, too. A sitcom. *All in the Family*. It is a show, essentially, about a man who is mystified by a world and culture and family that change all around him while he remains the same, a man who digs in his heels, who holds on to the past, refusing to be dragged into the present.

Henry was an extremely nice guy. Courteous, cheerful, compliant. He almost always took his meds when the nurses asked him to. On the rare occasions that he was stubborn and refused to take them, the nurses knew of an easy way to get him to cooperate. It was a trick passed down over the decades, from one nurse to another.

Henry, a nurse would say, Dr. Scoville insists that you take your meds right now!

And then, invariably, he would comply.

This strategy worked right through till the end, until Henry died, two years ago. The fact that Dr. William Beecher Scoville had died long before then made no difference. Dr. Scoville remained an undying authority figure in Henry's life because Henry's life never really progressed beyond the day in 1953 when Dr. Scoville, my grandfather, removed some small but important pieces of Henry's brain.

May 25, 1992:

> Researcher: How long have you had trouble remembering things?
> Henry: That I don't know myself. I can't tell you because I don't remember.
> R: Well, do you think it's days, or weeks, months, years?
> H: Well, see, I can't put it exactly on a day, week, or month, or year basis.
> R: But do you think it's been more than a year that you've had this problem?
> H: Well, I think it's about that. About a year. Or more. Because I believe I had an . . . this is just a thought that I'm having myself, well, I possibly have had an operation or something.
> R: Uh-huh. Tell me about that.
> H: And uh, I remember, I don't remember just where it was done, in, uh . . .
> R: Do you remember your doctor's name?
> H: No, I don't.
> R: Does the name Dr. Scoville sound familiar?
> H: Yes! That does.
> R: Tell me about Dr. Scoville.
> H: Well, he, he would, he did some traveling around. He did, well, medical research on people. All kinds of people. In Europe, too! And the wealthy. And out on the movie stars, too.
> R: That's right. Did you ever meet him?
> H: Yes, I think I did. Several times . . .
> R: Was it in the hospital?
> H: No, the first time I met him was in his office. Before I went to a hospital. And there, well . . . well, what he learned about me helped others, too. And I'm glad about that.

My grandfather's home was full of curious artifacts, each with its own story. He traveled a lot, all over the world, and always came home with something new and strange. It was a lifelong occupation: in the mid-1920s, he took what was in those days a very unusual trip to China and ended up paying some of his college tuition by selling the exotic jewelry he brought back to the States. Many of the items in his home looked as if they belonged in a cabinet of curiosities. There was a carved wooden totem hanging on one wall of his dining room, some sort of pagan king or god. It was maybe three feet tall and had a soulful, mournful expression. One Thanksgiving, when I asked where it came from, I was told he'd received it during a trip to South America, perhaps in gratitude for some operation he'd performed there. The carving had apparently once been an object of worship, owing mainly to the fact that it would, at times unpredictable, weep, drops of water streaming from its eyes. Did seasonal moisture variations and the way the wood responded to them cause the tears? Probably. That or magic. My grandfather had appreciated the totem's beauty but was unsentimental about its emotions. When he brought it home, he had someone polish it before he hung it on the wall. It never cried again.

Brain surgery is an ancient craft — there is a four-thousand-year-old hieroglyphic text describing successful operations — and among my grandfather's most interesting artifacts was a collection of premodern and tribal neurosurgical instruments. As a kid, I found those picks and blades fascinating and terrible to contemplate. It wasn't just the age of the tools, it was the acts they were intended for. Brain surgery, whatever the era, always requires at least two frightening qualities in its practitioners: the will to make forcible entry into another's skull and the hubris to believe you can fix the problems inside.

He was always a risk taker. During medical school he'd climbed, on a dare, one of the suspension cables of the still-under-construction George Washington Bridge and had spent the night shivering in a crate up top, waiting for the dawn. Once, while attending a neurosurgical conference in Spain, he visited a small bullring where toreadors were practicing their craft, threw off his jacket, and stepped into the ring himself. He loved cars, loved driving them as fast as they'd go. He loved tinkering with them, too. He always told people he'd have been a mechanic if he hadn't become a brain surgeon.

In 1939, a year after he completed a residency at Mass General, he cofounded the Department of Neurosurgery at Hartford Hospital. It was an interesting time for an ambitious young doctor to be entering his particular field, since a new kind of brain operation was on the verge of becoming a worldwide sensation. The operation had originated in Europe, pioneered by Egas Muniz, a Portuguese surgeon, who would go on to win a Nobel Prize for his efforts. Dr. Muniz called his operation leukotomy, but in the United States it came to be known by another name: lobotomy. The operation was crude—the surgeon would drive a skinny ice pick up through the thin bone behind his patient's eyeballs, then quickly swish the pick back and forth, cutting a messy swath through the frontal lobes—but it was inarguably effective at transforming otherwise difficult and uncontrollable individuals into placid, carefree creatures.

My grandfather had no problem with the basic idea of treating psychiatric problems through brain surgery. He believed that psychosurgical procedures might eventually, as he once wrote, relegate "psychoanalysis to that scientific limbo where perhaps it belongs." But he didn't like the lobotomy. It was too blunt, too imprecise, knocked everything out, the good with the bad. It makes people "easy to control and easy to handle," he said, "but, God forbid, at what a cost!" There had to be a better way. Some way to achieve the lobotomy's beneficial effects—the calming, the reduction in anxiety—without the lobotomy's attendant insidious, zombielike stupor.

So he tinkered. He began experimenting, at asylums in and around Hartford, with procedures he called "fractional lobotomies," attempting to target only the specific brain structures he believed were implicated in a particular patient's problems. Soon he was reporting a "most gratifying improvement in depressions, psycho-neuroses and tension states without any gross blunting of personality." And he mused about possible future advances and refinements to his approach, wondering whether his experiments in targeted brain lesioning might "bring us one blind step nearer" to the locations of the "fundamental mechanisms of mental disease and of epilepsy."

Around this time, Henry showed up in his office for a consultation.

When Henry Molaison is seven, a bicyclist collides with him, knocking him out cold for five minutes, leaving him with a deep gash in the left side of his forehead.

After that, the seizures begin.

They are minor at first but gradually increase in severity. At the age of sixteen, he experiences the first seizure of the type that used to be called grand mal and is now called tonic-clonic. He attends Windham High School in Willimantic, Connecticut, is smart, sharper than most. But the other students—watching him seize up ten or more times a day—tease him without mercy. Henry drops out of high school for a while but eventually goes back, struggles his way through to a diploma.

The seizures keep getting worse. He is in his early twenties, foggy-headed, living at home with his parents, no social life, barely holding down a job as a motor winder, when he has his first appointment with my grandfather.

My grandfather prescribes him Dilantin, phenobarbital, Tridione, Mesantoin, the best anticonvulsants of the day, in maximal doses. They don't work.

Henry's epilepsy is severe, horrible, intractable.

My grandfather suggests they try something else. Henry and his parents agree.

They are willing to try anything.

August 25, 1953. Henry lies on his back on an operating table in the Hartford Hospital neurosurgery suite. At the head of the table, flanked by scrub nurses and assistants, my grandfather leans over Henry with a trepan in his hand. Henry has been sedated and given a local anesthetic, and the flesh has been peeled down from his forehead, but he is conscious. A trepan is a sort of wide-mouthed serrated drill. The particular trepan he's using, like a lot of his surgical instruments, and like the operation itself, is of his own invention. To make this trepan, he bought a hole saw from a local auto-supply or machine shop for about a dollar, then attached it to a standard Hudson drill handle, the kind you crank by hand. Now he places the trepan down on Henry's exposed skull, just above one of his eye sockets, and bears down. It grinds through, extracting a button of bone roughly the size of a poker chip. He moves the trepan over a few inches, to the same spot above the other eye socket, and repeats the process. Scrub nurses flush the holes, exposing Henry's pale frontal lobes to the harsh light of my grandfather's headlamp. He puts the trepan aside, picks up a slightly curved metal retractor, inserts it into one of the holes, then levers

Henry's frontal lobe up and out of the way, so he can reach the hidden structures deeper in the recesses of Henry's brain.

Once he finds what he's looking for, the sea horse–shaped hippocampus and its adjacent organs—the hook-shaped uncus, the almond-shaped amygdala—he proceeds with the extraction. He has performed variations of this operation on a number of asylum residents but never on an epileptic. Still, he knows that the hippocampus—an organ whose precise function is a mystery to him, as it is to everyone else in 1953—has been implicated in some forms of epilepsy, and other surgeons have had success reducing seizures by removing one half of it, performing a so-called unilateral resection. With Henry, however, he decides to see what happens if he takes out both sides, not just one. He uses an electric cautery to cut the tissue and a skinny vacuum to suck it out, leaving behind just a negligible stump. Then he repeats the procedure on the other side. In all, taking both hemispheres into account, he removes several tablespoons of neuronal tissue.

Before he plugs the holes, he uses another tool to snap a few tiny metal clips onto the frayed far frontier of the fresh lesion. The operation is, as he will later write, a "frankly experimental" one, and these clips, visible in X-rays, will help him document its parameters.

Researcher: Do you know what you did yesterday?
Henry: No, I don't.
R: How about this morning?
H: I don't even remember that.
R: Could you tell me what you had for lunch today?
H: I don't know, to tell you the truth.

April 25, 1955, the night train from Montreal to Hartford. Brenda Milner, a young Cambridge-educated neuroscientist from McGill University, tries to sleep in the hurtling darkness. She's on her way to investigate a curious case of amnesia. Over the last year or so, she's begun to explore the mystery of how memory works. She knows that the scientific consensus is that memory is a diffuse phenomenon, one that can't be tracked to any particular part of the brain, but she's noticed that people with damage to their hippocampus sometimes have a particularly hard time remembering things. She has a hunch that the hippocampus plays some role in memory formation, and the patient she's going to see now will allow her to explore this hunch as never before: his hippocampus isn't just damaged, it's gone. From what she's been told, a surgeon, in a novel procedure, essentially removed it.

The next morning, in my grandfather's office, Milner meets Henry for the first time. He's a good-looking kid with hazel eyes, cordial, polite, quick with a joke, easygoing. They chat for a little while, and it strikes her that you might never know, after just a short conversation, that there was anything wrong with him. And then Milner steps out of the office for a cup of coffee, comes back a few minutes later, and meets Henry for the first time all over again.

Later my grandfather explains to Milner about the thirty or so other patients who've received variations of the operation he performed on Henry. These "severely deteriorated cases" were mostly schizophrenics and are scattered at various asylums around the state. He hasn't done much follow-up work on them and isn't sure if they've suffered the same sorts of memory deficits as Henry. He arranges a car and driver for Milner, so she can visit some of them. For the next few days, she descends into the back wards of one state institution after another. She has never been to an asylum before. She takes careful notes.

Patient A. Z.: "35-year-old woman, a paranoid schizophrenic . . . tense, assaultative, and sexually preoccupied . . ."

Patient I. S.: "54-year-old woman . . . auditory hallucinations and marked emotional lability . . ."

Patient A. L.: "31-year-old schizophrenic man . . . auditory and visual hallucinations . . ."

Patient M. B.: "55-year-old manic depressive woman . . . anxious, irritable, argumentative, and restless . . ."

And on and on, quick sketches of profound damage. Many of these cases interest Milner—particularly Patient D. C., a brilliant and psychotic medical doctor who attempted to kill his wife—but she returns from her asylum odyssey convinced of Henry's singular importance. Though some of the asylum patients clearly suffer from a similar amnesia to Henry's, Henry uniquely combines a near-complete resection of both hemispheres of the hippocampus with a mind not otherwise muddied by mental illness. The only other patient who received an operation identical to Henry's—a "radical bilateral medial temporal lobe excision (with the posterior limit of removal 8 cm from the temporal tips)"—is so deeply disturbed that nobody even noticed her inability to create new memories until nearly a year after her surgery.

Henry is the one.

The testing begins.

In 1848 an explosion drives a steel tamping bar through the skull of a twenty-five-year-old railroad foreman named Phineas Gage, obliterating a portion of his frontal lobes. He recovers and seems to possess all his earlier faculties, with one exception: the formerly mild-mannered Gage is now something of a hellion, an impulsive shit-starter. Ipso facto, the frontal lobes must play some function in regulating and restraining our more animalistic instincts.

In 1861 a French neurosurgeon named Pierre-Paul Broca announces that he has found the root of speech articulation in the brain. He bases his discovery on a patient of his, a man with damage to the left hemisphere of his inferior frontal lobe. The man comes to be known as Monsieur Tan, because, though he can understand what people say, "tan" is the only syllable he is capable of pronouncing.

Thirteen years later, Carl Wernicke, a German neurologist, describes a patient with damage to his posterior left temporal lobe, a man who speaks fluently but completely nonsensically, unable to form a logical sentence or understand the sentences of others. If Broca's area, as the damaged part of Monsieur Tan's brain came to be known, is responsible for speech articulation, then Wernicke's area must be responsible for language comprehension.

And so it goes. The broken illuminate the unbroken.

There is a word neuroscientists use to describe Henry. They call him pure. The purity in question doesn't have anything to do with morals or hygiene. It is entirely anatomical. My grandfather's resection produced a living, breathing test subject, Patient H. M., who will allow scientists to study how memory works with, as Brenda Milner will write, "the exactness of a planned experiment." And it is unlikely that a patient like Henry could arise without the willful act of a surgical procedure. Another scientist who later studies Henry discounts the possibility, for example, that a soldier might wind up with a brain similar to Henry's by being shot in the head:

"To get a pure one would be rare. Because think about what it would take to blow out both hippocampi. You'd be dead. I think it would be most compatible with not being alive."

Henry is the purest, and his purity makes him valuable.

Shortly after Milner's first visit with Henry, she and my grandfather begin coauthoring a paper, "Loss of Recent Memory After Bilateral Hippocampal Lesions," which appears in 1957 in the *Journal of Neurology, Neurosurgery and Psychiatry*. The paper, which formally introduces the scientific world to Patient H. M., reaches a deceptively straightforward conclusion: since my grandfather removed Henry's hippocampus and Henry no longer seems capable of creating new memories, the hippocampus must play an essential role in creating new memories.

That paper—which for the first time convincingly localizes memory function to a specific part of the brain—becomes the founding text of modern memory science.

But Milner continues riding the night train down to Hartford, spending days at a time with Henry. Her initial focus is on probing the depths of his memory dysfunction, the way experiences seem to slip *away* from him, leaving no trace. Then she begins investigating whether there are any experiences at all that he can remember.

One afternoon she sits Henry down at a desk and puts a piece of paper in front of him. The paper has a large drawing of a five-pointed star on it. There is a mirror angled at the star and a curtain over the paper so that Henry can no longer see the star directly but can see only its reflection in the mirror. She asks him to trace the star. It's a hard task for anyone, with any sort of brain, though after a while, with practice, people with normal brains tend to improve their results, mastering the necessary counterintuitive muscle movements. The first time Henry tries it, he performs poorly. But the funny thing is, the next time he tries, he does it a little better. And the next time better still. With each new attempt, he never remembers ever having attempted it before, but soon he's completing the task as well as anyone. Even Henry recognizes the strangeness of this.

"I thought this would be difficult," he says to Milner after tracing the star almost perfectly, "but I seem to have done this well."

In 1962 Milner publishes her account of the star-tracing experiment. That article arrives at another simple but revolutionary conclusion: since Henry is incapable of consciously remembering the events he's living through but seems to be able to unconsciously remember how to perform certain physical tasks, the human mind must contain at least two separate and independent memory systems. The world of memory science is upended again.

Less than a decade after his surgery, the study of Henry's incomplete brain has already spawned the century's two pivotal insights into how memory works. Milner has revealed him to be a sort of human Rosetta stone, an incarnate key to ancient mysteries.

Henry himself, of course, is oblivious to his own burgeoning importance. After the operation, he continues to live with his parents. The operation did succeed in reducing the frequency and severity of his seizures, and Henry is perfectly capable of helping out with tasks around the house. He can mow the lawn just fine, since the cut grass itself shows him what he has already cut and what he has not. His father, an electrician, dies in 1967, when Henry is forty-one years old. Henry goes back to work, packing lighters at a rehabilitation center designed to provide constructive occupations for mentally retarded people. A few years after that, he and his mother, whose health is failing, move into the home of an aunt. In 1977 Henry's mother moves into a nursing home. Henry lives alone with his aunt until 1980. When she becomes too ill to care for him, he moves into the Bickford Health Care Center. A conservator is appointed, a cousin of Henry's, and that conservator, like Henry's guardians before him, signs waivers allowing the scientists to continue their work.

Milner spends nearly two decades studying Henry. Then she moves on, pursues other research interests, lets other scientists pick up with Henry where she left off.

Researcher: Are you happy?
Henry: Yes. What I, well, the way I figure it is, what they find out about me helps them to help other people. And that's more important. That's what I thought about being, too. A doctor.
R: Really?
H: A brain surgeon.
R: Tell me about that.
H: And as soon as I started to wear glasses, in a way, I said no to myself.

R: Why?

H: Because [during] brain surgery you have bright lights, and I thought there would be an extra glare from the rim of your glasses that would go into your eye. And right at the moment you want to make the great reseverance, the most important severance, you just might move a little too far . . .

Jacopo Annese usually drives a Porsche, but today, summer 2006, he's a passenger in a sedan watching the desiccated red-brick husks of old paper mills glide by his window. Main Street, Windsor Locks, Connecticut, just a few blocks from the Bickford Health Care Center, where he's going to meet Henry for the first time. He's a little surprised, to be honest, that Suzanne Corkin, who's driving, didn't make him wear a bag over his head. Corkin, a former graduate student of Brenda Milner's at McGill, who runs the Behavioral Neuroscience Laboratory at MIT, took over as the lead investigator on all things Henry-related in the late 1970s. She is known to be fiercely protective of her prized test subject, vetting researchers exhaustively, demanding the signing of nondisclosure contracts, disallowing tape recorders, that sort of thing. She's built a good portion of her career on her special access to Henry, and she won't let just anybody in.

But when Annese requested that she set up this meeting, when he told her that he'd really like to see Henry at least once while Henry was still alive, she consented, arranged it, even chauffeured him.

Corkin first met Henry at Brenda Milner's lab in Montreal in 1962, and over the years, as the mining of his mind has continued, she's witnessed firsthand how Henry continues to give up riches, broadening our understanding of how memory works. But she's also keenly aware of Henry's enduring mysteries, has documented things about him that nobody can quite explain, not yet.

For example, Henry's inability to recall postoperative episodes, an amnesia that was once thought to be complete, has revealed itself over the years to have some puzzling exceptions. Certain things have managed, somehow, to make their way through, to stick and become memories. Henry knows a president was assassinated in Dallas, though Kennedy's motorcade didn't leave Love Field until more than a decade after Henry left my grandfather's operating room. Henry can hear the incomplete name of an icon— "Bob Dy . . ."—and complete it; even though in 1953 Robert Zimmer-

man was just a twelve-year-old chafing against the dead-end mo-
notony of small-town Minnesota. Henry can tell you that Archie
Bunker's son-in-law is named Meathead.

How is this possible?

And Corkin has discovered a number of other curious, anoma-
lous, presently inexplicable things about Henry. For instance, if
she holds a dolorimeter, a handheld, gun-shaped device housing
a hundred-watt bulb, to the underside of Henry's forearm, he is
able to tolerate the resulting pain for longer periods of time than
most people. Is he naturally, congenitally pain-resistant? Or did my
grandfather's removal of his amygdala, an organ that sits just in
front of the hippocampus and is suspected to mediate both emotion and pain, account for his toughness?

And why, as Corkin has also discovered, is Henry able to recognize the intensity of smells but not their provenance, unable to discriminate the fragrance of a rose from the stench of dog shit? Is
the hippocampus responsible for smell identification as well? Or
had my grandfather's "suction catheter" inadvertently damaged
Henry's nearby olfactory bulb as well?

You can spend a half century testing somebody, examining, pok-
ing, prodding, feeding—Henry will happily eat at least two full
dinners in a row if you give him a minute between removing the
first tray and replacing it with the second—and you can come up
with all sorts of theories to explain your findings. You can even
throw a person in an MRI machine, study the flickering images on
your computer screen. But the brain is the ultimate black box.
Eventually, to grasp the first cut, you'll have to make another.

The car pulls into the parking lot of the nursing home, noses
into an empty space. Annese and Corkin get out and walk inside
together. Henry's waiting. The three have lunch in the cafeteria.

He's an old man now, overweight, wheelchair-bound, largely in-
communicative. Lately, Henry's creeping decrepitude has itself
suggested some new experiments. During another meeting, Corkin quizzed Henry on how old he thought he was. He guessed that
he was perhaps in his thirties. Then she handed him a mirror.

"What do you think about how you look?" she asked while he
stared at himself.

"I'm not a boy," he said eventually.

Still, despite Henry's current condition, his lack of engagement,
Annese is glad to get a chance to meet him, to spend at least a little

time around him. Ever since graduate school, he's always found the anonymous cadavers the hardest. It makes it much easier, for some reason, if you know something about the person as a person before you deal with the person as a corpse.

On the way out, Annese notices a snapshot of Henry tacked to one of the bulletin boards near the entrance. Nobody's looking, and he has to resist the temptation to take it, to slip it into his pocket, to keep it as a sort of totem, something he could ponder in his off-hours, something he could use to help him imagine his way further into Henry's mind before the day he has to start digging into his brain.

Sometime around 1969, in the middle of taking some tests, Henry suddenly stopped and looked up at his examiner.

"Right now," he said, "I'm wondering, have I done or said something amiss? You see, at this moment, everything looks clear to me, but what happened just before? That's what worries me. It's like waking from a dream."

He might have said the same thing, felt the same way, in 1979, 1989, 1999. Every day, every waking hour. The world around him evolving, changing, growing, while the world within him cycles endlessly, fruitlessly. Memory is the traction of our lives. Without it, you can't move. You're nowhere at all, really. Like when you're just waking from a dream.

Some researchers once decided to find out what would happen if you actually did wake Henry from a dream. He spent several nights in a sleep laboratory, hooked up to sensors. Whenever he entered REM sleep, a researcher would shake him till his eyes opened and then ask him what he'd been dreaming about. He usually just talked about the same sorts of things he liked to talk about when he was awake — childhood memories of a road trip to Florida with his parents, of shooting targets in his backyard, of fishing with his dad. In the end, the researchers never published their dream studies, because nobody could decide whether Henry was dreaming at all, whether he was even capable of dreaming. But some of the transcripts survived.

    Henry, Henry, Henry!
    Oh!
    Were you dreaming?

Yeah.

What were you dreaming about?

I was having an argument with myself . . .

About what?

What I could have been . . . I dreamed of Pennsylvania. I dreamed of being a doctor. A brain surgeon. And it was all quick. Flashlike, being successful. And living down that way . . . tall straight trees.

My grandfather grew up in Pennsylvania, got his M.D. at U Penn. And yes, he was successful.

You could argue that Henry gives his career, like the careers of many other people, a boost. Gets his name out there, at the top of one of the most important scientific papers of the era and in the citations of thousands of others. He goes on to become clinical professor of neurosurgery emeritus at Yale and serves as president of both the American Academy of Neurological Surgery and the World Federation of Neurosurgical Societies.

Unlike Henry, he can move on. New patients, new procedures, a life steadily unfurling. A sustained narrative. Henry is an important episode in his past, but other episodes follow.

Still, he never performs another operation like the one he performed on Henry, and his characteristic hubris is, I like to think, tempered by a deeper appreciation for the dangers inherent in opening a man's skull. In 1973, during a conference about the ethics of brain surgery, he listens while a younger colleague of his, Dr. José Delgado, a professor of neurophysiology at Yale, advocates for the widespread use of corrective neural implants. Dr. Delgado, in an earlier publicity stunt, stopped a charging bull in its tracks by remotely activating electrodes he'd planted in its brain. "The question," Dr. Delgado declares, "rather than, What is man? should be, What kind of man are we going to construct?"

"With all due respect to Dr. Delgado," my grandfather responds, "I work almost wholly on humans, and we are more aware of the disastrous effects that sometimes occur in neurosurgery."

He died in a car crash on the New Jersey Turnpike eleven years later, in 1984. He was seventy-eight years old and still operating on patients at least once a week.

Winter 2008. New employees at the Bickford Health Care Center always receive a briefing on Henry and his special circumstances.

For example, they are directed never to speak to anyone outside the center about Henry, as the fact that he resides there is a closely guarded secret. If a stranger calls inquiring after Henry, the staff member receiving the call is supposed to give a noncommittal response neither confirming nor denying his presence and then immediately phone Henry's conservator, warning him about the snoop. The cloak of anonymity placed over Henry is effective: he has lived at the center for decades, and though he is the most famous patient in the history of neuroscience, no outsider has ever found him.

New staff are also briefed on the special rules that apply specifically to Henry's dying and his death. Suzanne Corkin drafted these rules, and the rules are printed out and always attached to Henry's chart.

So on the morning and afternoon of December 2, as Henry froths and gasps, fading from respiratory failure at age eighty-two, Corkin, as per protocol, receives periodic phone calls, keeping her abreast of the situation.

When his heart finally stops, a final call is placed to Corkin, and then someone at the center rushes to the freezer and digs out the flexible Cryopaks that were placed there in anticipation of this moment. By the time the hearse arrives, the ice blankets are wrapped securely around Henry's head, keeping his brain nice and cool.

Everything goes smoothly, according to plan, and a couple of hours and 106 miles later, the hearse pulls into the parking lot of Building 149 in the Charlestown Navy Yard, the Athinoula Martinos Center for Biomedical Imaging, in Boston, where Corkin is waiting.

The body bag is unzipped and the ice blankets unwrapped. Corkin has known Henry for forty-six years, having met him for the first time when she was still a graduate student in Brenda Milner's lab at McGill. Of course, her relationship with Henry, transactionally speaking, was just like Milner's: she knew Henry, not the other way around. Forty-six years of meeting someone for the first time, introducing yourself to an old friend. And now this last meeting that only she will remember.

During the night that follows, Corkin watches as Henry undergoes a series of high-resolution MRI scans. Then, the next morning, she attends the harvesting. She stands on a chair outside the autopsy room in the Mass General pathology department and peeks in through a window as Jacopo Annese and two other men saw off the top of Henry's skull and, with the care of obstetricians delivering a baby, pull his brain into the light. Corkin has spent the bulk of her career pondering the inner workings of Henry's brain. Now she finally gets to see it.

Later the brain sits for a while inside a bucket that's inside a cooler, steeping in a preservative solution, hanging upside down, suspended by a piece of kitchen twine looped through its basilar artery. Eventually, when the brain is firm enough to travel safely, Corkin rides to Logan Airport with the cooler. She accompanies it to the gate of a JetBlue flight from Boston to San Diego. There are cameramen following her. It's a self-consciously historic moment. She already has a book and movie deal. She puts the cooler down and Annese picks it up. Corkin thinks of Henry's brain as a treasure. She watches Annese walk down the ramp with it. She watches them disappear.

It's hard to let go.

There's a slim book sitting on a shelf right next to Annese's desk in his glass-walled office at the Brain Observatory in San Diego. Unlike a lot of the other books in this place—*A Study of Error, Serial Murder Syndrome, Man and Society in Calamity, Flesh in the Age of Reason, The Open and Closed Mind*—this one doesn't have a very lively title. But *Localization in the Cerebral Cortex,* by Korbinian Brodmann, is, to Annese, as vital a book as can be. It was originally published in 1909 and contains a series of meticulous hand-drawn maps of the human brain, divided up into fifty-two so-called Brodmann areas, each unique in its neuronal organization and, consequently, its function. Brodmann gleaned the borders of his areas through a rough and painstaking combination of microscopy and histology, and he did a great job, all things considered. He created an enduring Rand McNally road atlas of the mind, one that my grandfather used to direct his surgeries and one that most neuroscientists and neurosurgeons still use today.

As a fellow anatomist, Annese admires Brodmann's work immensely and recently wrote a glowing tribute to him that appeared in the journal *Nature.* But he will soon make Brodmann's old maps irrelevant.

That's what the Brain Observatory is all about, really.

If Korbinian Brodmann created the mind's Rand McNally, Jacopo Annese is creating its Google Maps.

A short walk from Annese's office, past an imported espresso machine and through a secure door, a number of tall, glass-fronted refrigerators stand against a wall. Many of these refrigerators contain plastic buckets, and though the plastic is murkier than the glass, you can still see what's inside. Most of the brains are human, but there is one from a dolphin, too. The dolphin brain is huge, significantly bigger than any of the human ones, though Annese cautions that it would be a mistake to read too much into size.

And what's true of individual brains is true of brain collections as well. With his Brain Observatory, Annese is setting out to create not the world's largest but the world's most useful collection of brains. Each specimen will, through a proprietary process developed by Annese, be preserved in both histological and digital form, at an unprecedented, neuronal level of resolution. Unlike Brodmann's hand-drawn sketches, Annese's maps will be three-dimensional and fully scalable, allowing future neuroscientists to zoom in from an overhead view of the hundred-billion-neuron forest all the way down to whatever intriguing thicket they like. And though each individual brain is by definition unique, as more and more brains come online, both the commonalities and differences between them should become increasingly apparent, allowing, Annese hopes, for the eventual synthesis of the holy grail of any neuroanatomist: a modern multidimensional atlas of the human mind, one that conclusively maps form to function. For the first time, we'll be able to meaningfully and easily compare large numbers of brains, perhaps finally understanding why one brain might be less empathetic or better at calculus or likelier to develop Alzheimer's than another. The Brain Observatory promises to revolutionize our understanding of how these three-pound hunks of tissue inside our skulls do what they do, which means, of course, that it promises to revolutionize our understanding of ourselves.

And what could be a better cornerstone for the Brain Observatory, what could be a better first volume for Annese's collection, than the infinitely pored-over brain of Patient H. M.? The boxes containing the cryogenic vials containing the slices of Henry's brain sit in their own freezers, to the left of the others, under lock and key. Precious cargo. San Diego is earthquake-prone, but there are backup generators and sensors that will automatically dial Annese's home and cell phones in the event of an emergency, so that wherever he is, he can jump into his Porsche, rush over, protect Henry.

Henry, just by being Henry, helps bring Annese's larger ambitions closer to reality. People who've read newspaper articles about Annese's work with Henry's brain have already called him up, made direct arrangements to donate their own. One of them, Bette, a feisty eighty-one-year-old who was one of the original flying monkeys in *The Wizard of Oz,* will be dropping by the Brain Observatory soon to get her second set of MRI scans. Annese knows the publicity will continue, hopes it will continue to inspire donations. He had wanted to get the brain of the guy *Rain Man* was based on, but that hadn't worked out. Eventually he'd like to get somebody really big, a household name, Bill Clinton, someone like that.

But Henry's brain is more than just an attention-garnering curiosity. And it's more than just a proof of concept, something Annese can use to demonstrate to the world the power of his methods.

It's an object—2,401 objects now—that contains enduring mysteries still waiting to be solved.

I remember following my grandfather up a hill. He was usually a sharp dresser—a *New York Times* reporter once described him as "almost unreal in his dashing appearance"—but on that particular morning I believe he wore a simple gray scarf, a floppy blue hat, and a threadbare ski suit. He was hauling a wooden toboggan behind him. It was either Christmas or Thanksgiving, one of the two holidays when the whole family would get together at his artifact-stuffed house in Farmington, Connecticut. I don't remember the slide down the hill, just the walk up.

I remember midway through one Christmas dinner, maybe his last one, when he pushed himself up from his chair at the head of the table, wandered back to his study, and came back a few minutes later with a crumpled bullet in his hand. He placed the slug down beside his plate, told us the story behind it. Stamford, Connecticut, turn of the century, a burglar breaks into the home of a young bachelor. The bachelor keeps a pistol by his bedside table but his pistol jams. The burglar's doesn't. A bullet enters the bachelor's chest, where it encounters a deflecting rib, skids away from a lucky heart. The bachelor survives and keeps the bullet as a memento. He eventually passes it down to his son.

The bullet just sat there for the rest of the dinner, beside my grandfather's plate, and like some of the other artifacts in his home, it was both fascinating and terrible to contemplate. Had it found its target, had its aim been true, then my grandfather, his children, his children's children, most of the people sitting around the table, myself included, would have never existed. It was a matter of centimeters, a fluke of aim, bone, ballistics, and it had made all the difference, its repercussions rippling down through generations.

An intact human brain sits on a small rectangle of green marble, under a sheet of saran wrap. Annese is wearing khakis and a black button-down shirt, plus a medical smock, blue rubber gloves, and safety glasses. Brains can be full of pathogens. He plucks off the saran wrap, picks up a scalpel, and begins to peel away the pia mater, a thin, sticky membrane that covers the brain.

He had never met the owner of this brain while she was alive, but he had performed the harvesting himself, so he knew what she had looked like, at least. She was a small lady, in her seventies, had reminded him a bit of a friend of his, and he likes to imagine that their personalities were similar.

He'd performed Henry's peeling on this same table, on this same slab of formaldehyde-slicked marble. He'd been alone then, music on—the Beatles, mostly—and everything had gone perfectly. He had removed the oxidized clips my grandfather had left behind, set them aside. Then the membranes, the blood vessels, all the obstructing tissue, stripping everything away, until he had been left, finally, with just Henry's naked brain. Peelings usually last three or four hours, but with Henry he took his time, made sure everything was just right, peeled for five hours straight.

He'd still been on a bit of a high then, still sort of pinching himself. He was part of the group that Corkin convened years ago to decide what to do with Henry's brain postmortem, and so of course Annese knew that the group had decided to give it to him as the cornerstone of his brain observatory—the most important brain of the twentieth century catalyzing a new era of brain research in the twenty-first. But despite all the planning, all the verbal agreements, he hadn't been absolutely sure he'd wind up with his prize until he actually boarded the flight with the cooler in hand. A part of him, the fatalistic Italian part, thought that something would happen at the last minute, that Corkin would change her mind, take Henry back.

But she didn't.

He presented the gate agent with two tickets, boarded the plane with time to spare. He gave Henry the window seat. Words he'd black-Sharpied on the cooler, along with his phone numbers and e-mail address: DIAGNOSTIC SPECIMEN. FRAGILE. IF FOUND, PLEASE DO NOT OPEN. CONTACT DR. ANNESE IMMEDI-ATELY.

And from then on, everything went just as smoothly as could be. After the peeling, he embedded Henry's naked brain in gelatin, froze it solid. The embedding was, he's not too modest to say, a masterpiece. And the embedding set the stage for the slicing, which, despite the pressure, the sleep deprivation, the four hundred thousand pairs of eyes watching the procedure over the Internet, went off without a hitch.

There's still lots more work to be done. The mounting of about two thousand more slides. The staining of those same slides. But he's getting close. Soon, within months, Annese will release a three-dimensional surface model of Henry's brain, built from the 2,401 high-res "block face" images taken during the slicing. This model will be at least ten times as detailed as anything one could possibly produce in an MRI machine and will have the additional benefit of being derived from images of the actual brain rather than a computerized interpretation of it. And then, bit by bit, he'll supplement that model with imagery of even greater resolution. A custom-built microscope scanner will digitize each of the mounted, stained slides at such a level of magnification that single neurons will be clearly visible. All of this, the resulting petabyte or two, will be accessible for free online to researchers worldwide. Over the last fifty-five years of his life, Henry was hidden away while a select coterie of scientists gathered more data about him, his abilities and his deficits, than any human in history. Now, after his death, Annese is poised to release Henry's brain into the wilds of the Internet, and the whole world will be able to reillumine that unprecedented volume of clinical data in the light of an unprecedented neuroanatomical map.

One of Annese's assistants pokes her head into the room, tells him that some more slides are ready for staining. A few minutes later, Annese holds one of the fresh-dipped seven-by-five-inch slides up to the light, letting a purplish dye drip off the glass. The dye has adhered to the slide's cross section of pale, almost invisible brain tissue, darkening it, developing it like a photograph.

A cross section of brain looks a lot like an inkblot, a Rorschach, and this one, at first glance, gives the impression of the head of a vaguely sinister goat. But then Annese starts guiding me through it.

"You can see here," he says, indicating a spot where the tissue looks darker, the neurons more cramped, "where your grandfather pushed up his frontal lobes."

We look at another slide, and he points to an area that would have sat a little below and back from Henry's frontal lobes, a portion of the slide where no dye stuck, since there was no tissue for it to adhere to. It's part of the lesion itself, the little bit of nothing that spawned everything. And though Annese doesn't want to go into too much detail, not before his findings are officially published, he tells me, sotto voce, that he's already discovered some surprising new things about what my grandfather destroyed in Henry's brain, and what he spared. For years, memory researchers have assumed that the stump that remained in Henry's brain was completely atrophic and nonfunctioning. According to Annese, however, that doesn't seem to be the case. The little that's left of Henry's hippocampus looks like it's in pretty good shape, actually.

It's the sort of revelation that could shake up the field of memory science yet again.

In 1953, when my grandfather closed a door in Henry's mind, did he leave it open just a crack? Does this explain the surprising exceptions to Henry's profound amnesia? So much of our understanding of how memory works is based on our understanding of how Henry's memory didn't work. But were we misunderstanding him, at least in part, all these years? These are the sorts of questions scientists will grapple with and argue over in the years to come, as the Brain Observatory goes online, as Henry's mind is preserved everywhere and nowhere at once, as his cells are counted and his final mysteries come to light.

Annese puts the slide on a rack to dry, and I look at it again, at the blank spot near the middle, the hole you can see right through. It's just a small emptiness, a tiny lacuna.

A matter of centimeters.

A beginning and an end.

JONATHAN FRANZEN

Emptying the Skies

FROM *The New Yorker*

THE SOUTHEASTERN CORNER of the Republic of Cyprus has
been heavily developed for foreign tourism in recent years. Large
medium-rise hotels, specializing in vacation packages for Germans
and Russians, overlook beaches occupied by sunbeds and umbrel-
las in orderly ranks, and the Mediterranean is nothing if not extremely blue. You can spend a very pleasant week here, driving the
modern roads and drinking the good local beer, without suspect-
ing that the area harbors the most intensive songbird-killing operations in the European Union.

On the last day of April, I went to the prospering tourist town
of Protaras to meet four members of a German bird-protection
organization, the Committee Against Bird Slaughter (CABS), that
runs seasonal volunteer "camps" in Mediterranean countries. Because the peak season for songbird trapping in Cyprus is autumn,
when southbound migrants are loaded up with fat from a northern
summer's feasting, I was worried that we might not see any action,
but the first orchard we walked into, by the side of a busy road, was
full of lime sticks: straight switches, about thirty inches long, that
are coated with the gluey gum of the Syrian plum and deployed
artfully, to provide inviting perches, in the branches of low trees.
The CABS team, which was led by a skinny, full-bearded young Italian named Andrea Rutigliano, fanned into the orchard, taking
down the sticks, rubbing them in dirt to neutralize the glue, and
breaking them in half. All the sticks had feathers on them. In a
lemon tree, we found a male collared flycatcher hanging upside
down like a piece of animal fruit, its tail and its legs and its black-

and-white wings stuck in glue. While it twitched and futilely turned its head, Rutigliano videoed it from multiple angles, and an older Italian volunteer, Dino Mensi, took still photographs. "The photos are important," said Alex Heyd, a sober-faced German who is the organization's general secretary, "because you win the war in the newspapers, not in the field."

In hot sunshine, the two Italians worked together to free the fly-catcher, gently liberating individual feathers, applying squirts of diluted soap to soften the resistant gum, and wincing when a feather was lost. Rutigliano then carefully groomed gum from the bird's tiny feet. "You have to get every bit of lime off," he said. "The first year I was doing this, I left a bit on the foot of one bird and saw it fly and get stuck again. I had to climb the tree." Rutigliano put the flycatcher in my hands, I opened them, and it flew off low through the orchard, resuming its northward journey.

We were surrounded by traffic noise, melon fields, housing de-velopments, hotel complexes. David Conlin, a beefy British mili-tary veteran, threw a bundle of disabled sticks into some weeds and said, "It's shocking—that you can stop anywhere around here and find these." I watched Rutigliano and Mensi work to free a second bird, a wood warbler, a lovely yellow-throated thing. It felt wrong to be seeing at such close range a species that ordinarily requires care-ful work with binoculars to get a decent view of. It felt literally dis-enchanting. I wanted to say to the wood warbler what St. Francis of Assisi is said to have said when he saw a captured wild animal: "Why did you let yourself be caught?"

As we were leaving the orchard, Rutigliano suggested that Heyd turn his CABS T-shirt inside out, so that we would look more like ordinary tourists taking a walk. In Cyprus it's permissible to enter any private land that isn't fenced, and all forms of songbird trap-ping have been criminal offenses since 1974, but what we were do-ing still felt to me high-handed and possibly dangerous. The team, in its black and drab clothing, looked more like commandos than like tourists. A local woman, perhaps the orchard's owner, watched without expression as we headed inland on a dirt lane. Then a man in a pickup truck passed us, and the team, fearing that he might be going ahead to take down lime sticks, followed him at a trot.

In the man's backyard, we found two pairs of twenty-foot-long metal pipes propped up in parallel on lawn chairs: a small-scale lime-stick factory of the sort that can provide good income for the mostly older Cypriot men who know the trade. "He's manufacturing them and keeping a few for himself," Rutigliano said. He and the others strolled brazenly around the man's chicken coop and rabbit cages, taking down a few empty sticks and laying them on the pipes. We then trespassed up a hillside and back down into an orchard crisscrossed by irrigation hoses and full of trapped birds. *"Questo giardino è un disastro!"* said Mensi, who spoke only Italian.

A female blackcap had torn most of its tail off and was stuck not only by both legs and both wings but also by the bill, which sprang open as soon as Rutigliano unglued it; it began to cry out furiously. When the bird was freed altogether, he squirted a little water in its mouth and set it on the ground. It fell forward and flopped piteously, pushing its head into the mud. "It's been hanging so long that its leg muscles are overstretched," he said. "We'll keep it tonight, and it can fly tomorrow."

"Even without a tail?" I said.

"Certainly." He scooped up the bird and stowed it in an outer pocket of his backpack.

Blackcaps are one of Europe's most common warblers and the traditional national delicacy of Cyprus, where they're known as *ambelopoulia*. They are the main target of Cypriot trappers, but the bycatch of other species is enormous: rare shrikes, other warblers, larger birds like cuckoos and golden orioles, even small owls and hawks. Stuck in lime in the second orchard were five collared flycatchers, a house sparrow, and a spotted flycatcher (formerly widespread, now becoming rare in much of northern Europe), as well as three more blackcaps. After the team members had sent them on their way, they wrangled about the tally of lime sticks at the site and settled on a figure of fifty-nine.

A little farther inland, in a dry and weedy grove with a view of the blue sea and the golden arches of a new McDonald's, we found one active lime stick with one living bird hanging from it. The bird was a thrush nightingale, a gray-plumaged species that I had seen only once before. It was deeply tangled in lime and had broken a wing. "The break is between two bones, so it cannot recover," Rutigliano said, palpating the joint through feathers. "Unfortunately, we need to kill this bird."

It seemed likely that the thrush nightingale had been caught on a stick overlooked by a trapper who had taken down his other sticks that morning. While Heyd and Conlin discussed whether to get up before dawn the next day and try to "ambush" the trapper, Rutigliano stroked the head of the thrush nightingale. "He's so beautiful," he said, like a little boy. "I can't kill him."

"What should we do?" Heyd said.

"Maybe give him a chance to hop around on the ground and die on his own."

"I don't think there's a good chance for it," Heyd said.

Rutigliano put the bird on the ground and watched as it scurried, looking more mouselike than birdlike, under a small thornbush. "Maybe in a few hours he can walk better," he said, unrealistically.

"Do you want me to make the decision?" Heyd said.

Rutigliano, without answering, wandered up the hill and out of sight.

"Where did it go?" Heyd asked me.

I pointed at the shrub. Heyd reached into it from two sides, captured the bird, held it gently in his hands, and looked up at me and Conlin. "Are we agreed?" he said, in German.

I nodded, and with a twist of his wrist he tore the bird's head off.

The sun had expanded its reach across the entire sky, killing its blue with whiteness. As we scouted for an approach from which to ambush the grove, it was already hard to say how many hours we'd been walking. Every time we saw a Cypriot in a truck or a field, we had to duck down and backtrack over rocks and pants-piercing thistles, for fear that somebody would alert the owner of the trapping site. There was nothing larger at stake here than a few songbirds, there were no land mines on the hillside, and yet the blazing stillness had a flavor of wartime menace.

Lime-stick trapping has been traditional and widespread in Cyprus since at least the sixteenth century. Migratory birds were an important seasonal source of protein in the countryside, and older Cypriots today remember being told by their mothers to go out to the garden and catch some dinner. In more recent decades, *ambelopoulia* became popular with affluent, urbanized Cypriots as a kind of nostalgic treat—you might bring a friend a jar of pickled birds as a house gift, or you might order a platter of them fried in a restaurant for a special occasion. By the mid-1990s, two decades after the country had outlawed all forms of bird trapping, as many as ten million songbirds a year were being killed. To meet the restaurant demand, traditional lime-stick trapping had been augmented by large-scale netting operations, and the Cypriot government, which was trying to clean up its act and win membership in the European Union, cracked down hard on the netters. By 2006 the annual take had fallen to around a million.

In the past few years, however, with Cyprus now comfortably ensconced in the EU, signs advertising illicit *ambelopoulia* have begun to reappear in restaurants, and the number of active trapping sites is rising. The Cypriot hunting lobby, which represents the republic's fifty thousand hunters, is this year supporting two parliamentary proposals to relax antipoaching laws. One would reduce lime-stick use to a misdemeanor; the other would decriminalize the use of electronic recordings to attract birds. Opinion polls show that while most Cypriots disapprove of bird trapping, most also don't think it's a serious issue, and many enjoy eating *ambelopoulia*. When the country's Game Fund organized raids on restaurants serving the birds, the media coverage was roundly negative, leading with an account of food being pulled from the hands of a pregnant female diner.

"Food is sacred here," said Martin Hellicar, the campaigns manager of BirdLife Cyprus, a local organization more averse to provocation than CABS is. "I don't think you'll ever get someone convicted for eating these things."

Hellicar and I had spent a day touring netting sites in the country's southeast corner. Any small olive grove can be used for netting, but the really big sites are in plantations of acacia, an alien species there's no reason to irrigate if you're not trapping birds. We saw these plantations everywhere. Long runners of cheap carpeting are laid down between rows of acacias; hundreds of meters of nearly invisible mist nets are strung from poles that are typically anchored in old car tires filled with concrete; then, in the night, birdsong is played at high volume to lure migrants to rest in the lush acacias. In the morning, at first light, the poachers throw handfuls of pebbles to startle the birds into the nets. (A telltale sign of trapping is a mound of these pebbles dumped by the side of the road.) Since it's a superstition among poachers that letting birds go free ruins a site, the unmarketable species are torn up and dropped on the ground or left to die in the nets. The marketable birds can fetch up to five euros apiece, and a well-run site can yield a thousand birds or more a day.

The worst area in Cyprus for poaching is the British military base on Cape Pyla. The British may be the bird-lovingest people in Europe, but the base, which leases its extensive firing ranges to Cypriot farmers, is in a delicate position diplomatically; after one recent enforcement sweep by the army, twenty-two Sovereign Base Area signs were torn down by angry locals. Off the base, enforcement is hampered by logistics and politics. Poachers employ lookouts and night guards and have learned to erect little shacks on their sites, because Game Fund officers are required to get a warrant to search any "domicile," and in the time it takes to do this the poachers can take down their nets and hide their electronic equipment. Because large-scale poachers are nowadays straight-up criminals, the officers are also afraid of violent attacks. "The biggest problem is that no one in Cyprus, not even the politicians, comes out and says that eating *ambelopoulia* is wrong," the director of the Game Fund, Pantelis Hadjigerou, told me. Indeed, the record holder for most *ambelopoulia* eaten in one sitting (fifty-four) was a popular politician in north Cyprus.

"Our ideal would be to find a well-known personality to come out and say, 'I don't eat *ambelopoulia*, it's wrong,'" the director of BirdLife Cyprus, Clairie Papazoglou, said to me. "But there's a little pact here that says that if anything bad happens it has to stay on the island, we can't look bad to the outside world."

"Before Cyprus joined the EU," Hellicar said, "the trappers said, We'll pull back for a while. Now, for eighteen- and nineteen-year-olds, there's a kind of patriotic machismo to poaching. It's a symbol of resistance to Big Brother EU."

What seemed Orwellian to me was Cyprus's internal politics. It's been thirty-six years since Turkey occupied the northern part of the island, and the ethnically Greek south has prospered immensely since then, but the national news is still dominated, seven days a week, by the Cyprus Problem. "Every other issue is swept under the carpet, everything else is insignificant," the Cypriot social anthropologist Yiannis Papadakis told me. "They say, 'How dare you take us to European Court for something as stupid as birds? We're taking Turkey to court!' There was never any serious debate about joining the EU—it was simply the means by which we were going to solve the Cyprus Problem."

The European Union's most powerful instrument of conservation is its landmark Birds Directive, which was issued in 1979 and requires member states to protect all European bird species and preserve sufficient habitat for them. Since joining the EU in 2004, Cyprus has received repeated warnings from the European Commission for infringement of the directive, but it has so far avoided judgments and fines; if a member state's environmental laws accord on paper with the directive, the commission is reluctant to interfere with sovereign enforcement.

Cyprus's nominally Communist ruling party ardently embraces private development. The tourism ministry is touting plans to build fourteen new residential golf complexes (the island currently has three), even though the country has very limited supplies of fresh water. Anyone who owns land reachable by road can build on it, and, as a result, the countryside is remarkably fragmented. I visited four of the southeast's most important nature preserves, all of them theoretically due special protection under EU regulations, and was uniformly depressed by their condition. The big seasonal lake at Paralimni, for example, near where I was patrolling with the CABS people, is a noisy dust bowl commandeered for an illegal shooting range and an illegal motocross course, carpeted with shotgun shells, and extensively littered with construction debris, discarded large appliances, and household trash.

And yet birds still come to Cyprus; they have no choice. Returning to town at some less white-skied hour, the CABS patrol stopped to admire a black-headed bunting, a jewel of gold and black and chestnut, that was singing from the top of a bush. For a moment our tension abated, and we were all just bird watchers exclaiming in our native languages.

"*Ah, che bello!*"

"Fantastic!"

"*Unglaublich schön!*"

Before we quit for the day, Rutigliano wanted to make one last stop, at an orchard where the previous year a CABS volunteer had been roughed up by trappers. As we were turning, in the team's rental car, off the main highway and up a dirt track, a red four-seater pickup truck was coming down the track, and its driver made a neck-slicing gesture at us. After the truck had moved onto the highway, two of its passengers leaned out of windows to give us the finger.

Heyd, the sober German, wanted to turn around and leave immediately, but the others argued that there was no reason to think the men were coming back. We proceeded up to the orchard and found it hung with four collared flycatchers and one wood warbler, which, because it couldn't get airborne, Rutigliano gave to me to put in my backpack. When all the lime sticks had been destroyed, Heyd again, more nervously, suggested that we leave. But there was another grove in the distance which the two Italians wanted to investigate. "I don't have a bad feeling," Rutigliano said.

"There's an English expression, 'Don't press your luck,'" Conlin said.

At that moment the red pickup sped back into sight, fifty yards down the slope from us, and stopped with a lurch. Three men jumped out and began running toward us, picking up baseball-size rocks and hurling them at us as they ran. I would have guessed that it was easy to dodge a few flying rocks, but it wasn't so easy, and Conlin and Heyd were hit by them. Rutigliano was shooting video, Mensi was taking pictures, and there was a lot of confused shouting — "Keep shooting, keep shooting!" "Call the police!" "What the hell is the number?" Mindful of the warbler in my backpack, and not eager to be mistaken for a CABS member, I followed Heyd as he retreated up the slope. From a not very safe distance, we stopped and watched two men attacking Mensi, trying to pull his backpack from his shoulders and his camera from his hands. The men, who were in their thirties and deeply suntanned, were shouting, "Why do you do this? Why do you make photos?" Mensi, keening terribly, his muscles bulging, was clutching the camera to his stomach. The men picked him up, threw him down, and fell on him; there ensued a blur of fighting. I couldn't see Rutigliano but later learned that he was being hit in the face, knocked to the ground, and kicked in the legs and the ribs. His video camera was smashed on a rock; Mensi was also hit in the head with it. Conlin was standing amid the fray with formidable military bearing, holding two cell phones and trying to dial the police. He said to me later that he'd told the attackers that he would drag them through every court in the country if they touched him.

Heyd had continued to retreat, which seemed to me a good idea. When I saw him look back and go pale and break into a dead run, I panicked, too.

Running from danger is like no other kind of running—it's hard to look where you're going. I jumped a stone fence and dashed through a field full of brambles, found myself stumbling into a ditch and getting hit in the chin by a piece of metal fencing, and decided: That's enough of that. I was worried about the warbler I was carrying. I saw Heyd running on up through a large garden, speaking to a middle-aged man, and then, looking frightened, continuing to run. I walked up to the garden's owner and tried to explain the situation, but he spoke only Greek. Seeming at once concerned and suspicious, he fetched his daughter, who was able to tell me, in English, that I'd blundered into the yard of the district director of Greenpeace. She gave me water and two plates of cookies and told my story to her father, who responded with one angry word. "Barbarians!" the daughter translated.

Back down by the rental car, under clouds threatening rain, Mensi was touching his ribs gently and dabbing at the cuts and abrasions that covered his arms; both his camera and his backpack had been stolen. Conlin showed me the smashed video camera, and Rutigliano, who had lost his glasses and was limping heavily, confessed to me, with matter-of-fact fanaticism, "I wanted something like this to happen. Just not this bad."

A second CABS team had arrived and was milling around with grim expressions. In its car was an empty wine carton into which, as a police cruiser was pulling up, I was able to transfer the wood warbler, which was looking subdued but no worse for the wear. I would have felt better about its rescue had there not been, on my cell phone, a new text message from a Cypriot friend of mine, confirming our clandestine date to eat *ambelopoulia* the following night. I was managing to half-convince myself that I could simply be a good journalistic observer and not personally have to eat one, but it wasn't at all clear how I could avoid it.

Every spring, some five billion birds come flooding up from Africa to breed in Eurasia, and every year as many as a billion are killed deliberately by humans, most notably on the migratory flyways of the Mediterranean. As its waters are fished clean by trawlers with sonar and efficient nets, its skies are vacuumed clean of migrants by the extremely effective technology of birdsong recordings. Since the 1970s, as a result of the Birds Directive and various other conservation treaties, the situation of some of the most endangered bird species has improved somewhat. But hunters throughout the Mediterranean are now seizing on this marginal improvement and pushing back. Cyprus recently experimented with a spring season on quail and turtledove; Malta, in April, opened its own spring season; and Italy's parliament, in May, passed a law that extends the fall season there. While Europeans may think of themselves as models of environmental enlightenment—they certainly lecture the United States and China on carbon emissions as if they were —the populations of many resident and migratory birds in Europe have been collapsing alarmingly in the past ten years. You don't have to be a bird watcher to miss the calling of the cuckoo, the circling of lapwings over fields, the singing of corn buntings from utility poles. A world of birds already battered by habitat loss and intensive agriculture is being hastened toward extinction by hunters and trappers. Spring in the Old World is liable to fall silent far sooner than in the New.

The Republic of Malta, which consists of several densely populated chunks of limestone with collectively less than twice the area of the District of Columbia, is the most savagely bird-hostile place in Europe. There are twelve thousand registered hunters (about 3 percent of the country's population), a large number of whom consider it their birthright to shoot any bird unlucky enough to migrate over Malta, regardless of the season or the bird's protection status. The Maltese shoot bee-eaters, hoopoes, golden orioles, shearwaters, storks, and herons. They stand outside the fences of the international airport and shoot swallows for target practice. They shoot from urban rooftops and from the sides of busy roads. They stand in closely spaced cliffside bunkers and mow down flocks of migrating hawks. They shoot endangered raptors, such as lesser spotted eagles and pallid harriers, that governments farther north in Europe are spending millions of euros to conserve. Rarities are stuffed and added to trophy collections; nonrarities are left on the ground or buried under rocks, so as not to incriminate their shooters. When bird watchers in Italy see a migrant that's missing a chunk of its wing or its tail, they call it "Maltese plumage."

In the 1990s, in the runup to Malta's accession to the EU, the government began to enforce an existing law against shooting nongame species, and Malta became a cause célèbre among groups as far-flung as the U.K.'s Royal Society for the Protection of Birds, which sent volunteers to assist with law enforcement. As a result, in the words of a British volunteer I spoke to, "the situation has gone from being diabolical to merely atrocious." But Maltese hunters, who argue that the country is too small for its shooting to make a meaningful dent in European bird populations, fiercely resent what they see as foreign interference in their "tradition." The national hunters' organization, the Federazzjoni Kaċċaturi Nassaba Konservazzjonisti, said in its April 2008 newsletter, "FKNK believes that the police's work should only be done by Maltese police and not by arrogant foreign extremists who think Malta is theirs because it's in the EU."

When, in 2006, the local bird group BirdLife Malta hired a Turkish national, Tolga Temuge, a former Greenpeace campaigns director, to launch an aggressive campaign against illegal hunting, hunters were reminded of Malta's siege by the Turks in 1565 and reacted with explosive rage. The FKNK's general secretary, Lino Farugia, inveighed against "the Turk" and his "Maltese lackeys," and there ensued a string of threats and attacks on BirdLife's property and personnel. A BirdLife member was shot in the face; three cars belonging to BirdLife volunteers were set on fire; and several thousand young trees were uprooted at a reforestation site that hunters resent for its competition with the main island's only other forest, which they control and shoot roosting birds in. As a widely read hunters' magazine explained in August 2008, "There is a limit to what extent one can expect to stretch the strong moral ties and values of Maltese families and stop their Latin blood from boiling over and expect them to give up their land and culture in a cowardly retreat."

And yet, in contrast to Cyprus, Maltese public opinion is strongly antihunting. Along with banking, tourism is Malta's main industry, and the newspapers frequently print angry letters from tourists who have been menaced by hunters or have witnessed avian atrocities. The Maltese middle class itself is unhappy that the country's very limited open space is overrun by trigger-happy hunters who post NO TRESPASSING signs on public land. Unlike BirdLife Cyprus, BirdLife Malta has succeeded in enlisting prominent citizens, including the owner of the Radisson Hotel group, in a media campaign called "Reclaiming YOUR Countryside."

Malta is a two-party country, however, and because its national elections are typically decided by a few thousand votes, neither the Labor Party nor the Nationalists can afford to alienate their hunting constituents so much that they stay away from the polls. Enforcement of hunting laws therefore continues to be lax: minimal manpower is devoted to it, many local police are friendly with hunters, and even the good police can be lethargic in responding to complaints. Even when offenders are prosecuted, Maltese courts have been reluctant to fine them more than a few hundred euros.

This year the Nationalist government opened the country's spring season on quail and turtledove in defiance of a European Court of Justice ruling last fall. The EU Birds Directive permits member states to apply "derogations" and allow the killing of small numbers of protected species for "judicious use," such as control of bird flocks around airports or subsistence hunting by traditional rural communities. The Maltese government had sought a derogation for continuing the "tradition" of spring hunting, which the directive normally forbids, and the Court had ruled that Malta's proposal failed three of four tests provided by the directive: strict enforcement, small numbers, and parity with other EU member states. Regarding the fourth test, however—whether an "alternative" exists—Malta presented evidence, in the form of bag counts, that autumn hunting of quail and turtledove was not a satisfactory alternative to spring hunting. Although the government was aware that the bag counts were unreliable (the FKNK's general secretary himself once publicly admitted that the actual bag might be ten times higher than the reported count), the European Commission has a policy of trusting the data presented by the governments of member states. Malta further argued that because quail and turtledove aren't globally threatened species (they're still plentiful in Asia), they didn't merit absolute protection, and the commission's lawyers failed to point out that what counted was the species' status within the EU, where, in fact, their populations are in serious decline. The Court, therefore, while ruling against Malta and forbidding a spring hunt, did allow that it had passed one of the four tests. And the government at home proclaimed a "victory" and proceeded, in early April, to authorize a hunt.

I joined Tolga Temuge, who looks like a Turkish David Foster Wallace, on an early-morning patrol on the first day of the season. We weren't expecting to see much shooting, because the FKNK, angered by the government's terms—the season would last only six half-days, instead of the traditional six to eight weeks, and only 2,500 licenses would be granted—had organized a boycott of the season, threatening to "name and shame" any hunter who applied for a license. "The European Commission *failed*," Temuge said as we drove the dark, dusty labyrinth of Malta's road system. "The European hunting organization and BirdLife International did a lot of hard work to arrive at sustainable hunting limits, and then Malta joins the EU, as the smallest member state, and threatens to bring down the whole edifice of the excellent Birds Directive. Malta's disregard for it is setting a bad precedent for other member states, especially in the Mediterranean, to behave the same way."

When the sky lightened, we stopped in a rough limestone lane, amid walled fields of golden hay, and listened for gunshots. I heard dogs barking, a cock crowing, trucks shifting gears, and, somewhere nearby, electronic quail song playing. Patrolling elsewhere on the island were six other of Temuge's teams, staffed mainly by foreign volunteers, with a few hired Maltese security men. As the sun came up, we began to hear distant gunshots, but not many; the country seemed essentially bird-free that morning. We proceeded through a village in which a couple of shots rang out—"Fucking unbelievable!" Temuge cried. "This is a residential area! Fucking unbelievable!"—and back into the stony maze of walls that passes for countryside in Malta. Further gunshots led us to a small field in which two men in their thirties were standing with a handheld radio. As soon as they saw us, they picked up hoes and began tending lush plantings of beans and onions. "Once you're in the area, they know," Temuge said. "Everybody knows. If they have radios, it's ninety percent sure they're hunters." It did indeed seem awfully early to be out doing hoe-work, and as long as we were standing by the field we heard no more shots. Four blazing male golden orioles flashed by, unlucky to have chosen Malta as a migratory stopover but lucky that we were standing there. In a low tree I spotted a female chaffinch, which is one of the most common birds in Europe and is all but absent in Malta, owing to the country's widespread illegal finch trapping. Temuge became very excited when I called it out. "A chaffinch!" he said. "That would be incredible, if we're starting to have breeding chaffinch here again." It was like somebody in North America being amazed to see a robin.

Maltese hunters are in the weak position of wanting something that would get Malta into real, punishable trouble with the EU: the legal right to shoot birds bound for their breeding grounds. Their leaders at the FKNK thus have little choice but to adopt uncompromising positions, such as this spring's boycott, which raises false hopes in the FKNK rank and file, fostering frustration and feelings of betrayal when, inevitably, the government disappoints them. I met with the FKNK's spokesman, Joseph Perici Calascione, a nervous but articulate man, at the organization's cramped, cluttered headquarters. "How could anybody, in their wildest imagination, expect us to be satisfied with a spring season that left eighty percent of hunters unable to get a license?" Perici Calascione said. "We've already gone two years without a season that was part of our tradition, part of our living. We weren't looking for a season as it was three years ago, but still a reasonable season, which the government had promised us in no uncertain terms before accession to the EU."

I brought up the matter of illegal shooting, and Perici Calascione offered me a Scotch. When I declined, he poured himself one. "We're completely against the illegal shooting of protected species," he said. "We're prepared to have hunting marshals in place to spot these individuals and take away their membership. And this would have been in place, had we been given a good season." Perici Calascione conceded that he was uncomfortable with the more incendiary statements of the FKNK's general secretary, but he himself became visibly distressed as he tried to convey how much hunting mattered to him; he sounded strangely like a victimized environmentalist. "Everybody is frustrated," he said, with a tremor in his voice. "Psychiatric incidents have increased, we've had suicides among our membership—our culture is threatened."

Just how much Maltese-style shooting is a "culture" and a "tradition" is debatable. While spring hunting and the killing and taxidermy of rare birds are unquestionably traditions of long standing, the phenomenon of indiscriminate slaughter seems not to have arisen until the 1960s, when Malta achieved its independence and began to prosper. Malta, indeed, represents a stark refutation of the theory that a society's affluence leads to better environmental stewardship. Affluence in Malta brought more sophisticated weapons, more money to pay taxidermists, and more cars and better roads, which made the countryside more easily accessible to hunters. Where hunting had once been a tradition handed down from father to son, it now became the pastime of young men who went out in unruly groups.

On a piece of land belonging to a hotel that hopes to build a golf course on it, I met with an old-fashioned hunter who is disgusted with his countrymen's bad behavior and with the FKNK's tolerance of it. He told me that undisciplined shooting is in the Maltese "blood" and that it was unreasonable to expect hunters to suddenly change after the country joined the EU. ("If you were born of a prostitute," he said, "you won't become a nun.") But he also put much of the blame on younger hunters and said that Malta's lowering of the hunting age from twenty-one to eighteen had made matters worse. "And now that they've changed the spring-hunting law," he said, "law-abiding people can't go out, but the indiscriminate shooters still go out, because there's not enough law enforcement. I've been in the country for three weeks this spring, and I've seen one police car."

Spring was always the main hunting season in Malta, and the hunter said that if the season is closed permanently he will probably keep hunting in the fall only as long as his two dogs live, and then quit and be just a bird watcher. "Something else is happening," he said. "Because where are the turtledoves? When I was young and going out with my father, we'd look up at the sky and see thousands of them. Now it's peak season, and I was out all day yesterday and saw twelve. I haven't seen a nightjar in two years. I haven't seen a rock thrush in five years. Last autumn, I went out every morning and afternoon looking for woodcock, with my dogs, and I saw three of them and didn't fire once. And that's part of the problem: people get frustrated. 'I don't find a woodcock, so let's shoot a kestrel.'"

Late on a Sunday afternoon, from a secluded height, Temuge and I used a telescope to spy on two men who were scanning the sky and fields with binoculars. "They're definitely hunters," Temuge said. "They keep their guns hidden until something comes by for them to shoot." But, as an hour passed and nothing came by, the men picked up rakes and began weeding a garden, only occasionally returning to their binoculars, and then another hour passed and they worked harder in the garden, because there were no birds.

Italy is a long, narrow gauntlet for a winged migrant to run. Poachers in Brescia, in the north, trap a million songbirds annually for sale to restaurants offering *pulenta e osei*—polenta with little birds. The woods of Sardinia are full of wire snares, the Venetian wetlands are a slaughtering ground for wintering ducks, and Umbria, the home of Saint Francis, has more registered hunters per capita than any other region. Hunters in Tuscany pursue their quotas of woodcock and wood pigeon and four legally shootable songbirds, including song thrush and skylark; but at dawn, in the mist, it's hard to distinguish legal from illegal quarry, and who's keeping track anyway? To the south, in Campania, much of which is controlled by the Camorra (the local mafia), the most inviting habitat for migratory waterfowl and waders is in fields flooded by the Camorra and rented to hunters for up to a thousand euros a day; songbird wholesalers from Brescia bring down refrigerated trucks to collect the take from small-time poachers; entire Campanian provinces are blanketed with traps for seven tuneful European finch species, and flush Camorristi pay handsomely for well-trained singers at the illegal bird markets there. Farther south, in Calabria and Sicily, the highly publicized springtime hunting of migrating honey buzzards has been reduced by intensive law enforcement and volunteer monitoring, but Calabria, especially, is still full of poachers who, if they can get away with it, will shoot anything that flies.

A curious old statute in Italy's civil code, enacted by the Fascists to encourage familiarity with firearms, gives hunters, and only hunters, the right to enter private property, regardless of who owns it, in pursuit of game. By the 1980s, there were more than two million licensed hunters running wild in the Italian countryside, which had emptied out as the population flowed into the cities. Most urban Italians dislike hunting, however, and in 1992 the Italian parliament passed one of Europe's more restrictive hunting laws, which included, most radically, a declaration that all wild

fauna belong exclusively to the Italian state, thereby reducing
hunting to a special concession. In the two decades since then,
the populations of some of Italy's most lovable megafauna, including wolves, have rebounded spectacularly, while the number of licensed hunters has fallen below eight hundred thousand. These
two trends have prompted Franco Orsi, a Ligurian senator from
Silvio Berlusconi's party, to propose a law that would liberalize the
use of decoy birds and expand the times and places in which hunting is permitted. A second, "communitary" law, intended to bring
Italy into compliance with the Birds Directive and thereby avoid
hundreds of millions of euros in fines pending against it, has just
been passed by the parliament and includes at least one clear victory for hunters: a shifting of the hunting season for certain bird
species into February.

I met with Orsi at his party's offices in Genoa, on the eve of regional elections that brought fresh gains for Berlusconi's coalition. Orsi, a handsome, soft-eyed man in his forties, is a passionate
hunter who chooses vacation destinations on the basis of what he
can shoot in them. His argument for updating the 1992 law is that
it has led to an explosive increase of harmful species; that Italian
hunters should be allowed to do whatever French and Spanish
hunters do; that private landowners could manage land for game
better than the state does; and that hunting is a socially and spiritu-
ally beneficial activity. He showed me a newspaper picture of wild
boar running down a Genoese street; he described the menace
posed by starlings at airports and in vineyards. But when I agreed
that controlling boar and starlings is a good idea, he went on to say
that hunters don't like killing boar in the season the authorities
want them to. "And, anyway, I can't accept that hunting is only for
wild boar, nutria, and starlings," he said. "That's something the
army can do."

I asked Orsi if he favored hunting every bird species to the maximum compatible with sustaining existing numbers.

"Let's imagine fauna as capital that every year produces interest," he said. "If I spend the interest, I can still keep the capital, and
the future of the species and of hunting will be preserved."

"But there's also the investment strategy of reinvesting part of
the interest, to grow the capital," I said.

"That depends on each species. There's an optimal density for each one, and some have a density that's larger than optimal, others smaller. So hunting has to regulate the balance."

My impression, from earlier visits to Italy, was that its avian populations are pretty much all suboptimal. Since Orsi didn't seem to share it, I asked him how he thought hunting harmless birds benefited society. To my surprise, he quoted Peter Singer, the author of *Animal Liberation,* to the effect that if every man had to kill the animals he eats, we would all be vegetarians. "In our urban society, we've lost the relationship between man and animal which has elements of violence," Orsi said. "When I was fourteen, my grandfather made me kill a chicken, which was the family tradition, and now every time I eat chicken I remember that it was an animal. To go back to Peter Singer, the overconsumption of animals in our society corresponds to an overconsumption of resources. Huge amounts of space are devoted to wasteful, industrialized farming, because we've lost a sense of rural identity. We shouldn't think that hunting is the only form of human violence against the environment. And hunting, in this sense, is educational."

I thought Orsi had a point, but to the Italian environmentalists I spoke with, his rhetoric proved only that he was skilled at handling journalists. Behind the national push to liberalize hunting laws, the *ambientalisti* all see the hand of Italy's large arms and munitions industry. As one of them said to me, "When somebody asks you what your business produces, do you say, 'Land mines that blow up Bosnian children,' or do you say, 'Traditional shotguns for people who enjoy waiting at dawn in a wetland for the ducks to come'?"

It's impossible to know how many birds are shot in Italy. The annual reported take of song thrushes, for example, ranges from three million to seven million, but Fernando Spina, a senior scientist at Italy's environmental-protection agency, considers these numbers "hugely conservative," since only the most conscientious hunters fill out their game cards correctly, local game authorities lack the manpower to police the hunters, the provincial databases are largely uncomputerized, and most local Italian hunting authorities routinely ignore requests for data. What is known is that Italy is a crucial migratory flyway. Banded birds have been recovered there from every country in Europe, thirty-eight countries in Africa, and six in Asia. And return migration begins in Italy very early, in some cases as early as late December. The EU's Birds Directive protects all birds on return migration, permitting hunting only within the limits of natural autumn mortality, and most responsible hunters therefore believe that the season should end on December 31. Italy's new communitary law goes the other way, however, and extends the season into February. Since early-return migrants tend to be the fittest of their species, the new law makes targets of precisely those birds with the best chance of breeding success. A longer season also protects poachers of protected species, because an illegal gunshot sounds just like a legal one. And without good data, nobody can say whether a region's annual bag limit on a species falls within natural mortality. "The bag limit is an arbitrary number, set by local officials," Spina said. "It has no relation to actual census numbers."

Although habitat loss is the biggest reason that European bird populations are collapsing, Italian-style hunting (*caccia selvaggia*, "wild hunting," its detractors call it) adds particular insult to the injury. When I asked Fulco Pratesi, a former big-game hunter who founded WWF Italy and who now considers hunting "a mania," why Italian hunters are so wild to kill birds, he cited his countrymen's love of weapons, their attachment to an "attitude of virility," their delight in breaking laws, and, strangely, their love of being in nature. "It's like a rapist who loves women but expresses it in a violent and perverse way," Pratesi said. "Birds that weigh twenty-two grams are being shot with thirty-two-gram ammunition." Italians, he added, more easily feel affection for "symbolic" animals like wolves and bears and have actually done a better job of protecting them than the rest of Europe has. "But birds are invisible," he said. "We don't see them, we don't hear them. In northern Europe, the arrival of migrating birds is visible and audible, and it moves people. Here, people live in cities and large housing complexes, and birds are literally up in the air."

For most of its history, Italy was visited every spring and fall by unimaginable numbers of packets of flying protein, and, unlike northern Europe, where people learned to see the correlation between overharvesting and diminishing returns, supplies seemed limitless in the Mediterranean. A poacher from Reggio di Calabria, still bitter about being forbidden to shoot honey buzzards, said to me, "We were only killing about twenty-five hundred a spring in Reggio, out of a total passage of sixty to a hundred thousand—it wasn't a big deal." The only way he could understand the banning of his sport was in terms of money. He told me, in all seriousness, that certain organizations that wanted to tap into state money had set themselves up as antipoachers, and that it was their need for poachers to oppose which had led to the writing of antipoaching laws. "And now these people are getting rich with money from the state," he explained.

In one of the southern provinces, I got to know an impishly boyish ex-poacher named Sergio. He'd been well into middle age before giving up poaching, feeling that he'd finally outgrown that stage of life, and he now tells stories of his "sins of youth" for comic effect. Going hunting at night was always illegal but never a problem, Sergio said, if your poaching companions were the parish priest and the brigadier of the local carabiniere. The brigadier was especially helpful in discouraging forest rangers from patrolling in their neighborhood. One night when Sergio was out hunting with him, they froze a barn owl in the headlights of the brigadier's Jeep. The brigadier told Sergio to shoot it. When Sergio demurred, the brigadier took out a shovel, walked around behind the owl, and whacked it on the head. Then he put it in the rear compartment of the Jeep.

"Why?" I asked Sergio. "Why did he want to kill the owl?"

"Because we were poaching!"

At the end of the night, when the brigadier opened the rear compartment, the owl, which had only been stunned, flew up and attacked him—Sergio spread his arms and made a ridiculously ferocious face to show me how.

For Sergio, the point of poaching had always been eating. He taught me a rhyme in his local dialect, which approximately translates: *For meat of the feather, eat a crow; for a heart that's kind, love a crone.* "You can cook crow for six days, and it's still tough," he told me. "But it's not bad in a broth. I also ate badger and fox—I ate everything." The only bird that no Italian seems interested in eating is the seagull. Even the honey buzzard, although southern families traditionally kept one specimen stuffed and mounted in the best room of their house (its local nickname is *adorno,* for "adornment"), was eaten as a springtime treat; the poacher in Reggio gave me his recipe for fricasseeing it with sugar and vinegar.

Italian wild hunters who, unlike Sergio, haven't outgrown the pursuit and who are frustrated by declining game populations and increasing state restrictions, have learned to go elsewhere in the Mediterranean for a thrill. On the Campanian seacoast, I spoke with a gap-toothed, gleefully unrepentant young-old poacher who, now that he can no longer set up a blind on the beach and shoot unlimited numbers of arriving migrants, contents himself with looking forward to vacations in Albania, where you can still shoot as much as you can find of whatever you want, whenever you want, for a very low fee. Although hunters from all nations go abroad, the Italians are widely considered to be the worst. The wealthiest of them go to Siberia to shoot woodcock during their springtime display flights or to Egypt, where, I was told, you can hire a local police officer to fetch your kills while you shoot ibises and globally threatened duck species until your arms are tired; there are pictures on the Internet of visiting hunters standing beside meter-high piles of bird carcasses.

The responsible hunters in Italy hate the wild ones; they hate Franco Orsi. "We have a culture clash in Italy between two visions of hunting," Massimo Canale, a young hunter in Reggio di Calabria, told me. "One side, Orsi's side, says, 'Let's just open it up.' On the other side are people with a sense of responsibility for where they live. To become a selective hunter, you need more than just a license. You need to study biology, physics, ballistics. You become selective for boar and deer—you have a role to play." Canale discovered his predatory instinct as a child, while hunting indiscriminately with his grandfather, and he feels fortunate to have met people who taught him a better way. "I don't mind not killing something on any given day," he said, "but killing is the goal, and I'd be lying if I said it wasn't. I have a conflict between my predatory instinct and my rationality, and my way of trying to tame my instinct is through selective hunting. In my opinion, it's the only way to hunt in 2010. And Orsi doesn't know or care about it."

The two visions of hunting correspond broadly to Italy's two faces. There's the frankly criminal Italy of the Camorra and its allies and the quasi-criminal Italy of Berlusconi's cronies, but there is also, still, *l'Italia che lavora*— "the Italy that works." The Italians who combat poaching are motivated by disgust with their country's lawlessness, and they rely heavily on tips from responsible hunters like Canale, who become frustrated when, for example, they're unable to find quail to shoot because all the birds have been attracted to illegal recordings. In Salerno, the least disorderly of Campania's provinces, I joined a squad of WWF guards who took me out to an artificial pond, now drained, where they had recently stalked the president of a regional hunters' association and caught him illegally using electronic recordings to attract birds. Looming near the pond, amid fields rendered desolate by white plastic crop covers, was a disintegrating mountain of "ecoballs"—shrink-wrapped bales of Neapolitan garbage that had been dumped all over the Campanian countryside and become a symbol of Italy's environmental crisis. "It was the second time in two years that we'd caught the guy," the squad leader said. "He was part of the committee that regulates hunting in the region, and he'd remained president in spite of having been charged. There are other regional presidents who do the same thing but are harder to catch."

One shining example of the Italy that works has been the suppression of honey-buzzard poaching at the Strait of Messina. Every year since 1985, the national forest police have assigned an extra team with helicopters to patrol the Calabrian side of the strait. Although the Calabrian situation has lately deteriorated somewhat —this year's team was smaller than in the past and stayed for fewer days, and the estimated death toll was four hundred, double the number in recent years—the Sicilian side of the strait is the domain of a famous crusader, Anna Giordano, and remains essentially free of poachers. Beginning as a fifteen-year-old in 1981, Giordano undertook surveillance of the concrete blinds from which raptors were being shot by the thousands as they sailed in low over the mountains above Messina. Unlike the Calabrians, who ate the buzzards, the Sicilians shot purely for the sake of tradition, for competition with one another, and for trophies. Some of them shot everything; others restricted themselves to honey buzzard ("The Bird," they called it) unless they saw a real rarity, like golden eagle. Giordano hurried from the blinds to the nearest pay phone, from which she summoned the forest police, and then back to the blinds. Although her cars were vandalized, and although she was constantly threatened and vilified, she was never physically harmed, probably because she was a young woman. (The Italian word for "bird," *uccello,* is also slang for "penis" and lent itself to dirty jibes about her, but a poster I saw on the wall of her office flipped these jibes around: "Your Virility? A Dead Bird.") With increasing success, especially after the advent of cell phones, Giordano compelled the forest police to crack down on the poachers, and her growing fame brought media attention and legions of volunteers. In recent years her teams have reported seasonal gunshot totals in the single digits.

"In the early years," Giordano said when I joined her on a hilltop to look at passing hawks, "we didn't even dare raise our binoculars when we were counting raptors, because the poachers would watch us and start shooting if they saw us looking at something. Our logs from back then show lots of 'unidentified raptors.' And now we can stand up here all afternoon, comparing the markings of first-calendar-year female harriers and not hear a shot. A couple of years ago, one of the worst poachers, a violent, stupid, vulgar guy who'd always been in our face wherever we went, drove up to me and asked if we could talk. I was, like, 'Heh-heh-heh-heh, okay.' He asked me if I remembered what I'd said to him twenty-five years ago. I said I couldn't remember what I said yesterday. He said, 'You said the day would come when I would love the birds instead of killing them. I just came up here to tell you you were right. I used to say to my son when we were going out, Have you got the gun? Now I say, Have you got the binoculars?' And I handed him my own binoculars—to a poacher!—so he could see a honey buzzard that was flying over."

Giordano is small, dark, and zealous. She has lately been attacking the local government for failing to regulate housing development around Messina, and, as if to insure that she has too much to do, she also helps operate a wildlife rescue center. I'd already visited one Italian animal hospital, on the grounds of a shuttered psychiatric hospital in Naples, and seen an X-ray of a hawk heavily dotted with lead shot, several recovering raptors in large cages, and a seagull whose left leg was blackened and shriveled from having stepped in acid. At Giordano's center, on a hill behind Messina, I watched her feed scraps of raw turkey to a small eagle that had been blinded by a shotgun pellet. She grasped the eagle's taloned legs in one hand and cradled the bird against her belly. Its tail feathers sadly bedraggled, its gaze stern but impotent, it suffered her to open its bill and stuff in meat until its gullet bulged. The bird seemed to me at once all eagle and no longer an eagle at all. I didn't know what it was.

Like most Cypriot restaurants that serve *ambelopoulia*, the one I went to with a friend and a friend of his (I'll call them Takis and

Demetrios) had a small private dining room in which the little birds could be consumed discreetly. We walked through the main room, in which a TV was blaring one of the Brazilian soap operas that are popular in Cyprus, and sat down to an onslaught of Cypriot specialties: smoked pork, fried cheese, pickled caper twigs, wild asparagus and mushrooms with eggs, wine-soaked sausage, couscous. The proprietor also brought us three fried song thrushes, which we hadn't asked for, and hovered by our table as if to make sure I ate mine. I thought of Saint Francis, who had set aside his sympathy for animals once a year on Christmas and eaten meat. I thought of a kid named Woody, who, on a backpacking trip I'd taken as a teenager, had given me a bite of fried robin. I thought of a prominent Italian conservationist who'd admitted to me that song thrushes are "bloody tasty." The conservationist was right. The meat was dark and richly flavorful, and the bird was enough bigger than an *ambelopoulia* that I could think of it as ordinary restaurant food, more or less, and of myself as an ordinary consumer.

After the proprietor went away, I asked Takis and Demetrios what kind of Cypriots like to eat *ambelopoulia*.

"The people who do it a lot," Demetrios said, "are the same ones who go to cabarets, the lounges where there's pole dancing and Eastern European girls who make themselves available. In other words, people with not a high level of morality. Which is to say, most Cypriots. There's a saying here, 'Whatever you can stuff your mouth with, whatever your ass can grab—'"

"I.e., because life is short," Takis said.

"People come to Cyprus and think they're in a European country because we belong to the EU," Demetrios said. "In fact, we're a Middle Eastern country that's part of Europe by accident."

The night before, at the Paralimni police station, I'd given a statement to a young detective who seemed to want me to say that the attackers of the CABS team had only been trying to get the team to stop taking pictures and video of them. "For people here," the detective explained when we were done, "it's a tradition to trap birds, and you can't change that overnight. Trying to talk to them and explain why it's wrong is more helpful than the aggressive approach of CABS." He may have been right, but I'd been hearing the same plea for patience all over the Mediterranean, and it was sounding to me like a version of modern consumerism's more general plea regarding nature: Just wait until we've used up everything, and then you nature-lovers can have what's left.

While Takis and Demetrios and I waited for the dozen *ambelopoulia* that were coming, we argued about who was going to eat them. "Maybe I'll take one small bite," I said.

"I don't even like *ambelopoulia*," Takis said.

"Neither do I," Demetrios said.

"Okay," I said. "How about if I take two and you each take five?" They shook their heads.

Dismayingly soon, the proprietor returned with a plate. In the room's harsh light, the *ambelopoulia* looked like a dozen little gleaming yellowish gray turds. "You're the first American I've ever served," the proprietor said. "I've had lots of Russians, but never an American." I put one on my plate, and the proprietor told me that eating it was the same as taking two Viagras.

When we were alone again, my field of vision shrank to a few inches, the way it had when I'd dissected a frog in ninth-grade biology. I made myself eat the two almond-size breast muscles, which were the only obvious meat; the rest was greasy cartilage and entrails and tiny bones. I couldn't tell if the meat's bitterness was real or the product of emotion, the killing of a blackcap's enchantment. Takis and Demetrios were making short work of their eight birds, taking clean bones from their mouths and exclaiming that *ambelopoulia* were much better than they remembered; were rather good, in fact. I trashed a second bird and then, feeling somewhat sick, wrapped my remaining two in a paper napkin and put them in my pocket. The proprietor returned and asked if I'd enjoyed the birds.

"Mm!" I said.

"If you hadn't asked for them"—this in a regretful tone—"I think you really would have liked the lamb tonight."

I made no reply, but now, as if satisfied by my complicity, the proprietor became talkative: "Young kids today don't like to eat them. It used to start young, and you'd get used to the taste. My toddler can eat ten at a time."

Takis and Demetrios exchanged skeptical glances.

"It's a shame they've been outlawed," the proprietor went on, "because they used to be a great tourist attraction. Now it's become almost like the drug trade. A dozen of them cost me sixty euros.

These damned foreigners come and take down the nets and destroy them, and we've surrendered to them. Trapping *ambelopoulia* used to be one of the few ways people around here could make a good living."

Outside, by the edge of the restaurant parking lot, near some bushes in which I'd earlier heard *ambelopoulia* singing, I knelt down and scraped a hole in the dirt with my fingers. The world was feeling especially empty of meaning, and the best I could do to fight this feeling was to unwrap the two dead birds from the napkin, put them in the hole, and tamp some dirt down on them. Then Takis led me to a nearby tavern with medium-sized birds grilling on charcoal outside. It was a sort of poor man's cabaret, and as soon as we'd ordered beers at the bar one of the hostesses, a heavy-legged blonde from Moldova, pulled up a stool behind us.

The blue of the Mediterranean isn't pretty to me anymore. The clarity of its water, prized by vacationers, is the clarity of a sterile swimming pool. There are few smells on its beaches, and few birds, and its depths are on their way to being empty; much of the fish now consumed in Europe comes illegally, no questions asked, from the ocean west of Africa. I look at the blue and see not a sea but a postcard, paper thin.

And yet it is the Mediterranean, specifically Italy, that gave us the poet Ovid, who in the *Metamorphoses* deplored the eating of animals, and the vegetarian Leonardo da Vinci, who envisioned a day when the life of an animal would be valued as highly as that of a person, and Saint Francis, who once petitioned the Holy Roman Emperor to scatter grain on fields on Christmas Day and give the crested larks a feast. For Saint Francis, the crested larks, whose drab brown plumage and peaked head feathers resemble the hooded brown robes of his Friars Minor, his Little Brothers, were a model for his order: wandering, as light as air, and saving up nothing, just gleaning their daily minimum of food, and always singing, singing. He addressed them as his Sister Larks. Once, by the side of an Umbrian road, he preached to the local birds, which are said to have gathered around him quietly and listened with a look of understanding, and then chastised himself for not having thought to preach to them sooner. Another time, when he wanted to preach to human beings, a flock of swallows was chattering noisily, and he

said to them, either angrily or politely—the sources are unclear—
"Sister Swallows, you've had your say. Now be quiet and let me have
my say." According to the legend, the swallows immediately fell silent.

I visited the site of the Sermon to the Birds with a Franciscan
friar, Guglielmo Spirito, who is also a passionate amateur Tolkien
scholar. "Even as a child," Guglielmo said, "I knew that if I ever
joined the church it would be as a Franciscan. The main thing that
attracted me, when I was young, was his relationship with animals.
To me the lesson of Saint Francis is the same as that of fairy tales:
that oneness with nature is not only desirable but possible. He's
an example of wholeness regained, wholeness actually within our
reach." There was no intimation of wholeness at the little shrine,
across a busy road from a Vulcangas station, that now commemo-
rates the Sermon to the Birds; I could hear a few crows cawing and
tits twittering, but mostly just the roar of passing cars and trucks
and farm equipment. Back in Assisi, however, Guglielmo took me
to two other Franciscan sites that felt more enchanted. One was
the Sacred Hut, the crude stone building in which Saint Francis
and his first followers had lived in voluntary poverty and invented a
brotherhood. The other was the tiny chapel of Santa Maria degli
Angeli, outside which, in the night, as Saint Francis lay dying, his
sister larks are said to have circled and sung. Both structures are
now entirely enclosed by later, larger, more ornate churches; one
of the architects, some pragmatic Italian, had seen fit to plant a fat
marble column in the middle of the Sacred Hut.

Nobody since Jesus has lived a life more radically in keeping with
his gospel than Saint Francis did; and Saint Francis, unburdened
by the weight of being the Messiah, went Jesus one better and extended his gospel to all creation. It seemed to me that if wild birds
survive in modern Europe, it will be in the manner of those ancient small Franciscan buildings sheltered by the structures of a
vain and powerful church: as beloved exceptions to its rule.

IAN FRAZIER

Fish Out of Water

FROM *The New Yorker*

IN THE SHEDD AQUARIUM, on the lakefront in downtown Chicago, there's a video display that makes visitors laugh until they're falling down. The video is in an area of the aquarium devoted to invasive species, and it shows silver carp *(Hypophthalmichthys molitrix)*, a fish originally from China and eastern Siberia, jumping in the Illinois River near Peoria. A peculiarity of silver carp is that when they are alarmed by potential predators they leap from the water, sometimes rocketing fifteen feet into the air. In the video, several people are cruising in a small motorboat below the spillway of a lock or a dam while fish fly all around. The people get hit in the arms, the back, the sides. They're ducking, they're yelling, the silver carp are flying, the boat is swerving. Aquarium visitors whoop and wipe the tears away and watch the video again.

The invasion of Asian carp into the waters of the South and the Midwest differs from other ongoing environmental problems in that it slaps you in the head. Videos like the one in the Shedd are the reason a lot of people know about Asian carp. Not only are the newcomers upsetting the balance in midcountry ecosystems; they are knocking boaters' glasses off and breaking their noses and chipping their teeth and leaving body bruises in the shape of fish. So far, apparently, there have been no fatalities. And while threats to the environment tend to be ignorable (if only in the short run), this one is not, because millions of people go boating, and the novelty of being hit by a fish wears off fast.

Right now there are actually two kinds of Asian carp to worry about: silver carp and their nonjumping companions, bighead carp *(Hypophthalmichthys nobilis)*. Bigheads, which can grow to a hundred pounds, are bigger than silvers. Neither really has the appearance of a carp, because their mouths are not the downward-pointing mouths of bottom-feeding fish. Unlike the common carp, which we think of as an American fish although it was introduced here in the 1880s, silver carp and bighead carp feed not on the bottom but in the top few feet of the water column. These carp eat only plankton, which they filter from the water with rakers in their gills. They are highly efficient feeders, outconsuming other fish and leaving less for the fry of such game fish as bass, crappies, and walleyes. The fear is that when they get into a lake or river you will soon have nothing else.

In the United States, the Asian carp started their journey from a place of formerly ominous reputation: Down the River. As long ago as the 1970s, bigheads and silvers escaped into the lower Mississippi River from waste-treatment plants and commercial catfish ponds in Arkansas and Mississippi. Down South they were worker fish, imported to clean up enclosed areas by eating algae. Presumably, Mississippi River floods gave them the chance to get away. Once at large, the carp headed north, eventually turning up in the Missouri, the Tennessee, the Ohio, the Des Moines, the Wabash, the Illinois. For the long term, they seem to have their sights set on Canada. Today, just a few decades after their escape, they are almost there.

Not to get too sentimental about it, but the Mississippi River is us, and vice versa. It's our bloodstream. Last summer I was driving along the river in western Illinois thinking how horrible the Mississippi had been lately, with its outsized floods and its destruction of New Orleans, and I noted the recent flooding still in progress along the Illinois shore — the miles of roads and fields submerged, and the ferry landing at Golden Eagle, Illinois, now separated from dry land by seventy feet of mud and water, and low-lying parking lots full of river mud cracked like pieces of a jigsaw puzzle curling in the sun. In the sprawl of standing water over parts of the landscape, no actual river could be found. Then the road I was on descended from a ridge to the mostly unflooded river town of Hamburg, Illinois, and the Mississippi itself was running fast beside the main street, and just across the shining expanse were the houses and church steeples on the Missouri side. An old and powerful

emotion hit me; my blood leaned with the current and I let the re-
criminations go by.

The fact that Asian carp are now in this river and many others,
sucking in plankton and growing big and reproducing and waiting
to smack a Jet Skier's face, is really not good. Possibly these carp
will change large parts of our national watersheds forever. We may
be infected with a virus for which there is no cure.

Among Asian-carp-infested rivers, the Illinois has it worst of all.
This river is formed by the junction of the Kankakee and the Des
Plaines about fifty miles southwest of downtown Chicago. It runs at
a diagonal partway across the state and then turns due south, meeting the Mississippi north of St. Louis. Via the Des Plaines, most of
the treated wastewater of Chicago flows south into the Illinois. It's
the main industrial river of the state. The fields of corn and soybeans through which it passes are the factory floor, the river is the
conveyor. If there's any stretch of this river that doesn't hum and
throb—with barges, tugs, grain elevators, power plants, coal de-
pots, refineries—I didn't see it.

One morning in August I was fishing in the Illinois by the boat-
launch ramp in a riverfront park in Havana, Illinois, a town whose
full name is pleasant to say. Not much water traffic was passing at
this early hour. A light breeze interfered with the deep green reflection of the trees on the far shore, and the echoes of car tires
rolling on the brick ramp bounced off some barges parked over
there. Fish were rising near the bank in the brown current. I cast to
them with a dry fly, but they wouldn't bite. Later, a guy in a sporting goods store explained that these were Asian carp feeding on
grain dust from the elevators upstream. In fact, because of the tini-
ness of what they eat, Asian carp are almost impossible to catch
with hook and line.

I quit fishing as the sun rose above the trees and the day became
stifling. Two fishermen came to the boat launch from upriver in an
eighteen-foot bass boat with a twenty-five-horsepower outboard.
They had on camo hats and blue-jean bib overalls. As they were
winching their boat onto the trailer, one of them fanned himself
with his hat and yelled to me, "Hey, turn down the heat!" I went
over and said hello and asked how they had done. They opened
their cooler and showed me a beautiful catch of crappies and a

medium-sized striped bass. I expressed surprise at their success and said I had heard that the Asian carp had depleted this fishery. Not necessarily, the guys said, adding that there were so many little Asian carp now that the other fish had more prey to feed on. I asked if they thought Asian carp were a serious problem.

"Oh, hell yes, they're a problem," the smaller of the two guys said. "They jump up and hit you all the time, and they get slime on you, and they shit all over your boat. And *bleed*—oh, they bleed like no sonsabitches you've ever seen. Yesterday a carp hit Junior in the chest—I thought it would go into his bib—and it left a trail of shit and slime and blood all down the front of him."

"Hey, look—there's my nephew!" Junior interrupted. A blue-and-white-striped tugboat was going by with a young fellow in jeans and a T-shirt standing in the wheelhouse at the top. "My nephew's the youngest pilot on the river—became a pilot right out of high school," Junior said. He and the other guy waved and the pilot waved back, the happiest man in Illinois. A large American flag fluttered at the tug's stern. All around the boat, from the roiled wake and from the curls of foam at the bow, carp of mint-bright silver were leaping in the sun.

In the parking lot of the local field station of the Illinois Natural History Survey, just up from the boat launch, two big outboards with twin motors sat dripping on their trailers. After a thorough hosing down, the boats still smelled of disinfectant, and a few fish scales clung to the aluminum structures in their bows. At a desk in the field-station office, Matt O'Hara, a tall, broad-shouldered fish biologist who had just come in from the river, sat and talked with me, while at an adjoining desk a colleague talked to a film crew from an outdoor channel. Another reporter and a film crew from ESPN were expected shortly. During a recent fish survey, Matt O'Hara said, so many fish jumped out of the river at the first jolt of electroshocking that a camera filming from the nearby shore was unable to see the boat. In fact, the visual craziness of the leaping-carp phenomenon, propagated in Internet videos, had been drawing TV crews from all over. Dealing with crews from every major channel, from cable shows, and from Canada, Russia, England, France, and Japan had become part of these biologists' job.

Their actual and more important job is keeping track of what's in the river. North America's two largest watersheds, the Mississippi and the Great Lakes, connect in one place, Chicago. Boats travel from the Atlantic Ocean into the St. Lawrence Seaway, go through the lakes, turn into the Chicago Sanitary and Ship Canal, continue through the canal into the Des Plaines River, enter the Illinois, and head onward from there to the Mississippi and the Gulf of Mexico. This means, of course, that invasive species can travel the same route in either direction. All of them must pass by the Havana field station's door, and Matt O'Hara and others at the station check the river continually to see if they do. These days the watchers' eyes are on the Asian carp, whose dense population in the Illinois—eight thousand or more silver carp per river mile—exerts a seemingly inexorable pressure northward, toward Chicago and the huge, Asian-carp-free (probably) watershed beyond.

Like other scientists who grasp the threat, Matt O'Hara becomes severe when he talks about it, though in a quiet and midwestern way. "I don't know if silver carp and bighead carp would necessarily thrive in the Great Lakes," he told me. "Lake Erie is one of the biggest fisheries in the world—the lakes together are a seven- to ten-billion-dollar fishery annually, and most of that catch is from Lake Erie—and it does look as if the carp would do well in that particular lake, and probably trash it. What I'm even more concerned about in the shorter run is the rivers. The Illinois became seriously infested just in the last seven years. A study has identified twenty-two rivers in the Great Lakes system that might be as vulnerable as the Illinois. Some of these are major rivers with important salmon runs, like the Pere Marquette and the Manistee, in Michigan. Asian carp in those rivers could become a disaster really fast.

"You can eat Asian carp, and they're good. Score the fillets crossways with a sharp knife and cook them in hot oil, and it dissolves a lot of the bones. But given what they do to an ecosystem, I can't say I see any advantage at all with these fish, definitely not if they get in the lakes. They're terrible for the aesthetics, and they certainly make people leery of going out in a boat. Even today, it is still legal to import bighead carp, although silvers are now illegal. And there's another kind of carp, the black carp, that's also in southern catfish ponds and is legal to import and could be an additional disaster if it moved north. Black carp eat snails and mussels and would probably strip our native mollusk and shellfish populations, with all kinds of consequences. That's a danger people should think about, too. We knew fifteen years ago the silvers and the bigheads were going to be a problem, but nothing was done about them. Until recently, nobody paid attention to carp."

Well, okay—let's eat them. This solution has already occurred to the state of Illinois. Americans in general are not keen on eating carp, so, looking elsewhere, the governor of Illinois recently announced an agreement to sell local carp to the Chinese. Big River Fish Corporation, of Pearl, Illinois, would harvest, package, and ship carp to the Beijing Zhouchen Animal Husbandry Company, while the state would invest $2 million in Big River to improve its facilities and processing capacity. The Chinese had already inspected the product and announced their complete satisfaction with "the wild Asian carp" of Illinois. Zhouchen said it would buy 30 million pounds, and possibly more, by the end of 2011.

Pearl, Illinois, by the Illinois River in the west-central part of the state, is on Route 100 and has 187 residents. A small post office makes up about a quarter of its downtown. Big River Fish, around a bend on Route 100, announces itself with two hand-painted wooden signs, one facing north and one south, against the trunk of a maple tree. The signs show the profile of a long-whiskered catfish done all in black and shades of gray except for the hypnotic yellow eye. Rick Smith, Big River's president, was not in his office the morning I drove the eighty-some miles down from Havana. He had just left for the big motorcycle rally in Sturgis, South Dakota, towing his fried-fish cook shack. Rick Smith belongs to the very small number of motorcycle-rally food vendors who also ink multi-million-dollar deals with the Chinese.

Mike Houston, a Big River employee, was locking up as I arrived. The company would close for about a week until the boss returned, he said. Mike Houston had a red ponytail, gray hair at the temples, blue eyes, a ginger-colored beard, a red Ohio State baseball cap, and a T-shirt with a saying about beer. Accompanying him was another employee, a wiry bald man with tattoos and a walrus mustache. "Yeah, we've already sold our first load of Asian carp to the Chinese," Mike Houston said. "They're also buyin' yellow carp, grass carp, catfish. Thirty million pounds is a lot of fish, but they say it will feed one city. We get the fish in the round—that means whole—and then we gut 'em and power-wash the cavity. That cavity's got to be completely clean before you can ship 'em. The Chinese want the carp with the tails, heads, and scales still on. We flash-freeze 'em in our forty-below-zero freezing chamber and then bag 'em individually and put 'em in big plastic shipping bags that hold twelve or fifteen hundred pounds of fish.

"We're expanding our operation, but right now we can move a hell of a lot of fish out of here. We can operate year-round, because there's no restrictions and nobody cares how many carp you catch. Guttin' these fish is hard work. Some of the Asian carp go fifty, sixty pounds. Just liftin' 'em is hard. I used to be a chef, I can work a knife, and I've had days when I've gutted and washed twenty thousand pounds of fish. These Asian carp, what they eat, basically, is muck. You'd be surprised, though—their meat is all white meat, and it's good. I've eaten it. People say it's difficult to find Asian carp in China because they're all fished out. I like to think, sellin' silver carp and bighead carp to the Chinese, that we're sendin' their own product back to 'em. And I'll tell you, even with all the fish we move, we ain't makin' a dent in the Asian carp that's out there. There's commercial fishermen in Havana and in Beardstown that we can call up and say we need a hundred thousand pounds of fish right away, and those boys can get a hundred thousand pounds to us in a day or two. And a day or two later they can have a hundred thousand more. No, we ain't makin' a dent."

The town of Bath, nine miles south of Havana on a long slough of the Illinois, takes a sporting approach to the problem. Every year in August, the Boat Tavern, a waterfront bar in Bath, holds a fishing tournament in which teams of anglers catch silver carp. It's called the Redneck Fishing Tournament; the first was in 2004. Prizes are given for the most fish and the best costumes. This year, 105 boats (at a fifty-dollar entry fee apiece) competed before a crowd of about two thousand in the event's two days. The method of fishing was straightforward: flush the carp from the water with boat engines and snatch them from the air with nets. There was also barbecuing, beer drinking, karaoke singing, games for children and teenagers, bluegrass bands, booths selling T-shirts, and so on. Among the crowd, T-shirt adages were on the order of FRIENDS DON'T LET FRIENDS FISH SOBER and WHAT HAPPENS IN THE BARN STAYS IN THE BARN.

From the top of the boat ramp leading to the event, the Redneck Fishing Tournament smelled like ketchup and mud. This was on Saturday afternoon, the event's peak. Spectators milled, media swarmed. At occurrences of even slight interest, a forest of boom mikes converged, while video cameras pointed here and there promiscuously. A notice on a telephone pole announced that just by entering the premises you were granting a production company called Left/Right the permission to use your voice, words, and image in a project temporarily called "Untitled Carp" anywhere on earth or in the universe in perpetuity.

Tall cottonwoods, ash trees, and maples shaded the shore, which was rutted black mud firmed up in places with heaps of new sand. Crushed blue-and-white Busch beer cans disappeared into the mud, crinkling underfoot. Aluminum johnboats, some camo, some not, lined the riverfront in fleets. Fishing costumes involved headgear: army helmets, football helmets with face guards or antlers or buffalo horns, octopus-tentacle hats, pirate bandannas, Viking helmets with horns and fur, devil hats with upward-pointing horns, a hat like a giant red-and-white fishing bobber, a Burger King crown. Competitors had their faces painted camo colors or gold or red or zebra-striped. Bath, Illinois, was first surveyed by Abraham Lincoln, and on August 16, 1858, while campaigning against Stephen Douglas in the race for the U.S. Senate, Lincoln delivered his famous "House Divided" speech to a large crowd in Bath. He took as his text the New Testament verse "A house divided against itself cannot stand." One hundred and fifty-two years later, the Confederate-flag halter tops mingling with the American flags among the tournament crowd would have puzzled him; likewise, the pirate flags.

The competition took place in heats that lasted two hours. Before each start, boats put off from shore with their engines idling and waited for Betty DeFord, of the Boat Tavern, to give the signal. This honor belonged to her as the inventor and organizer of the tournament. At the air-horn blast, the boats raced off, net-men and -women holding extra-large dip nets at the ready. Soon you could see people plucking fish out of the air. Most of the boats then careered away and out of sight. Two hours later they eased back to the riverbank, many of them heavily loaded. Grinning competitors carried big, heavy-duty plastic barrels of silver carp, a man on each side, to the counting station, where festival officials counted the fish one by one and threw them into the tarpaulin-lined bed of a pickup truck. Among participants and onlookers, a cheerful giddiness prevailed. I had never seen so many fish. It was like an old-time dream of frontier bounty.

Randy Stockham of Havana won first prize, $1,400. He and his team caught 188 fish. Second prize of $1,100 went to Ron St. Germain of Michigan, with 186. Mike Mamer and his Sushi Slayers, of Washington, Illinois, were third, at 153 fish ($800). Top prizes for costumes were awarded to devils from Greenview, Illinois, and cavemen from Michigan. The tournament also donated $1,500 to help local children suffering from developmental disorders and cancer. A total of 3,239 fish were caught, all of them silver carp. They ended up as fertilizer on Betty DeFord's thirty-acre farm.

"I'm just a grandmother and a bartender," Betty DeFord said when I called her at the Boat Tavern a week later for a final recap. "I don't even own this bar. I started the Redneck Fishing Tournament because I want to warn other parts of the country about what these carp will do. I remember when this water had no Asian carp, and you could go frog gigging with a flashlight and a trolling motor on a summer night. The way these fish attack, that's impossible now. We just finished our tournament, and the carp are jumping more out there than they ever were. I want everybody to know: these fish need to be gone."

As you proceed northward toward Chicago on the Illinois, the industrial noise along its banks increases. Turning into the Des Plaines River south of Joliet, you might feel you're in the clanking, racketing final ascent of a roller coaster's highest hill. After Joliet, the machinery of greater Chicago multiplies along the riverbank until, in the municipality of Romeoville, white refinery towers in ranks send white clouds spiraling skyward, and a mesa of coal like a geological feature stretches for a third of a mile, and empty semi-trucks make hollow, drumlike sounds as they cross railroad tracks, and other vehicles beep, backing up. The place is a no man's land. Here the channelized Des Plaines and the Chicago Sanitary and Ship Canal, which have already met up some miles to the south, run parallel, about three hundred yards apart. The water in both of them is the color of old lead.

Romeoville is where the Army Corps of Engineers and the state of Illinois think they can stop the carp. At two locations on the ship canal, electric barriers zap the water. The canal, a rectilinear rock-

wall ditch 160 feet wide and about 25 feet deep, passes through
Romeoville behind chainlink fences topped with barbed wire and
hung with signs: "Danger: Electric Fish Barrier" and "Electric
Charge in Water/Do Not Stop, Anchor, or Fish" and "Caution:
This water is not suitable for wading, swimming, jet skiing, water
skiing/tubing, or any body contact." At the fish barriers, steel cables or bars on the bottom of the canal pulse low-voltage direct
current. The electric charge—according to the navy, whose divers
tested it somehow—is strong enough, under some circumstances,
to cause muscle paralysis, inability to breathe, and ventricular fib-
rillation in human beings.

What it does to fish is less clear. Entering Chicago, the plucky
Asian carp swims into a whole cityful of complications, and it continues to swim single-mindedly while questions of politics, bureaucracy, urban hydrology, and interstate commerce work themselves
out. To put it another way: Chicago is a swamp. Various water-
courses, only temporarily subdued, thread throughout the metro
area. The engineers who caused local rivers to run backward a
hundred years ago so as to send Chicago sewage south to the Mississippi rather than into the city's front yard—Lake Michigan—
did not have too hard a job, because the underlying swamp could
do the same thing itself when in flood. Given enough rainfall, the
waters of swampy Chicago become one. The Des Plaines River
needs no electric barrier because its Chicago section does not connect to the lake; however, when the Des Plaines floods it sloshes
into the ship canal at points above the canal's electric barriers. And
the Des Plaines is very likely to have Asian carp.

Meditating on the complications of the carp's presence in Chicago can disable the brain. Do the electric barriers actually stop all
the carp, or are the little ones able to get through? Do stunned fish
sometimes wash through in the occasional reverses of flow in the
canal? Does the electric charge remain strong and uniform when
ships and barges go by? (Probably; instruments in tollbooth-like
buildings beside the canal adjust the charge when necessary.) Does
the electric field have weak spots where fish can pass? Does a win-
tertime influx of road salt in the water cause the charge to fluctu-
ate? What about when the current must be turned off for mainte-
nance of the bars or cables? Is the rotenone chemical fish killer
that is administered when the current is off effective without fail?

The Chicago waterway system has several locks—why not simply

close these and be done with it? Would closing the locks have a
negligible effect on the economy of Illinois, as a study commis-
sioned by the state of Michigan has claimed, or damage its economy irreparably, as demonstrated in a study preferred by Illinois?
Would closing the locks risk flooding thousands of Chicago base-
ments, as local officials say? How about the part of the waterway
where there is no lock at all? Wouldn't the carp simply go that way?
Will the new thirteen-mile barricade of concrete and special wire
mesh designed to keep carp from swerving out of the Des Plaines
and into the ship canal during flood times actually work? Will un-
informed anglers introduce Asian-carp minnows while using them
for bait? And what about immigrant communities from Asia who
are known to perform ritual releases of fish and other animals during certain religious ceremonies? Might they have performed such
rituals involving Asian carp already? Might they do so in the future?
Will all this bring prevention to naught?

The Illinois Department of Natural Resources addresses itself to
many of these questions, as does the U.S. Army Corps of Engineers
(Great Lakes and Ohio River Division), the Metropolitan Water
Reclamation District of Greater Chicago, the U.S. Fish and Wildlife Service, the Senate Great Lakes Task Force, the Congressional
Great Lakes Task Force, the White House Council on Environmental Quality, the interagency Asian Carp Rapid Response Team, and
the interagency Asian Carp Regional Coordinating Committee's
Asian Carp Control Framework. The Illinois Chamber of Commerce, the American Waterways Operators, and the Chemical Industry Council of Illinois put in a word for industry, while the Natural Resources Defense Council, Freshwater Future, the Sierra
Club, Healing Our Waters Great Lakes Coalition, and the Alliance
for the Great Lakes offer perspective from the environmental side.
The Chicago Department of the Environment, the Great Lakes
Fishery Commission, the Great Lakes and St. Lawrence Cities Ini-
tiative, and the University of Toledo's Lake Erie Center are heard
from as well. For legal guidance, concerned parties refer to the
Nonindigenous Aquatic Nuisance Prevention and Control Act of
1990 (revised as the National Invasive Species Act of 1996, revised
as the National Aquatic Invasive Species Act of 2007); also, to the
Asian Carp Prevention and Control Act of 2006.

Man proposes, carp disposes. Through waving weedbeds of bureaucracy and human cross-purposes, the fish swims. Starting in 2009, tests for Asian-carp DNA in Chicago waters indicated that the fish might have moved beyond the barriers. In February 2010 the state of Illinois, hoping to calm its neighbors, began an intensive program of fishing for actual fish with electroshocking and nets, while the DNA tests continued. These efforts turned up more DNA but no fish. In March the Illinois DNR announced that six weeks of searching had found no Asian carp throughout the entire Chicago Area Waterway System. Then, on June 22, a commercial fisherman working for the DNR netted a nineteen-pound-six-ounce bighead carp in Lake Calumet, thirty miles beyond the electric barrier and six miles from Lake Michigan. Between this well-fed, healthy male Asian carp and the Great Lakes, no obstacle intervened.

If the neighbors had been worried before, they began to sweat and hyperventilate now. Outcries of alarm came from officialdom in Minnesota, Wisconsin, Michigan, Ohio, Pennsylvania, and New York, as well as from Canada. Michigan's attorney general, Mike Cox, brought suit in federal district court to force the Corps of Engineers and the city of Chicago to close the locks immediately. Four other Great Lakes states, but not New York, joined on Michigan's side; two previous lawsuits for the same purpose had already failed. The Senate Energy and Natural Resources Subcommittee on Water and Power held hearings on the federal response to Asian carp. The governor and the attorney general of Ohio called on the president to convene a White House Asian Carp Emergency Summit. Dave Camp, a congressman from Michigan, proposed a piece of legislation he called the CARP ACT, for Close All Routes and Prevent Asian Carp Today.

Senator Dick Durbin of Illinois announced that he had asked the president to appoint a Coordinated Response Commander for Asian Carp, and the president had agreed. This so-called carp czar was to be chosen within thirty days. The president made clear that he had a "zero tolerance" policy for invasive species. Senator Kirsten Gillibrand of New York announced her support for another anti-carp bill, the Permanent Prevention of Asian Carp Act. Michigan's Mike Cox accused Obama of not doing enough about the problem because he sympathized with his own state of Illinois. (Cox, a Republican, is running for governor.) Senator Carl Levin, Democrat of Michigan, sent Obama a letter asking him to act quickly to protect the Great Lakes from Asian carp, and thirty-two representatives and fourteen senators added their names.

Unlike Romeoville, the place where the apocalyptic bighead was caught is quiet. To some it might even be paradise. The part of Lake Calumet in which the commercial fisherman netted this carp doubles as the water hazard beside the concluding holes of a luxury golf course called Harborside, rated by *Golfweek* as the third-best municipal golf course in the country. One afternoon I walked some of the course. To play eighteen at Harborside on a Saturday or a Sunday costs $95; new SUVs filled the parking lot. Several of the raised tees provided a panoramic view of the Chicago skyline, while the surrounding Rust Belt ruins and ghetto neighborhoods of south Chicago just beyond the fence seemed far away. I asked employees at the clubhouse golf shop, waiters in the restaurant, a bartender, and a man tending the greens, but none had heard of the nation-shaking carp netted here two months before.

The reason the fisherman happened to be fishing there in the first place was that a team of scientists from Notre Dame had already found bighead-carp DNA in the water. Working for the Corps of Engineers, the university's Center for Aquatic Conservation has been doing DNA tests for more than a year. As part of the increased tracking of Asian carp, the Notre Dame scientists collected samples from bodies of water all over Chicago. David M. Lodge, a professor of biology, heads the center. I stopped by South Bend to see him on my way back from Illinois.

Professor Lodge is a tall and genial man in his fifties, with a southern accent, blue eyes, and a D.Phil. in zoology from Oxford University, which he attended on a Rhodes. He wore a yellow tennis shirt, with his ballpoint pen and mechanical pencil stuck neatly in the button part of the neck, an innovation I admired, because I was wearing the same kind of shirt and had compensated for its lack of breast pocket by putting my pens in my pants pockets, always an awkward deal. He asked if I was writing something funny on carp. I said I probably was. "I know you can't not laugh when you see the silver carp jumping all over the place, but it's really not funny," he said. "It's a tragic thing, and people are wrong to trivialize it. We should focus on these fish's potential environmental and economic impact. In the Great Lakes—just as we're seeing now in south Louisiana—the environment *is* the economy. Look how the degrading of Lake Erie in the sixties and seventies contributed to the decline of Detroit and Cleveland and Buffalo. To people who say this is a question of jobs versus the environment, I say it's not either-or.

"The bigger issue is how we as a country protect ourselves against invasive species. At the moment, we are not very good at preventing invasions. We're constantly reacting after it's too late. Most invasions, if detected early, can be stopped, because establishing an organism so it's viable in a new environment is not automatic. Our current approach is more or less open-door. Right now the canal-and-river passage across Illinois from the lake to the Mississippi is a highway for the dispersal of organisms. The Great Lakes is a hot spot for aquatic invasions. In the lakes there are a hundred and six species nonnative to North America that are not in the Mississippi, while there are only fifty in the Mississippi that are not in the Great Lakes. An even greater threat, really, is of invasions going in the opposite direction from the carp's—that is, going from the lakes to the Mississippi. The Mississippi system holds the richest heritage of biodiversity in North America. The electric barrier at Romeo-ville was built originally to stop a small invasive fish called the round goby from coming south—too late, as it turned out, because today the round gobies are established in the Illinois. A later invasive, the tubenose goby, does appear to have been stopped. So the barrier may have helped. But over time it will not be able to stop everything.

"As for the Asian carp and the lakes, worst case is they're already established there. We wouldn't necessarily know. Usually, the first sign we have that organisms are invading is that members of the public see them. But people aren't underwater, and netting and electroshocking are not good tools for finding out what's going on in a body of water, especially not in one as big as Lake Michigan."

To explain more about how DNA testing for Asian carp works, David Lodge led me to the office of his colleague Christopher Jerde, a brown-haired South Dakotan who helped develop the technique. I did not follow all the science of it. Essentially, the technique uses DNA found in water samples to determine what species might have been present, just as DNA evidence can suggest that a person was at a crime scene. The DNA sequences in water samples are not followed to the point where individual fish are singled out, but that could be done, too. Christopher Jerde said that people who wanted to ignore the carp problem kept pooh-poohing Notre Dame's DNA findings before the confirming bighead carp was caught. They argued that the DNA could have been carried in on a duck's feathers or something. But in places where his team got multiple positive hits, he knew the fish had to be there. When he was proved right, he took no pleasure. The carp invasion only makes him mourn.

Some planners who take a long view believe that the Midwest's invasive-species problem requires a bold solution: the complete separation, or reseparation, of the Great Lakes from the Mississippi at Chicago. This huge infrastructure project would consist of concrete dikes, new shipping terminals, new water-treatment facilities, maybe barge lifts to transfer freight, maybe the rereversing of Chicago's wastewater flow so that it no longer goes south. Business interests tend to hate this idea. I asked Jerde about it, and he said, "Right here is where you could put something like that," and called up a Google Earth photo of Chicago's South Side on his computer. "This empty area here is a Rust Belt nowhere left over from old Chicago," he said. "The main hub of a new shipping and hydrological arrangement for Chicago could go right here." As it happened, the brown and empty region he indicated on the satellite photo was not far from the green oases of the Harborside golf course.

"Well, we don't have opinions on policy," he continued. "We just provide the data. We give it to the Corps or the DNR, and they present it to the people who make the decisions, and it's out of our hands. When people were coming up with all those supposed rebuttals to our findings, we couldn't defend ourselves. It's frustrating. People say we're either wrong or exaggerating, so there's no problem and there's nothing we need to do, or else they turn around and say it's too late, the carp are already here, so there's nothing we *can* do. I don't know whether we can stop these fish, but if we do nothing I can guarantee this problem and plenty of others will get worse.

"Now, let me just show you a few of the places where we've had positive DNA hits for Asian carp." He touched a button on his computer, and little yellow thumbtack-like circles appeared on the

Google Earth photo. They followed the Ship Canal, they appeared in the Grand Calumet and Little Calumet rivers, they clustered at the south end of the Romeoville electric barrier—proof that it was stopping fish, he said. "Last summer we had a positive hit here, in Calumet Harbor, which is separated from the lake only by a breakwater," he said. Then he moved in to a closer view of Chicago's downtown. "We also got a positive right here." He pointed to a yellow thumbtack at the Chicago River Lock and Dam, under Lake Shore Drive, about a mile and a half north of the Shedd Aquarium.

DAVID H. FREEDMAN

Lies, Damned Lies, and Medical Science

FROM *The Atlantic*

IN 2001 RUMORS were circulating in Greek hospitals that surgery residents, eager to rack up scalpel time, were falsely diagnosing hapless Albanian immigrants with appendicitis. At the University of Ioannina medical school's teaching hospital, a newly minted doctor named Athina Tatsioni was discussing the rumors with colleagues when a professor who had overheard asked her if she'd like to try to prove whether they were true — he seemed to be almost daring her. She accepted the challenge and, with the professor's and other colleagues' help, eventually produced a formal study showing that for whatever reason the appendices removed from patients with Albanian names in six Greek hospitals were more than three times as likely to be perfectly healthy as those removed from patients with Greek names. "It was hard to find a journal willing to publish it, but we did," recalls Tatsioni. "I also discovered that I really liked research." Good thing, because the study had actually been a sort of audition. The professor, it turned out, had been putting together a team of exceptionally brash and curious young clinicians and PhDs to join him in tackling an unusual and controversial agenda.

Last spring I sat in on one of the team's weekly meetings on the medical school's campus, which is plunked crazily across a series of sharp hills. The building in which we met, like most at the school, had the look of a barracks and was festooned with political graffiti. But the group convened in a spacious conference room that would have been at home at a Silicon Valley start-up. Sprawled around a large table were Tatsioni and eight other youngish Greek researchers and physicians who, in contrast to the pasty younger staff frequently seen in U.S. hospitals, looked like the casually glamorous cast of a television medical drama. The professor, a dapper and soft-spoken man named John Ioannidis, loosely presided.

One of the researchers, a biostatistician named Georgia Salanti, fired up a laptop and projector and started to take the group through a study she and a few colleagues were completing that asked this question: were drug companies manipulating published research to make their drugs look good? Salanti ticked off data that seemed to indicate they were, but the other team members almost immediately started interrupting. One noted that Salanti's study didn't address the fact that drug-company research wasn't measuring critically important "hard" outcomes for patients, such as survival versus death, and instead tended to measure "softer" outcomes, such as self-reported symptoms ("my chest doesn't hurt as much today"). Another pointed out that Salanti's study ignored the fact that when drug-company data seemed to show patients' health improving, the data often failed to show that the drug was responsible or that the improvement was more than marginal.

Salanti remained poised, as if the grilling were par for the course, and gamely acknowledged that the suggestions were all good — but a single study can't prove everything, she said. Just as I was getting the sense that the data in drug studies were endlessly malleable, Ioannidis, who had mostly been listening, delivered what felt like a coup de grâce: wasn't it possible, he asked, that drug companies were carefully selecting the topics of their studies — for example, comparing their new drugs against those already known to be inferior to others on the market — so that they were ahead of the game even before the data juggling began? "Maybe sometimes it's the questions that are biased, not the answers," he said, flashing a friendly smile. Everyone nodded. Though the results of drug studies often make newspaper headlines, you have to wonder whether they prove anything at all. Indeed, given the breadth of the potential problems raised at the meeting, can *any* medical-research studies be trusted?

That question has been central to Ioannidis's career. He's what's known as a meta-researcher, and he's become one of the world's foremost experts on the credibility of medical research. He and his team have shown, again and again, and in many different ways, that much of what biomedical researchers conclude in published studies—conclusions that doctors keep in mind when they prescribe antibiotics or blood-pressure medication, or when they advise us to consume more fiber or less meat, or when they recommend surgery for heart disease or back pain—is misleading, exaggerated, and often flat-out wrong. He charges that as much as 90 percent of the published medical information that doctors rely on is flawed. His work has been widely accepted by the medical community; it has been published in the field's top journals, where it is heavily cited; and he is a big draw at conferences. Given this exposure, and the fact that his work broadly targets everyone else's work in medicine, as well as everything that physicians do and all the health advice we get, Ioannidis may be one of the most influential scientists alive. Yet for all his influence, he worries that the field of medical research is so pervasively flawed, and so riddled with conflicts of interest, that it might be chronically resistant to change —or even to publicly admitting that there's a problem.

The city of Ioannina is a big college town a short drive from the ruins of a 20,000-seat amphitheater and a Zeusian sanctuary built at the site of the Dodona oracle. The oracle was said to have issued pronouncements to priests through the rustling of a sacred oak tree. Today a different oak tree at the site provides visitors with a chance to try their own hands at extracting a prophecy. "I take all the researchers who visit me here, and almost every single one of them asks the tree the same question," Ioannidis tells me as we contemplate the tree the day after the team's meeting. "'Will my research grant be approved?'" He chuckles, but Ioannidis (pronounced yo-NEE-dees) tends to laugh not so much in mirth as to soften the sting of his attack. And sure enough, he goes on to suggest that an obsession with winning funding has gone a long way toward weakening the reliability of medical research.

He first stumbled on the sorts of problems plaguing the field, he explains, as a young physician-researcher in the early 1990s at Harvard. At the time he was interested in diagnosing rare diseases, for which a lack of case data can leave doctors with little to go on other than intuition and rules of thumb. But he noticed that doctors seemed to proceed in much the same manner even when it came to cancer, heart disease, and other common ailments. Where were the hard data that would back up their treatment decisions? There was plenty of published research, but much of it was remarkably unscientific, based largely on observations of a small number of cases. A new "evidence-based medicine" movement was just starting to gather force, and Ioannidis decided to throw himself into it, working first with prominent researchers at Tufts University and then taking positions at Johns Hopkins University and the National Institutes of Health. He was unusually well armed: he had been a math prodigy of near-celebrity status in high school in Greece, and had followed his parents, who were both physician-researchers, into medicine. Now he'd have a chance to combine math and medicine by applying rigorous statistical analysis to what seemed a surprisingly sloppy field. "I assumed that everything we physicians did was basically right, but now I was going to help verify it," he says. "All we'd have to do was systematically review the evidence, trust what it told us, and then everything would be perfect."

It didn't turn out that way. In poring over medical journals, he was struck by how many findings of all types were refuted by later findings. Of course, medical-science "never minds" are hardly secret. And they sometimes make headlines, as when in recent years large studies or growing consensuses of researchers concluded that mammograms, colonoscopies, and PSA tests were far less useful cancer-detection tools than we had been told; or when widely prescribed antidepressants such as Prozac, Zoloft, and Paxil were revealed to be no more effective than a placebo for most cases of depression; or when we learned that staying out of the sun entirely can actually increase cancer risks; or when we were told that the advice to drink lots of water during intense exercise was potentially fatal; or when, last April, we were informed that taking fish oil, exercising, and doing puzzles doesn't really help fend off Alzheimer's disease, as had long been claimed. Peer-reviewed studies have come to opposite conclusions on whether using cell phones can cause brain cancer, whether sleeping more than eight hours a night is healthful or dangerous, whether taking aspirin every day is more likely to save your life or cut it short, and whether routine angioplasty works better than pills to unclog heart arteries.

But beyond the headlines, Ioannidis was shocked at the range and reach of the reversals he was seeing in everyday medical research. "Randomized controlled trials," which compare how one group responds to a treatment against how an identical group fares without the treatment, had long been considered nearly unshakable evidence, but they, too, ended up being wrong some of the time. "I realized even our gold-standard research had a lot of problems," he says. Baffled, he started looking for the specific ways in which studies were going wrong. And before long he discovered that the range of errors being committed was astonishing: from what questions researchers posed, to how they set up the studies, to which patients they recruited for the studies, to which measurements they took, to how they analyzed the data, to how they presented their results, to how particular studies came to be published in medical journals.

This array suggested a bigger, underlying dysfunction, and Ioannidis thought he knew what it was. "The studies were biased," he says. "Sometimes they were overtly biased. Sometimes it was difficult to see the bias, but it was there." Researchers headed into their studies wanting certain results—and, lo and behold, they were getting them. We think of the scientific process as being objective, rigorous, and even ruthless in separating out what is true from what we merely wish to be true, but in fact it's easy to manipulate results, even unintentionally or unconsciously. "At every step in the process, there is room to distort results, a way to make a stronger claim or to select what is going to be concluded," says Ioannidis. "There is an intellectual conflict of interest that pressures researchers to find whatever it is that is most likely to get them funded."

Perhaps only a minority of researchers were succumbing to this bias, but their distorted findings were having an outsize effect on published research. To get funding and tenured positions, and often merely to stay afloat, researchers have to get their work published in well-regarded journals, where rejection rates can climb above 90 percent. Not surprisingly, the studies that tend to make the grade are those with eye-catching findings. But while coming up with eye-catching theories is relatively easy, getting reality to bear them out is another matter. The great majority collapse under the weight of contradictory data when studied rigorously. Imagine, though, that five different research teams test an interesting theory that's making the rounds, and four of the groups correctly prove the idea false, while the one less cautious group incorrectly "proves" it true through some combination of error, fluke, and clever selection of data. Guess whose findings your doctor ends up reading about in the journal and you end up hearing about on the evening news? Researchers can sometimes win attention by refuting a prominent finding, which can help to at least raise doubts about results, but in general it is far more rewarding to add a new insight or exciting-sounding twist to existing research than to retest its basic premises—after all, simply re-proving someone else's results is unlikely to get you published, and attempting to undermine the work of respected colleagues can have ugly professional repercussions.

In the late 1990s, Ioannidis set up a base at the University of Ioannina. He pulled together his team, which remains largely intact today, and started chipping away at the problem in a series of papers that pointed out specific ways certain studies were getting misleading results. Other meta-researchers were also starting to spotlight disturbingly high rates of error in the medical literature. But Ioannidis wanted to get the big picture across, and to do so with solid data, clear reasoning, and good statistical analysis. The project dragged on, until finally he retreated to the tiny island of Sikinos in the Aegean Sea, where he drew inspiration from the relatively primitive surroundings and the intellectual traditions they recalled. "A pervasive theme of ancient Greek literature is that you need to pursue the truth, no matter what the truth might be," he says. In 2005 he unleashed two papers that challenged the foundations of medical research.

He chose to publish one paper, fittingly, in the online journal *PLoS Medicine,* which is committed to running any methodologically sound article without regard to how "interesting" the results may be. In the paper, Ioannidis laid out a detailed mathematical proof that, assuming modest levels of researcher bias, typically imperfect research techniques, and the well-known tendency to focus on exciting rather than highly plausible theories, researchers will come up with wrong findings most of the time. Simply put, if you're attracted to ideas that have a good chance of being wrong, and if you're motivated to prove them right, and if you have a little wiggle room in how you assemble the evidence, you'll probably succeed in proving wrong theories right. His model predicted, in different fields of medical research, rates of wrongness roughly corresponding to the observed rates at which findings were later convincingly refuted: 80 percent of nonrandomized studies (by far the most common type) turn out to be wrong, as do 25 percent of supposedly gold-standard randomized trials, and as much as 10 percent of the platinum-standard large randomized trials. The article spelled out his belief that researchers were frequently manipulating data analyses, chasing career-advancing findings rather than good science, and even using the peer-review process—in which journals ask researchers to help decide which studies to publish—to suppress opposing views. "You can question some of the details of John's calculations, but it's hard to argue that the essential ideas aren't absolutely correct," says Doug Altaian, an Oxford University researcher who directs the Centre for Statistics in Medicine.

Still, Ioannidis anticipated that the community might shrug off his findings: sure, a lot of dubious research makes it into journals, but we researchers and physicians know to ignore it and focus on the good stuff, so what's the big deal? The other paper headed off that claim. He zoomed in on forty-nine of the most highly regarded research findings in medicine over the previous thirteen years, as judged by the science community's two standard measures: the papers had appeared in the journals most widely cited in research articles, and the forty-nine articles themselves were the most widely cited articles in these journals. These were articles that helped lead to the widespread popularity of treatments such as the use of hormone-replacement therapy for menopausal women, vitamin E to reduce the risk of heart disease, coronary stents to ward off heart attacks, and daily low-dose aspirin to control blood pressure and prevent heart attacks and strokes. Ioannidis was putting his contentions to the test not against run-of-the-mill research, or even merely well-accepted research, but against the absolute tip of the research pyramid. Of the forty-nine articles, forty-five claimed to have uncovered effective interventions. Thirty-four of these claims had been retested, and fourteen of these, or 41 percent, had been convincingly shown to be wrong or significantly exaggerated. If between a third and a half of the most acclaimed research in medicine was proving untrustworthy, the scope and impact of the problem were undeniable. That article was published in the *Journal of the American Medical Association*.

Driving me back to campus in his smallish SUV—after insisting, as he apparently does with all his visitors, on showing me a nearby lake and the six monasteries situated on an islet within it—Ioannidis apologized profusely for running a yellow light, explaining with a laugh that he didn't trust the truck behind him to stop. Considering his willingness, even eagerness, to slap the face of the medical-research community, Ioannidis comes off as thoughtful, upbeat, and deeply civil. He's a careful listener, and his frequent grin and semi-apologetic chuckle can make the sharp prodding of his arguments seem almost good-natured. He is as quick, if not quicker, to question his own motives and competence as anyone else's. A neat and compact forty-five-year-old with a trim mustache, he presents as a sort of dashing nerd—Giancarlo Giannini with a bit of Mr. Bean.

The humility and graciousness seem to serve him well in getting across a message that is not easy to digest or, for that matter, believe: that even highly regarded researchers at prestigious institutions sometimes churn out attention-grabbing findings rather than findings likely to be right. But Ioannidis points out that obviously questionable findings cram the pages of top medical journals, not to mention the morning headlines. Consider, he says, the endless stream of results from nutritional studies in which researchers follow thousands of people for some number of years, tracking what they eat and what supplements they take, and how their health changes over the course of the study. "Then the researchers start asking, What did vitamin E do? What did vitamin C or D or A do? What changed with calorie intake, or protein or fat intake? What happened to cholesterol levels? Who got what type of cancer?" he says. "They run everything through the mill, one at a time, and they start finding associations, and eventually conclude that vitamin X lowers the risk of cancer Y, or this food helps with the risk of that disease." In a single week this fall, Google's news page offered these headlines: "More Omega-3 Fats Didn't Aid Heart Patients"; "Fruits, Vegetables Cut Cancer Risk for Smokers"; "Soy May Ease Sleep Problems in Older Women"; and dozens of similar stories.

When a five-year study of 10,000 people finds that those who take more vitamin X are less likely to get cancer Y, you'd think you have pretty good reason to take more vitamin X, and physicians routinely pass these recommendations on to patients. But these studies often sharply conflict with one another. Studies have gone back and forth on the cancer-preventing powers of vitamins A, D, and E; on the heart-health benefits of eating fat and carbs; and even on the question of whether being overweight is more likely to extend or shorten your life. How should we choose among these dueling high-profile nutritional findings? Ioannidis suggests a simple approach: ignore them all.

For starters, he explains, the odds are that in any large database of many nutritional and health factors, there will be a few apparent connections that are in fact merely flukes, not real health effects —it's a bit like combing through long, random strings of letters and claiming there's an important message in any words that happen to turn up. But even if a study managed to highlight a genuine health connection to some nutrient, you're unlikely to benefit much from taking more of it, because we consume thousands of nutrients that act together as a sort of network, and changing your intake of just one of them is bound to cause ripples throughout the network that are far too complex for these studies to detect and that may be as likely to harm you as help you. Even if changing that one factor does bring on the claimed improvement, there's still a good chance that it won't do you much good in the long run, because these studies rarely go on long enough to track the decades-long course of disease and ultimately death. Instead, they track easily measurable health "markers" such as cholesterol levels, blood pressure, and blood-sugar levels, and meta-experts have shown that changes in these markers often don't correlate as well with long-term health as we have been led to believe.

On the relatively rare occasions when a study does go on long enough to track mortality, the findings frequently upend those of the shorter studies. (For example, though the vast majority of studies of overweight individuals link excess weight to ill health, the longest of them haven't convincingly shown that overweight people are likely to die sooner, and a few of them have seemingly demonstrated that moderately overweight people are likely to live *longer*.) And these problems are aside from ubiquitous measurement errors (for example, people habitually misreport their diets in studies), routine misanalysis (researchers rely on complex software capable of juggling results in ways they don't always understand), and the less common, but serious, problem of outright fraud (which has been revealed, in confidential surveys, to be much more widespread than scientists like to acknowledge).

If a study somehow avoids every one of these problems and finds a real connection to long-term changes in health, you're still not guaranteed to benefit, because studies report average results that typically represent a vast range of individual outcomes. Should you be among the lucky minority that stands to benefit, don't expect a noticeable improvement in your health, because studies usually detect only modest effects that merely tend to whittle your chances of succumbing to a particular disease from small to somewhat smaller. "The odds that anything useful will survive from any of these studies are poor," says Ioannidis—dismissing in a breath a good chunk of the research into which we sink about $100 billion a year in the United States alone.

And so it goes for all medical studies, he says. Indeed, nutritional studies aren't the worst. Drug studies have the added corruptive force of financial conflict of interest. The exciting links between genes and various diseases and traits that are relentlessly hyped in the press for heralding miraculous around-the-corner treatments for everything from colon cancer to schizophrenia have in the past proved so vulnerable to error and distortion, Ioannidis has found, that in some cases you'd have done about as well by throwing darts at a chart of the genome. (These studies seem to have improved somewhat in recent years, but whether they will hold up or be useful in treatment are still open questions.) Vioxx, Zelnorm, and Baycol were among the widely prescribed drugs found to be safe and effective in large randomized controlled trials before the drugs were yanked from the market as unsafe or not so effective or both.

"Often the claims made by studies are so extravagant that you can immediately cross them out without needing to know much about the specific problems with the studies," Ioannidis says. But of course it's that very extravagance of claim (one large randomized controlled trial even proved that secret prayer by unknown parties can save the lives of heart-surgery patients, while another proved that secret prayer can harm them) that helps gets these findings into journals and then into our treatments and lifestyles, especially when the claim builds on impressive-sounding evidence. "Even when the evidence shows that a particular research idea is wrong, if you have thousands of scientists who have invested their careers in it, they'll continue to publish papers on it," he says. "It's like an epidemic, in the sense that they're infected with these wrong ideas, and they're spreading it to other researchers through journals."

Though scientists and science journalists are constantly talking up the value of the peer-review process, researchers admit among themselves that biased, erroneous, and even blatantly fraudulent studies easily slip through it. *Nature,* the grande dame of science journals, stated in a 2006 editorial, "Scientists understand that peer review per se provides only a minimal assurance of quality, and that the public conception of peer review as a stamp of authentication is far from the truth." What's more, the peer-review process often pressures researchers to shy away from striking out in genuinely new directions and instead to build on the findings of their colleagues (that is, their potential reviewers) in ways that only *seem* like breakthroughs—as with the exciting-sounding gene linkages (autism genes identified!) and nutritional findings (olive oil lowers blood pressure!) that are really just dubious and conflicting variations on a theme.

Most journal editors don't even claim to protect against the problems that plague these studies. University and government research overseers rarely step in to directly enforce research quality, and when they do, the science community goes ballistic over the outside interference. The ultimate protection against research error and bias is supposed to come from the way scientists constantly retest each other's results—except they don't. Only the most prominent findings are likely to be put to the test, because there's likely to be publication payoff in firming up the proof or contradicting it.

But even for medicine's most influential studies, the evidence sometimes remains surprisingly narrow. Of those forty-five supercited studies that Ioannidis focused on, eleven had never been retested. Perhaps worse, Ioannidis found that even when a research error is outed, it typically persists for years or even decades. He looked at three prominent health studies from the 1980s and 1990s that were each later soundly refuted and discovered that researchers continued to cite the original results as correct more often than as flawed—in one case for at least twelve years after the results were discredited.

Doctors may notice that their patients don't seem to fare as well with certain treatments as the literature would lead them to expect, but the field is appropriately conditioned to subjugate such anecdotal evidence to study findings. Yet much, perhaps even most, of what doctors do has never been formally put to the test in credible studies, given that the need to do so became obvious to the field only in the 1990s, leaving it playing catch-up with a century or more of non-evidence-based medicine, and contributing to Ioannidis's shockingly high estimate of the degree to which medical knowledge is flawed. That we're not routinely made seriously ill by this shortfall, he argues, is due largely to the fact that most medical interventions and advice don't address life-and-death situations but rather aim to leave us marginally healthier or less unhealthy, so we usually neither gain nor risk all that much.

Medical research is not especially plagued with wrongness. Other meta-research experts have confirmed that similar issues distort research in all fields of science, from physics to economics (where the highly regarded economists J. Bradford DeLong and Kevin Lang once showed how a remarkably consistent paucity of strong evidence in published economics studies made it unlikely that *any* of them were right). And needless to say, things only get worse when it comes to the pop expertise that endlessly spews at us from diet, relationship, investment, and parenting gurus and pundits. But we expect more of scientists, and especially of medical scientists, given that we believe we are staking our lives on their results. The public hardly recognizes how bad a bet this is. The medical community itself might still be largely oblivious to the scope of the problem, if Ioannidis hadn't forced a confrontation when he published his studies in 2005.

Ioannidis initially thought the community might come out fighting. Instead, it seemed relieved, as if it had been guiltily waiting for someone to blow the whistle, and eager to hear more. David Gorski, a surgeon and researcher at Detroit's Barbara Ann Karmanos Cancer Institute, noted in his prominent medical blog that when he presented Ioannidis's paper on highly cited research at a professional meeting, "not a single one of my surgical colleagues was the least bit surprised or disturbed by its findings." Ioannidis offers a theory for the relatively calm reception. "I think that people didn't feel I was only trying to provoke them, because I showed that it was a community problem, instead of pointing fingers at individual examples of bad research," he says. In a sense, he gave scientists an opportunity to cluck about the wrongness without having to acknowledge that they themselves succumb to it—it was something everyone else did.

To say that Ioannidis's work has been embraced would be an understatement. His *PLoS Medicine* paper is the most downloaded in the journal's history, and it's not even Ioannidis's most-cited work —that would be a paper he published in *Nature Genetics* on the problems with gene-link studies. Other researchers are eager to work with him: he has published papers with 1,328 different coauthors at 538 institutions in forty-three countries, he says. Last year he received, by his estimate, invitations to speak at a thousand conferences and institutions around the world, and he was accepting an average of about five invitations a month until a case last year of excessive-travel-induced vertigo led him to cut back. Even so, in the weeks before I visited him he had addressed an AIDS conference in San Francisco, the European Society for Clinical Investigation, Harvard's School of Public Health, and the medical schools at Stanford and Tufts.

The irony of his having achieved this sort of success by accusing the medical-research community of chasing after success is not lost on him, and he notes that it ought to raise the question of whether he himself might be pumping up his findings. "If I did a study and the results showed that in fact there wasn't really much bias in research, would I be willing to publish it?" he asks. "That would create a real psychological conflict for me." But his bigger worry, he says, is that while his fellow researchers seem to be getting the message, he hasn't necessarily forced anyone to do a better job. He fears he won't in the end have done much to improve anyone's health. "There may not be fierce objections to what I'm saying," he explains. "But it's difficult to change the way that everyday doctors, patients, and healthy people think and behave."

As helter-skelter as the University of Ioannina Medical School campus looks, the hospital abutting it looks reassuringly stolid. Athina Tatsioni has offered to take me on a tour of the facility, but we make it only as far as the entrance when she is greeted—accosted, really—by a worried-looking older woman. Tatsioni, normally a bit reserved, is warm and animated with the woman, and the two have a brief but intense conversation before embracing and saying goodbye. Tatsioni explains to me that the woman and her husband were patients of hers years ago; now the husband has been admitted to the hospital with abdominal pains, and Tatsioni has promised she'll stop by his room later to say hello. Recalling the appendicitis story, I prod a bit, and she confesses she plans to do her own exam. She needs to be circumspect, though, so she won't appear to be second-guessing the other doctors.

Tatsioni doesn't so much fear that someone will carve out the man's healthy appendix. Rather, she's concerned that, like many patients, he'll end up with prescriptions for multiple drugs that will do little to help him and may well harm him. "Usually what happens is that the doctor will ask for a suite of biochemical tests —liver fat, pancreas function, and so on," she tells me. "The tests could turn up something, but they're probably irrelevant. Just having a good talk with the patient and getting a close history is much more likely to tell me what's wrong." Of course, the doctors have all been trained to order these tests, she notes, and doing so is a lot quicker than a long bedside chat. They're also trained to ply the patient with whatever drugs might help whack any errant test numbers back into line. What they're not trained to do is to go back and look at the research papers that helped make these drugs the standard of care. "When you look the papers up, you often find the drugs didn't even work better than a placebo. And no one tested how they worked in combination with the other drugs," she says. "Just taking the patient off everything can improve their health right away." But not only is checking out the research another time-consuming task, patients often don't even *like* it when they're taken off their drugs, she explains; they find their prescriptions reassuring.

Later, Ioannidis tells me he makes a point of having several clinicians on his team. "Researchers and physicians often don't understand each other; they speak different languages," he says. Knowing that some of his researchers are spending more than half their time seeing patients makes him feel the team is better positioned to bridge that gap; their experience informs the team's research with firsthand knowledge and helps the team shape its papers in a way more likely to hit home with physicians. It's not that he envisions doctors making all their decisions based solely on solid evidence—there's simply too much complexity in patient treatment to pin down every situation with a great study. "Doctors need to rely on instinct and judgment to make choices," he says. "But these choices should be as informed as possible by the evidence. And if the evidence isn't good, doctors should know that, too. And so should patients."

In fact, the question of whether the problems with medical research should be broadcast to the public is a sticky one in the meta-research community. Already feeling that they're fighting to keep patients from turning to alternative medical treatments such as homeopathy, or misdiagnosing themselves on the Internet, or simply neglecting medical treatment altogether, many researchers and physicians aren't eager to provide even more reason to be skeptical of what doctors do—not to mention how public disenchantment with medicine could affect research funding. Ioannidis dismisses these concerns. "If we don't tell the public about these problems, then we're no better than nonscientists who falsely claim they can heal," he says. "If the drugs don't work and we're not sure how to treat something, why should we claim differently? Some fear that there may be less funding because we stop claiming we can prove we have miraculous treatments. But if we can't really provide those miracles, how long will we be able to fool the public anyway? The scientific enterprise is probably the most fantastic achievement in human history, but that doesn't mean we have a right to overstate what we're accomplishing."

We could solve much of the wrongness problem, Ioannidis says, if the world simply stopped expecting scientists to be right. That's because being wrong in science is fine, and even necessary—as long as scientists recognize that they blew it, report their mistake openly instead of disguising it as a success, and then move on to the next thing, until they come up with the very occasional genuine breakthrough. But as long as careers remain contingent on producing a stream of research that's dressed up to seem more right than it is, scientists will keep delivering exactly that.

"Science is a noble endeavor, but it's also a low-yield endeavor," he says. "I'm not sure that more than a very small percentage of medical research is ever likely to lead to major improvements in clinical outcomes and quality of life. We should be very comfortable with that fact."

ATUL GAWANDE

Letting Go

FROM *The New Yorker*

SARA THOMAS MONOPOLI was pregnant with her first child when her doctors learned that she was going to die. It started with a cough and a pain in her back. Then a chest X-ray showed that her left lung had collapsed, and her chest was filled with fluid. A sample of the fluid was drawn off with a long needle and sent for testing. Instead of an infection, as everyone had expected, it was lung cancer, and it had already spread to the lining of her chest. Her pregnancy was thirty-nine weeks along, and the obstetrician who had ordered the test broke the news to her as she sat with her husband and her parents. The obstetrician didn't get into the prognosis—she would bring in an oncologist for that—but Sara was stunned. Her mother, who had lost her best friend to lung cancer, began crying.

The doctors wanted to start treatment right away, and that meant inducing labor to get the baby out. For the moment, though, Sara and her husband, Rich, sat by themselves on a quiet terrace off the labor floor. It was a warm Monday in June 2007. She took Rich's hands, and they tried to absorb what they had heard. Monopoli was thirty-four. She had never smoked or lived with anyone who had. She exercised. She ate well. The diagnosis was bewildering. "This is going to be okay," Rich told her. "We're going to work through this. It's going to be hard, yes. But we'll figure it out. We can find the right treatment." For the moment, though, they had a baby to think about.

"So Sara and I looked at each other," Rich recalled, "and we said, 'We don't have cancer on Tuesday. It's a cancer-free day. We're having a baby. It's exciting. And we're going to enjoy our baby.'" On Tuesday, at 8:55 P.M., Vivian Monopoli, seven pounds nine ounces, was born. She had wavy brown hair like her mom, and she was perfectly healthy.

The next day Sara underwent blood tests and body scans. Dr. Paul Marcoux, an oncologist, met with her and her family to discuss the findings. He explained that she had a non-small-cell lung cancer that had started in her left lung. Nothing she had done had brought this on. More than 15 percent of lung cancers—more than people realize—occur in nonsmokers. Hers was advanced, having metastasized to multiple lymph nodes in her chest and its lining. The cancer was inoperable. But there were chemotherapy options, notably a relatively new drug called Tarceva, which targets a gene mutation commonly found in lung cancers of female nonsmokers. Eighty-five percent respond to this drug, and, Marcoux said, "some of these responses can be long-term."

Words like "respond" and "long-term" provide a reassuring gloss on a dire reality. There is no cure for lung cancer at this stage. Even with chemotherapy, the median survival is about a year. But it seemed harsh and pointless to confront Sara and Rich with this now. Vivian was in a bassinet by the bed. They were working hard to be optimistic. As Sara and Rich later told the social worker who was sent to see them, they did not want to focus on survival statistics. They wanted to focus on "aggressively managing" this diagnosis.

Sara was started on the Tarceva, which produced an itchy, acne-like facial rash and numbing tiredness. She also underwent a surgical procedure to drain the fluid around her lung; when the fluid kept coming back, a thoracic surgeon eventually placed a small, permanent tube in her chest, which she could drain whenever fluid accumulated and interfered with her breathing. Three weeks after the delivery, she was admitted to the hospital with severe shortness of breath from a pulmonary embolism—a blood clot in an artery to the lungs, which is dangerous but not uncommon in cancer patients. She was started on a blood thinner. Then test results showed that her tumor cells did not have the mutation that Tarceva targets. When Marcoux told Sara that the drug wasn't going to work, she had an almost violent physical reaction to the news, bolting to the bathroom in mid-discussion with a sudden bout of diarrhea.

Dr. Marcoux recommended a different, more standard chemotherapy, with two drugs called carboplatin and paclitaxel. But the paclitaxel triggered an extreme, nearly overwhelming allergic response, so he switched her to a regimen of carboplatin plus gemcitabine. Response rates, he said, were still very good for patients on this therapy.

She spent the remainder of the summer at home, with Vivian and her husband and her parents, who had moved in to help. She loved being a mother. Between chemotherapy cycles, she began trying to get her life back.

Then in October, a CT scan showed that the tumor deposits in her left lung and chest and lymph nodes had grown substantially. The chemotherapy had failed. She was switched to a drug called pemetrexed. Studies found that it could produce markedly longer survival in some patients. In reality, however, only a small percentage of patients gained very much. On average, the drug extended survival by only two months—from eleven months to thirteen months—and that was in patients who, unlike Sara, had responded to first-line chemotherapy.

She worked hard to take the setbacks and side effects in stride. She was upbeat by nature, and she managed to maintain her optimism. Little by little, however, she grew sicker—increasingly exhausted and short of breath. By November she didn't have the wind to walk the length of the hallway from the parking garage to Marcoux's office; Rich had to push her in a wheelchair.

A few days before Thanksgiving, she had another CT scan, which showed that the pemetrexed—her third drug regimen—wasn't working, either. The lung cancer had spread: from the left chest to the right; to the liver; to the lining of her abdomen; and to her spine. Time was running out.

This is the moment in Sara's story that poses a fundamental question for everyone living in the era of modern medicine: What do we want Sara and her doctors to do now? Or, to put it another way, if you were the one who had metastatic cancer—or, for that matter, a similarly advanced case of emphysema or congestive heart failure—what would you want your doctors to do?

The issue has become pressing in recent years, for reasons of expense. The soaring cost of health care is the greatest threat to the country's long-term solvency, and the terminally ill account for a lot of it. Twenty-five percent of all Medicare spending is for the 5 percent of patients who are in their final year of life, and most of that money goes for care in their last couple of months that is of little apparent benefit.

Spending on a disease like cancer tends to follow a particular pattern. There are high initial costs as the cancer is treated, and then, if all goes well, these costs taper off. Medical spending for a breast-cancer survivor, for instance, averaged an estimated $54,000 in 2003, the vast majority of it for the initial diagnostic testing, surgery, and, where necessary, radiation and chemotherapy. For a patient with a fatal version of the disease, though, the cost curve is U-shaped, rising again toward the end—to an average of $63,000 during the last six months of life with an incurable breast cancer. Our medical system is excellent at trying to stave off death with $8,000-a-month chemotherapy, $3,000-a-day intensive care, $5,000-an-hour surgery. But ultimately, death comes, and no one is good at knowing when to stop.

The subject seems to reach national awareness mainly as a question of who should "win" when the expensive decisions are made: the insurers and the taxpayers footing the bill or the patient battling for his or her life. Budget hawks urge us to face the fact that we can't afford everything. Demagogues shout about rationing and death panels. Market purists blame the existence of insurance: if patients and families paid the bills themselves, those expensive therapies would all come down in price. But they're debating the wrong question. The failure of our system of medical care for people facing the end of their life runs much deeper. To see this, you have to get close enough to grapple with the way decisions about care are actually made.

Recently, while seeing a patient in an intensive-care unit at my hospital, I stopped to talk with the critical-care physician on duty, someone I'd known since college. "I'm running a warehouse for the dying," she said bleakly. Out of the ten patients in her unit, she said, only two were likely to leave the hospital for any length of time. More typical was an almost eighty-year-old woman at the end of her life, with irreversible congestive heart failure, who was in the ICU for the second time in three weeks, drugged to oblivion and tubed in most natural orifices and a few artificial ones. Or the seventy-year-old with a cancer that had metastasized to her lungs and bone, and a fungal pneumonia that arises only in the final phase of the illness. She had chosen to forgo treatment, but her oncologist pushed her to change her mind, and she was put on a ventilator and antibiotics. Another woman, in her eighties, with end-stage respiratory and kidney failure, had been in the unit for two weeks. Her husband had died after a long illness with a feeding tube and a tracheotomy, and she had mentioned that she didn't want to die that way. But her children couldn't let her go and asked to proceed with the placement of various devices: a permanent tracheotomy, a feeding tube, and a dialysis catheter. So now she just lay there tethered to her pumps, drifting in and out of consciousness.

Almost all these patients had known, for some time, that they had a terminal condition. Yet they—along with their families and doctors—were unprepared for the final stage. "We are having more conversation now about what patients want for the end of their life, by far, than they have had in all their lives to this point," my friend said. "The problem is that's way too late." In 2008 the national Coping with Cancer project published a study showing that terminally ill cancer patients who were put on a mechanical ventilator, given electrical defibrillation or chest compressions, or admitted, near death, to intensive care had a substantially worse quality of life in their last week than those who received no such interventions. And six months after their death, their caregivers were three times as likely to suffer major depression. Spending one's final days in an ICU because of terminal illness is for most people a kind of failure. You lie on a ventilator, your every organ shutting down, your mind teetering on delirium and permanently beyond realizing that you will never leave this borrowed, fluorescent place. The end comes with no chance for you to have said goodbye or "It's okay" or "I'm sorry" or "I love you."

People have concerns besides simply prolonging their lives. Surveys of patients with terminal illness find that their top priorities include, in addition to avoiding suffering, being with family, having the touch of others, being mentally aware, and not becoming a burden to others. Our system of technological medical care has utterly failed to meet these needs, and the cost of this failure is measured in far more than dollars. The hard question we face, then, is not how we can afford this system's expense. It is how we can build a health-care system that will actually help dying patients achieve what's most important to them at the end of their lives.

*

For all but our most recent history, dying was typically a brief process. Whether the cause was childhood infection, difficult childbirth, heart attack, or pneumonia, the interval between recognizing that you had a life-threatening ailment and death was often just a matter of days or weeks. Consider how our presidents died before the modern era. George Washington developed a throat infection at home on December 13, 1799, that killed him by the next evening. John Quincy Adams, Millard Fillmore, and Andrew Johnson all succumbed to strokes and died within two days. Rutherford Hayes had a heart attack and died three days later. Some deadly illnesses took a longer course: James Monroe and Andrew Jackson died from the months-long consumptive process of what appears to have been tuberculosis; Ulysses Grant's oral cancer took a year to kill him; and James Madison was bedridden for two years before dying of "old age." But as the end-of-life researcher Joanne Lynn has observed, people usually experienced life-threatening illness the way they experienced bad weather—as something that struck with little warning—and you either got through it or you didn't.

Dying used to be accompanied by a prescribed set of customs. Guides to *ars moriendi*, the art of dying, were extraordinarily popular; a 1415 medieval Latin text was reprinted in more than a hundred editions across Europe. Reaffirming one's faith, repenting one's sins, and letting go of one's worldly possessions and desires were crucial, and the guides provided families with prayers and questions for the dying in order to put them in the right frame of mind during their final hours. Last words came to hold a particular place of reverence.

These days, swift catastrophic illness is the exception; for most people, death comes only after a long medical struggle with an incurable condition—advanced cancer, progressive organ failure (usually the heart, kidney, or liver), or the multiple debilities of very old age. In all such cases, death is certain, but the timing isn't. So everyone struggles with this uncertainty—with how and when to accept that the battle is lost. As for last words, they hardly seem to exist anymore. Technology sustains our organs until we are well past the point of awareness and coherence. Besides, how do you attend to the thoughts and concerns of the dying when medicine has made it almost impossible to be sure who the dying even are? Is someone with terminal cancer, dementia, incurable congestive heart failure dying exactly?

I once cared for a woman in her sixties who had severe chest and abdominal pain from a bowel obstruction that had ruptured her colon, caused her to have a heart attack, and put her into septic shock and renal failure. I performed an emergency operation to remove the damaged length of colon and give her a colostomy. A cardiologist stented her coronary arteries. We put her on dialysis, a ventilator, and intravenous feeding, and stabilized her. After a couple of weeks, though, it was clear that she was not going to get much better. The septic shock had left her with heart and respiratory failure as well as dry gangrene of her foot, which would have to be amputated. She had a large, open abdominal wound with leaking bowel contents, which would require twice-a-day cleaning and dressing for weeks in order to heal. She would not be able to eat. She would need a tracheotomy. Her kidneys were gone, and she would have to spend three days a week on a dialysis machine for the rest of her life.

She was unmarried and without children. So I sat with her sisters in the ICU family room to talk about whether we should proceed with the amputation and the tracheotomy. "Is she dying?" one of the sisters asked me. I didn't know how to answer the question. I wasn't even sure what the word "dying" meant anymore. In the past few decades, medical science has rendered obsolete centuries of experience, tradition, and language about our mortality, and created a new difficulty for mankind: how to die.

One Friday morning this spring, I went on patient rounds with Sarah Creed, a nurse with the hospice service that my hospital system operates. I didn't know much about hospice. I knew that it specialized in providing "comfort care" for the terminally ill, sometimes in special facilities, though nowadays usually at home. I knew that in order for a patient of mine to be eligible, I had to write a note certifying that he or she had a life expectancy of less than six months. And I knew few patients who had chosen it, except maybe in their very last few days, because they had to sign a form indicating that they understood their disease was incurable and that they were giving up on medical care to stop it. The picture I had of hospice was of a morphine drip. It was not of this brown-haired and blue-eyed former ICU nurse with a stethoscope, knocking on Lee Cox's door on a quiet street in Boston's Mattapan neighborhood.

"Hi, Lee," Creed said when she entered the house.

"Hi, Sarah," Cox said. She was seventy-two years old. She'd had several years of declining health due to congestive heart failure from a heart attack and pulmonary fibrosis, a progressive and irreversible lung disease. Doctors tried slowing the disease with steroids, but they didn't work. She had cycled in and out of the hospital, each time in worse shape. Ultimately, she accepted hospice care and moved in with her niece for support. She was dependent on oxygen and unable to do the most ordinary tasks. Just answering the door, with her thirty-foot length of oxygen tubing trailing after her, had left her winded. She stood resting for a moment, her lips pursed and her chest heaving.

Creed took Cox's arm gently as we walked to the kitchen to sit down, asking her how she had been doing. Then she asked a series of questions, targeting issues that tend to arise in patients with terminal illness. Did Cox have pain? How was her appetite, thirst, sleeping? Any trouble with confusion, anxiety, or restlessness? Had her shortness of breath grown worse? Was there chest pain or heart palpitations? Abdominal discomfort? Trouble with bowel movements or urination or walking?

She did have some new troubles. When she walked from the bedroom to the bathroom, she said, it now took at least five minutes to catch her breath, and that frightened her. She was also getting chest pain. Creed pulled a stethoscope and a blood-pressure cuff from her medical bag. Cox's blood pressure was acceptable, but her heart rate was high. Creed listened to her heart, which had a normal rhythm, and to her lungs, hearing the fine crackles of her pulmonary fibrosis but also a new wheeze. Her ankles were swollen with fluid, and when Creed asked for her pillbox she saw that Cox was out of her heart medication. She asked to see Cox's oxygen equipment. The liquid-oxygen cylinder at the foot of the neatly made bed was filled and working properly. The nebulizer equipment for her inhaler treatments, however, was broken.

Given the lack of heart medication and inhaler treatments, it was no wonder that she had worsened. Creed called Cox's pharmacy to confirm that her refills had been waiting, and had her arrange for her niece to pick up the medicine when she came home from work. Creed also called the nebulizer supplier for same-day emergency service.

She then chatted with Cox in the kitchen for a few minutes.

Cox's spirits were low. Creed took her hand. Everything was going to be all right, she said. She reminded her about the good days she'd had—the previous weekend, for example, when she'd been able to go out with her portable oxygen cylinder to shop with her niece and get her hair colored.

I asked Cox about her previous life. She had made radios in a Boston factory. She and her husband had two children and several grandchildren.

When I asked her why she had chosen hospice care, she looked downcast. "The lung doctor and heart doctor said they couldn't help me anymore," she said. Creed glared at me. My questions had made Cox sad again.

"It's good to have my niece and her husband helping to watch me every day," she said. "But it's not my home. I feel like I'm in the way."

Creed gave her a hug before we left and one last reminder. "What do you do if you have chest pain that doesn't go away?" she asked.

"Take a nitro," Cox said, referring to the nitroglycerin pill that she can slip under her tongue.

"And?"

"Call you."

"Where's the number?"

She pointed to the twenty-four-hour hospice call number that was taped beside her phone.

Outside, I confessed that I was confused by what Creed was doing. A lot of it seemed to be about extending Cox's life. Wasn't the goal of hospice to let nature take its course?

"That's not the goal," Creed said. The difference between standard medical care and hospice is not the difference between treating and doing nothing, she explained. The difference was in your priorities. In ordinary medicine, the goal is to extend life. We'll sacrifice the quality of your existence now—by performing surgery, providing chemotherapy, putting you in intensive care—for the chance of gaining time later. Hospice deploys nurses, doctors, and social workers to help people with a fatal illness have the fullest possible lives right now. That means focusing on objectives like freedom from pain and discomfort, or maintaining mental awareness for as long as possible, or getting out with family once in a while. Hospice and palliative-care specialists aren't much concerned about whether that makes people's lives longer or shorter.

Like many people, I had believed that hospice care hastens death, because patients forgo hospital treatments and are allowed high-dose narcotics to combat pain. But studies suggest otherwise. In one, researchers followed 4,493 Medicare patients with either terminal cancer or congestive heart failure. They found no difference in survival time between hospice and non-hospice patients with breast cancer, prostate cancer, and colon cancer. Curiously, hospice care seemed to extend survival for some patients; those with pancreatic cancer gained an average of three weeks, those with lung cancer gained six weeks, and those with congestive heart failure gained three months. The lesson seems almost Zen: you live longer only when you stop trying to live longer. When Cox was transferred to hospice care, her doctors thought that she wouldn't live much longer than a few weeks. With the supportive hospice therapy she received, she had already lived for a year.

Creed enters people's lives at a strange moment—when they have understood that they have a fatal illness but have not necessarily acknowledged that they are dying. "I'd say only about a quarter have accepted their fate when they come into hospice," she said. When she first encounters her patients, many feel that they have simply been abandoned by their doctors. "Ninety-nine percent understand they're dying, but one hundred percent hope they're not," she says. "They still want to beat their disease." The initial visit is always tricky, but she has found ways to smooth things over. "A nurse has five seconds to make a patient like you and trust you. It's in the whole way you present yourself. I do not come in saying, 'I'm so sorry.' Instead, it's 'I'm the hospice nurse, and here's what I have to offer you to make your life better. And I know we don't have a lot of time to waste.'"

That was how she started with Dave Galloway, whom we visited after leaving Lee Cox's home. He was forty-two years old. He and his wife, Sharon, were both Boston firefighters. They had a three-year-old daughter. He had pancreatic cancer, which had spread; his upper abdomen was now solid with tumor. During the past few months, the pain had become unbearable at times, and he was admitted to the hospital several times for pain crises. At his most recent admission, about a week earlier, it was found that the tumor had perforated his intestine. There wasn't even a temporary fix for this problem. The medical team started him on intravenous nutrition and offered him a choice between going to the intensive-care unit and going home with hospice. He chose to go home.

"I wish we'd gotten involved sooner," Creed told me. When she and the hospice's supervising doctor, Dr. JoAnne Nowak, evaluated Galloway upon his arrival at home, he appeared to have only a few days left. His eyes were hollow. His breathing was labored. Fluid swelled his entire lower body to the point that his skin blistered and wept. He was almost delirious with abdominal pain.

They got to work. They set up a pain pump with a button that let him dispense higher doses of narcotic than he had been allowed. They arranged for an electric hospital bed, so that he could sleep with his back raised. They also taught Sharon how to keep Dave clean, protect his skin from breakdown, and handle the crises to come. Creed told me that part of her job is to take the measure of a patient's family, and Sharon struck her as unusually capable. She was determined to take care of her husband to the end, and, perhaps because she was a firefighter, she had the resilience and the competence to do so. She did not want to hire a private-duty nurse. She handled everything, from the IV lines and the bed linens to orchestrating family members to lend a hand when she needed help.

Creed arranged for a specialized "comfort pack" to be delivered by FedEx and stored in a mini-refrigerator by Dave's bed. It contained a dose of morphine for breakthrough pain or shortness of breath, Ativan for anxiety attacks, Compazine for nausea, Haldol for delirium, Tylenol for fever, and atropine for drying up the upper-airway rattle that people can get in their final hours. If any such problem developed, Sharon was instructed to call the twenty-four-hour hospice nurse on duty, who would provide instructions about which rescue medications to use and, if necessary, come out to help.

Dave and Sharon were finally able to sleep through the night at home. Creed or another nurse came to see him every day, sometimes twice a day; three times that week, Sharon used the emergency hospice line to help her deal with Dave's pain crises or hallucinations. After a few days, they were even able to go out to a favorite restaurant; he wasn't hungry, but they enjoyed just being there, and the memories it stirred.

The hardest part so far, Sharon said, was deciding to forgo the two-liter intravenous feedings that Dave had been receiving each day. Although they were his only source of calories, the hospice staff encouraged discontinuing them because his body did not seem to be absorbing the nutrition. The infusion of sugars, proteins, and fats made the painful swelling of his skin and his shortness of breath worse—and for what? The mantra was live for now. Sharon had balked, for fear that she'd be starving him. The night before our visit, however, she and Dave decided to try going without the infusion. By morning, the swelling was markedly reduced. He could move more, and with less discomfort. He also began to eat a few morsels of food, just for the taste of it, and that made Sharon feel better about the decision.

When we arrived, Dave was making his way back to bed after a shower, his arm around his wife's shoulders and his slippered feet taking one shuffling step at a time.

"There's nothing he likes better than a long hot shower," Sharon said. "He'd live in the shower if he could."

Dave sat on the edge of his bed in fresh pajamas, catching his breath, and then Creed spoke to him as his daughter, Ashley, ran in and out of the room in her beaded pigtails, depositing stuffed animals in her dad's lap.

"How's your pain on a scale of one to ten?" Creed asked.

"A six," he said.

"Did you hit the pump?"

He didn't answer for a moment. "I'm reluctant," he admitted.

"Why?" Creed asked.

"It feels like defeat," he said.

"Defeat?"

"I don't want to become a drug addict," he explained. "I don't want to need this."

Creed got down on her knees in front of him. "Dave, I don't know anyone who can manage this kind of pain without the medication," she said. "It's not defeat. You've got a beautiful wife and daughter, and you're not going to be able to enjoy them with the pain."

"You're right about that," he said, looking at Ashley as she gave him a little horse. And he pressed the button.

Dave Galloway died one week later—at home, at peace, and surrounded by family. A week after that, Lee Cox died, too. But, as if to show just how resistant to formula human lives are, Cox had never reconciled herself to the incurability of her illnesses. So when her family found her in cardiac arrest one morning they followed her wishes and called 911 instead of the hospice service. The emergency medical technicians and firefighters and police rushed in. They pulled off her clothes and pumped her chest, put a tube in her airway and forced oxygen into her lungs, and tried to see if they could shock her heart back. But such efforts rarely succeed with terminal patients, and they did not succeed with her.

Hospice has tried to offer a new ideal for how we die. Although not everyone has embraced its rituals, those who have are helping to negotiate an *ars moriendi* for our age. But doing so represents a struggle—not only against suffering but also against the seemingly unstoppable momentum of medical treatment.

Just before Thanksgiving of 2007, Sara Monopoli, her husband, Rich, and her mother, Dawn Thomas, met with Dr. Marcoux to discuss the options she had left. By this point Sara had undergone three rounds of chemotherapy with limited, if any, effect. Perhaps Marcoux could have discussed what she most wanted as death neared and how best to achieve those wishes. But the signal he got from Sara and her family was that they wished to talk only about the next treatment options. They did not want to talk about dying.

Recently I spoke to Sara's husband and her parents. Sara knew that her disease was incurable, they pointed out. The week after she was given the diagnosis and delivered her baby, she spelled out her wishes for Vivian's upbringing after she was gone. She had told her family on several occasions that she did not want to die in the hospital. She wanted to spend her final moments peacefully at home. But the prospect that those moments might be coming soon, that there might be no way to slow the disease, "was not something she or I wanted to discuss," her mother said.

Her father, Gary, and her twin sister, Emily, still held out hope for a cure. The doctors simply weren't looking hard enough, they felt. "I just couldn't believe there wasn't something," Gary said. For Rich, the experience of Sara's illness had been disorienting: "We had a baby. We were young. And this was so shocking and so odd. We never discussed stopping treatment."

Marcoux took the measure of the room. With almost two decades of experience treating lung cancer, he had been through many of these conversations. He has a calm, reassuring air and a native Minnesotan's tendency to avoid confrontation or overintimacy. He tries to be scientific about decisions.

"I know that the vast majority of my patients are going to die of their disease," he told me. The data show that after failure of second-line chemotherapy, lung-cancer patients rarely get any added survival time from further treatments and often suffer significant side effects. But he, too, has his hopes.

He told them that at some point "supportive care" was an option for them to think about. But, he went on, there were also experimental therapies. He told them about several that were under trial. The most promising was a Pfizer drug that targeted one of the mutations found in her cancer's cells. Sara and her family instantly pinned their hopes on it. The drug was so new that it didn't even have a name, just a number—PF0231006—and this made it all the more enticing.

There were a few hovering issues, including the fact that the scientists didn't yet know the safe dose. The drug was only in a Phase I trial—that is, a trial designed to determine the toxicity of a range of doses, not whether the drug worked. Furthermore, a test of the drug against her cancer cells in a petri dish showed no effect. But Marcoux didn't think that these were decisive obstacles—just negatives. The critical problem was that the rules of the trial excluded Sara because of the pulmonary embolism she had developed that summer. To enroll, she would need to wait two months, in order to get far enough past the episode. In the meantime he suggested trying another conventional chemotherapy, called Navelbine. Sara began the treatment the Monday after Thanksgiving.

It's worth pausing to consider what had just happened. Step by step, Sara ended up on a *fourth* round of chemotherapy, one with a minuscule likelihood of altering the course of her disease and a great likelihood of causing debilitating side effects. An opportunity to prepare for the inevitable was forgone. And it all happened because of an assuredly normal circumstance: a patient and family unready to confront the reality of her disease.

I asked Marcoux what he hopes to accomplish for terminal lung-cancer patients when they first come to see him. "I'm thinking, Can I get them a pretty good year or two out of this?" he said.

"Those are *my* expectations. For me, the long tail for a patient like her is three to four years." But this is not what people want to hear. "They're thinking ten to twenty years. You hear that time and time again. And I'd be the same way if I were in their shoes."

You'd think doctors would be well equipped to navigate the shoals here, but at least two things get in the way. First, our own views may be unrealistic. A study led by the Harvard researcher Nicholas Christakis asked the doctors of almost five hundred terminally ill patients to estimate how long they thought their patient would survive, and then followed the patients. Sixty-three percent of doctors overestimated survival time. Just 17 percent underestimated it. The average estimate was 530 percent too high. And, the better the doctors knew their patients, the more likely they were to err.

Second, we often avoid voicing even these sentiments. Studies find that although doctors usually tell patients when a cancer is not curable, most are reluctant to give a specific prognosis, even when pressed. More than 40 percent of oncologists report offering treatments that they believe are unlikely to work. In an era in which the relationship between patient and doctor is increasingly miscast in retail terms — "the customer is always right" — doctors are especially hesitant to trample on a patient's expectations. You worry far more about being overly pessimistic than you do about being overly optimistic. And talking about dying is enormously fraught. When you have a patient like Sara Monopoli, the last thing you want to do is grapple with the truth. I know, because Marcoux wasn't the only one avoiding that conversation with her. I was, too.

Earlier that summer, a PET scan had revealed that, in addition to her lung cancer, she also had thyroid cancer, which had spread to the lymph nodes of her neck, and I was called in to decide whether to operate. This second, unrelated cancer was in fact operable. But thyroid cancers take years to become lethal. Her lung cancer would almost certainly end her life long before her thyroid cancer caused any trouble. Given the extent of the surgery that would have been required, and the potential complications, the best course was to do nothing. But explaining my reasoning to Sara meant confronting the mortality of her lung cancer, something that I felt ill prepared to do.

Sitting in my clinic, Sara did not seem discouraged by the discovery of this second cancer. She seemed determined. She'd read about the good outcomes from thyroid-cancer treatment. So she was geared up, eager to discuss when to operate. And I found myself swept along by her optimism. Suppose I was wrong, I wondered, and she proved to be that miracle patient who survived metastatic lung cancer?

My solution was to avoid the subject altogether. I told Sara that the thyroid cancer was slow-growing and treatable. The priority was her lung cancer, I said. Let's not hold up the treatment for that. We could monitor the thyroid cancer and plan surgery in a few months.

I saw her every six weeks and noted her physical decline from one visit to the next. Yet even in a wheelchair, Sara would always arrive smiling, makeup on and bangs bobby-pinned out of her eyes. She'd find small things to laugh about, like the tubes that created strange protuberances under her dress. She was ready to try anything, and I found myself focusing on the news about experimental therapies for her lung cancer. After one of her chemotherapies seemed to shrink the thyroid cancer slightly, I even raised with her the possibility that an experimental therapy could work against both her cancers, which was sheer fantasy. Discussing a fantasy was easier—less emotional, less explosive, less prone to misunderstanding—than discussing what was happening before my eyes.

Between the lung cancer and the chemo, Sara became steadily sicker. She slept most of the time and could do little out of the house. Clinic notes from December describe shortness of breath, dry heaves, coughing up blood, severe fatigue. In addition to the drainage tubes in her chest, she required needle-drainage procedures in her abdomen every week or two to relieve the severe pressure from the liters of fluid that the cancer was producing there.

A CT scan in December showed that the lung cancer was spreading through her spine, liver, and lungs. When we met in January, she could move only slowly and uncomfortably. Her lower body had become swollen. She couldn't speak more than a sentence without pausing for breath. By the first week of February, she needed oxygen at home to breathe. Enough time had elapsed since her pulmonary embolism, however, that she could start on Pfizer's experimental drug. She just needed one more set of scans for clearance. These revealed that the cancer had spread to her brain, with at least nine metastatic growths across both hemi-spheres. The experimental drug was not designed to cross the blood-brain barrier. PF0231006 was not going to work.

And still Sara, her family, and her medical team remained in bat-tle mode. Within twenty-four hours, Sara was scheduled to see a radiation oncologist for whole-brain radiation to try to reduce the metastases. On February 12, she completed five days of radiation treatment, which left her immeasurably fatigued, barely able to get out of bed. She ate almost nothing. She weighed twenty-five pounds less than she had in the fall. She confessed to Rich that for the past two months she had experienced double vision and was unable to feel her hands.

"Why didn't you tell anyone?" he asked her.

"I just didn't want to stop treatment," she said. "They would make me stop."

She was given two weeks to recover her strength after the radia-tion. Then she would be put on another experimental drug from a small biotech company. She was scheduled to start on February 25. Her chances were rapidly dwindling. But who was to say they were zero?

In 1985 the paleontologist and writer Stephen Jay Gould pub-lished an extraordinary essay entitled "The Median Isn't the Mes-sage," after he had been given a diagnosis, three years earlier, of abdominal mesothelioma, a rare and lethal cancer usually associ-ated with asbestos exposure. He went to a medical library when he got the diagnosis and pulled out the latest scientific articles on the disease. "The literature couldn't have been more brutally clear: mesothelioma is incurable, with a median survival of only eight months after discovery," he wrote. The news was devastating. But then he began looking at the graphs of the patient-survival curves.

Gould was a naturalist and more inclined to notice the varia-tion around the curve's middle point than the middle point it-self. What the naturalist saw was remarkable variation. The patients were not clustered around the median survival but instead fanned out in both directions. Moreover, the curve was skewed to the right, with a long tail, however slender, of patients who lived many years longer than the eight-month median. This is where he found sol-ace. He could imagine himself surviving far out in that long tail. And he did. Following surgery and experimental chemotherapy, he lived twenty more years before dying, in 2002, at the age of sixty, from a lung cancer that was unrelated to his original disease.

"It has become, in my view, a bit too trendy to regard the acceptance of death as something tantamount to intrinsic dignity," he wrote in his 1985 essay. "Of course I agree with the preacher of Ecclesiastes that there is a time to love and a time to die—and when my skein runs out I hope to face the end calmly and in my own way. For most situations, however, I prefer the more martial view that death is the ultimate enemy—and I find nothing reproachable in those who rage mightily against the dying of the light."

I think of Gould and his essay every time I have a patient with a terminal illness. There is almost always a long tail of possibility, however thin. What's wrong with looking for it? Nothing, it seems to me, unless it means we have failed to prepare for the outcome that's vastly more probable. The trouble is that we've built our medical system and culture around the long tail. We've created a multitrillion-dollar edifice for dispensing the medical equivalent of lottery tickets—and have only the rudiments of a system to prepare patients for the near-certainty that those tickets will not win. Hope is not a plan, but hope is our plan.

For Sara there would be no miraculous recovery, and, when the end approached, neither she nor her family was prepared. "I always wanted to respect her request to die peacefully at home," Rich later told me. "But I didn't believe we could make it happen. I didn't know how."

On the morning of Friday, February 22, three days before she was to start her new round of chemo, Rich awoke to find his wife sitting upright beside him, pitched forward on her arms, eyes wide, struggling for air. She was gray, breathing fast, her body heaving with each open-mouthed gasp. She looked as if she were drowning. He tried turning up the oxygen in her nasal tubing, but she got no better.

"I can't do this," she said, pausing between each word. "I'm scared."

He had no emergency kit in the refrigerator. No hospice nurse to call. And how was he to know whether this new development was fixable?

We'll go to the hospital, he told her. When he asked if they should drive, she shook her head, so he called 911 and told her mother, Dawn, who was in the next room, what was going on. A few minutes later, firemen swarmed up the stairs to her bedroom, sirens wailing outside. As they lifted Sara into the ambulance on a stretcher, Dawn came out in tears.

"We're going to get ahold of this," Rich told her. This was just another trip to the hospital, he said to himself. The doctors would figure this out.

At the hospital Sara was diagnosed with pneumonia. That troubled the family, because they thought they'd done everything to keep infection at bay. They'd washed hands scrupulously, limited visits by people with young children, even limited Sara's time with baby Vivian if she showed the slightest sign of a runny nose. But Sara's immune system and her ability to clear her lung secretions had been steadily weakened by the rounds of radiation and chemotherapy as well as by the cancer.

In another way, the diagnosis of pneumonia was reassuring, because it was just an infection. It could be treated. The medical team started Sara on intravenous antibiotics and high-flow oxygen through a mask. The family gathered at her bedside, hoping for the antibiotics to work. This could be reversible, they told one another. But that night and the next morning her breathing only grew more labored.

"I can't think of a single funny thing to say," Emily told Sara as their parents looked on.

"Neither can I," Sara murmured. Only later did the family realize that those were the last words they would ever hear from her. After that she began to drift in and out of consciousness. The medical team had only one option left: to put her on a ventilator. Sara was a fighter, right? And the next step for fighters was to escalate to intensive care.

This is a modern tragedy, replayed millions of times over. When there is no way of knowing exactly how long our skeins will run — and when we imagine ourselves to have much more time than we do — our every impulse is to fight, to die with chemo in our veins or a tube in our throats or fresh sutures in our flesh. The fact that we may be shortening or worsening the time we have left hardly seems to register. We imagine that we can wait until the doctors tell us that there is nothing more they can do. But rarely is there *nothing* more that doctors can do. They can give toxic drugs of unknown efficacy, operate to try to remove part of the tumor, put in a feeding tube if a person can't eat: there's always something. We want these choices. We don't want anyone—certainly not bureaucrats or the marketplace—to limit them. But that doesn't mean we are eager to make the choices ourselves. Instead, most often, we make no choice at all. We fall back on the default, and the default is: Do Something. Is there any way out of this?

In late 2004, executives at Aetna, the insurance company, started an experiment. They knew that only a small percentage of the terminally ill ever halted efforts at curative treatment and enrolled in hospice and that when they did, it was usually not until the very end. So Aetna decided to let a group of policyholders with a life expectancy of less than a year receive hospice services *without* forgoing other treatments. A patient like Sara Monopoli could continue to try chemotherapy and radiation and go to the hospital when she wished—but also have a hospice team at home focusing on what she needed for the best possible life now and for that morning when she might wake up unable to breathe. A two-year study of this "concurrent care" program found that enrolled patients were much more likely to use hospice: the figure leaped from 26 percent to 70 percent. That was no surprise, since they weren't forced to give up anything. The surprising result was that they did give up things. They visited the emergency room almost half as often as the control patients did. Their use of hospitals and ICUs dropped by more than two-thirds. Overall costs fell by almost a quarter.

This was stunning and puzzling: it wasn't obvious what made the approach work. Aetna ran a more modest concurrent-care program for a broader group of terminally ill patients. For these patients the traditional hospice rules applied—in order to qualify for home hospice, they had to give up attempts at curative treatment. But either way, they received phone calls from palliative-care nurses who offered to check in regularly and help them find services for anything from pain control to making out a living will. For these patients, too, hospice enrollment jumped to 70 percent, and their use of hospital services dropped sharply. Among elderly patients, use of intensive-care units fell by more than 85 percent. Satisfaction scores went way up. What was going on here? The program's leaders had the impression that they had simply given patients someone experienced and knowledgeable to talk to about their daily needs. And somehow that was enough—just talking.

The explanation strains credibility, but evidence for it has grown in recent years. Two-thirds of the terminal-cancer patients in the Coping with Cancer study reported having had no discussion with their doctors about their goals for end-of-life care, despite being, on average, just four months from death. But the third who did were far less likely to undergo cardiopulmonary resuscitation or be put on a ventilator or end up in an intensive-care unit. Two-thirds enrolled in hospice. These patients suffered less, were physically more capable, and were better able, for a longer period, to interact with others. Moreover, six months after the patients died, their family members were much less likely to experience persistent major depression. In other words, people who had substantive discussions with their doctor about their end-of-life preferences were far more likely to die at peace and in control of their situation, and to spare their family anguish.

Can mere discussions really do so much? Consider the case of La Crosse, Wisconsin. Its elderly residents have unusually low end-of-life hospital costs. During their last six months, according to Medicare data, they spend half as many days in the hospital as the national average, and there's no sign that doctors or patients are halting care prematurely. Despite average rates of obesity and smoking, their life expectancy outpaces the national mean by a year.

I spoke to Dr. Gregory Thompson, a critical-care specialist at Gundersen Lutheran Hospital, while he was on ICU duty one recent evening, and he ran through his list of patients with me. In most respects the patients were like those found in any ICU—terribly sick and living through the most perilous days of their lives. There was a young woman with multiple organ failure from a devastating case of pneumonia, a man in his mid-sixties with a ruptured colon that had caused a rampaging infection and a heart attack. Yet these patients were completely different from those in other ICUs I'd seen: none had a terminal disease; none battled the final stages of metastatic cancer or untreatable heart failure or dementia.

To understand La Crosse, Thompson said, you had to go back to 1991, when local medical leaders headed a systematic campaign to get physicians and patients to discuss end-of-life wishes. Within a few years, it became routine for all patients admitted to a hospital, nursing home, or assisted-living facility to complete a multiple-choice form that boiled down to four crucial questions. At this moment in your life, the form asked:

1. Do you want to be resuscitated if your heart stops?
2. Do you want aggressive treatments such as intubation and mechanical ventilation?
3. Do you want antibiotics?
4. Do you want tube or intravenous feeding if you can't eat on your own?

By 1996, 85 percent of La Crosse residents who died had written advanced directives, up from 15 percent, and doctors almost always knew of and followed the instructions. Having this system in place, Thompson said, has made his job vastly easier. But it's not because the specifics are spelled out for him every time a sick patient arrives in his unit.

"These things are not laid out in stone," he told me. Whatever the yes/no answers people may put on a piece of paper, one will find nuances and complexities in what they mean. "But instead of having the discussion when they get to the ICU, we find many times it has already taken place."

Answers to the list of questions change as patients go from entering the hospital for the delivery of a child to entering for complications of Alzheimer's disease. But in La Crosse the system means that people are far more likely to have talked about what they want and what they don't want before they and their relatives find themselves in the throes of crisis and fear. When wishes aren't clear, Thompson said, "families have also become much more receptive to having the discussion." The discussion, not the list, was what mattered most. Discussion had brought La Crosse's end-of-life costs down to just over half the national average. It was that simple —and that complicated.

One Saturday morning last winter, I met with a woman I had operated on the night before. She had been undergoing a procedure for the removal of an ovarian cyst when the gynecologist who was operating on her discovered that she had metastatic colon cancer. I was summoned, as a general surgeon, to see what could be done. I removed a section of her colon that had a large cancerous mass, but the cancer had already spread widely. I had not been able to get it all. Now I introduced myself. She said a resident had told her that a tumor was found and part of her colon had been excised.

Yes, I said. I'd been able to take out "the main area of involvement." I explained how much bowel was removed, what the recovery would be like — everything except how much cancer there was. But then I remembered how timid I'd been with Sara Monopoli, and all those studies about how much doctors beat around the bush. So when she asked me to tell her more about the cancer, I explained that it had spread not only to her ovaries but also to her lymph nodes. I said that it had not been possible to remove all the disease. But I found myself almost immediately minimizing what I'd said. "We'll bring in an oncologist," I hastened to add. "Chemotherapy can be very effective in these situations."

She absorbed the news in silence, looking down at the blankets drawn over her mutinous body. Then she looked up at me. "Am I going to die?"

I flinched. "No, no," I said. "Of course not."

A few days later, I tried again. "We don't have a cure," I explained. "But treatment can hold the disease down for a long time." The goal, I said, was to "prolong your life" as much as possible.

I've seen her regularly in the months since, as she embarked on chemotherapy. She has done well. So far the cancer is in check. Once I asked her and her husband about our initial conversations. They don't remember them very fondly. "That one phrase that you used — 'prolong your life' — it just . . ." She didn't want to sound critical.

"It was kind of blunt," her husband said.

"It sounded harsh," she echoed. She felt as if I'd dropped her off a cliff.

I spoke to Dr. Susan Block, a palliative-care specialist at my hospital who has had thousands of these difficult conversations and is a nationally recognized pioneer in training doctors and others in managing end-of-life issues with patients and their families. "You have to understand," Block told me. "A family meeting is a procedure, and it requires no less skill than performing an operation."

One basic mistake is conceptual. For doctors, the primary purpose of a discussion about terminal illness is to determine what people want—whether they want chemo or not, whether they want to be resuscitated or not, whether they want hospice or not. They focus on laying out the facts and the options. But that's a mistake, Block said.

"A large part of the task is helping people negotiate the overwhelming anxiety—anxiety about death, anxiety about suffering, anxiety about loved ones, anxiety about finances," she explained. "There are many worries and real terrors." No one conversation can address them all. Arriving at an acceptance of one's mortality and a clear understanding of the limits and the possibilities of medicine is a process, not an epiphany.

There is no single way to take people with terminal illness through the process, but, according to Block, there are some rules. You sit down. You make time. You're not determining whether they want treatment X versus Y. You're trying to learn what's most important to them under the circumstances—so that you can provide information and advice on the approach that gives them the best chance of achieving it. This requires as much listening as talking. If you are talking more than half of the time, Block says, you're talking too much.

The words you use matter. According to experts, you shouldn't say, "I'm sorry things turned out this way," for example. It can sound like pity. You should say, "I wish things were different." You don't ask, "What do you want when you are dying?" You ask, "If time becomes short, what is most important to you?"

Block has a list of items that she aims to cover with terminal patients in the time before decisions have to be made: what they understand their prognosis to be; what their concerns are about what lies ahead; whom they want to make decisions when they can't; how they want to spend their time as options become limited; what kinds of trade-offs they are willing to make.

Ten years ago her seventy-four-year-old father, Jack Block, a professor emeritus of psychology at the University of California at Berkeley, was admitted to a San Francisco hospital with symptoms from what proved to be a mass growing in the spinal cord of his neck. She flew out to see him. The neurosurgeon said that the procedure to remove the mass carried a 20 percent chance of leaving him quadriplegic, paralyzed from the neck down. But without it he had a 100 percent chance of becoming quadriplegic.

The evening before surgery, father and daughter chatted about friends and family, trying to keep their minds off what was to come, and then she left for the night. Halfway across the Bay Bridge, she recalled, "I realized, 'Oh, my God, I don't know what he really wants.'" He'd made her his health-care proxy, but they had talked about such situations only superficially. So she turned the car around.

Going back in "was really uncomfortable," she said. It made no difference that she was an expert in end-of-life discussions. "I just felt awful having the conversation with my dad." But she went through her list. She told him, "'I need to understand how much you're willing to go through to have a shot at being alive and what level of being alive is tolerable to you.' We had this quite agonizing conversation where he said—and this totally shocked me—'Well, if I'm able to eat chocolate ice cream and watch football on TV, then I'm willing to stay alive. I'm willing to go through a lot of pain if I have a shot at that.'

"I would never have expected him to say that," Block went on. "I mean, he's a professor emeritus. He's never watched a football game in my conscious memory. The whole picture—it wasn't the guy I thought I knew." But the conversation proved critical, because after surgery he developed bleeding in the spinal cord. The surgeons told her that in order to save his life, they would need to go back in. But he had already become nearly quadriplegic and would remain severely disabled for many months and possibly forever. What did she want to do?

"I had three minutes to make this decision, and, I realized, he had already made the decision." She asked the surgeons whether, if her father survived, he would still be able to eat chocolate ice cream and watch football on TV. Yes, they said. She gave the okay to take him back to the operating room.

"If I had not had that conversation with him," she told me, "my instinct would have been to let him go at that moment, because it just seemed so awful. And I would have beaten myself up. Did I let him go too soon?" Or she might have gone ahead and sent him to surgery, only to find—as occurred—that he survived only to go through what proved to be a year of "very horrible rehab" and disability. "I would have felt so guilty that I condemned him to that," she said. "But there was no decision for me to make." He had decided.

During the next two years, he regained the ability to walk short distances. He required caregivers to bathe and dress him. He had difficulty swallowing and eating. But his mind was intact and he had partial use of his hands—enough to write two books and more than a dozen scientific articles. He lived for ten years after the operation. This past year, however, his difficulties with swallowing advanced to the point where he could not eat without aspirating food particles, and he cycled between hospital and rehabilitation facilities with the pneumonias that resulted. He didn't want a feeding tube. And it became evident that the battle for the dwindling chance of a miraculous recovery was going to leave him unable ever to go home again. So this past January he decided to stop the battle and go home.

"We started him on hospice care," Block said. "We treated his choking and kept him comfortable. Eventually, he stopped eating and drinking. He died about five days later."

Susan Block and her father had the conversation that we all need to have when the chemotherapy stops working, when we start needing oxygen at home, when we face high-risk surgery, when the liver failure keeps progressing, when we become unable to dress ourselves. I've heard Swedish doctors call it a "breakpoint discussion," a systematic series of conversations to sort out when they need to switch from fighting for time to fighting for the other things that people value—being with family or traveling or enjoying chocolate ice cream. Few people have this discussion, and there is good reason for anyone to dread these conversations. They can unleash difficult emotions. People can become angry or overwhelmed. Handled poorly, the conversations can cost a person's trust. Handled well, they can take real time.

I spoke to an oncologist who told me about a twenty-nine-year-old patient she had recently cared for who had an inoperable brain tumor that continued to grow through second-line chemotherapy. The patient elected not to attempt any further chemotherapy, but getting to that decision required hours of discussion—for this was not the decision he had expected to make. First, the oncologist said, she had a discussion with him alone. They reviewed the story of how far he'd come, the options that remained. She was frank. She told him that in her entire career she had never seen third-line chemotherapy produce a significant response in his type of brain tumor. She had looked for experimental therapies, and none were truly promising. And, although she was willing to proceed with chemotherapy, she told him how much strength and time the treatment would take away from him and his family.

He did not shut down or rebel. His questions went on for an hour. He asked about this therapy and that therapy. And then, gradually, he began to ask about what would happen as the tumor got bigger, the symptoms he'd have, the ways they could try to control them, how the end might come.

The oncologist next met with the young man together with his family. That discussion didn't go so well. He had a wife and small children, and at first his wife wasn't ready to contemplate stopping chemo. But when the oncologist asked the patient to explain in his own words what they'd discussed, she understood. It was the same with his mother, who was a nurse. Meanwhile, his father sat quietly and said nothing the entire time.

A few days later, the patient returned to talk to the oncologist. "There should be something. There *must* be something," he said. His father had shown him reports of cures on the Internet. He confided how badly his father was taking the news. No patient wants to cause his family pain. According to Block, about two-thirds of patients are willing to undergo therapies they don't want if that is what their loved ones want.

The oncologist went to the father's home to meet with him. He had a sheaf of possible trials and treatments printed from the Internet. She went through them all. She was willing to change her opinion, she told him. But either the treatments were for brain tumors that were very different from his son's or else he didn't qualify. None were going to be miraculous. She told the father that he needed to understand: time with his son was limited, and the young man was going to need his father's help getting through it.

The oncologist noted wryly how much easier it would have been for her just to prescribe the chemotherapy. "But that meeting with the father was the turning point," she said. The patient and the family opted for hospice. They had more than a month together before he died. Later the father thanked the doctor. That last month, he said, the family simply focused on being together, and it proved to be the most meaningful time they'd ever spent.

Given how prolonged some of these conversations have to be, many people argue that the key problem has been the financial incentives: we pay doctors to give chemotherapy and to do surgery, but not to take the time required to sort out when doing so is unwise. This certainly is a factor. (The new health-reform act was to have added Medicare coverage for these conversations, until it was deemed funding for "death panels" and stripped out of the legislation.) But the issue isn't merely a matter of financing. It arises from a still unresolved argument about what the function of medicine really is—what, in other words, we should and should not be paying for doctors to do.

The simple view is that medicine exists to fight death and disease, and that is, of course, its most basic task. Death is the enemy. But the enemy has superior forces. Eventually, it wins. And in a war that you cannot win, you don't want a general who fights to the point of total annihilation. You don't want Custer. You want Robert E. Lee, someone who knew how to fight for territory when he could and how to surrender when he couldn't, someone who understood that the damage is greatest if all you do is fight to the bitter end.

More often these days, medicine seems to supply neither Custers nor Lees. We are increasingly the generals who march the soldiers onward, saying all the while, "You let me know when you want to stop." All-out treatment, we tell the terminally ill, is a train you can get off at any time—just say when. But for most patients and their families this is asking too much. They remain riven by doubt and fear and desperation; some are deluded by a fantasy of what medical science can achieve. But our responsibility in medicine is to deal with human beings as they are. People die only once. They have no experience to draw upon. They need doctors and nurses who are willing to have the hard discussions and say what they have seen, who will help people prepare for what is to come—and to escape a warehoused oblivion that few really want.

Sara Monopoli had had enough discussions to let her family and her oncologist know that she did not want hospitals or ICUs at the end—but not enough to have learned how to achieve this. From the moment she arrived in the emergency room that Friday morning in February, the train of events ran against a peaceful ending.

There was one person who was disturbed by this, though, and who finally decided to intercede—Chuck Morris, her primary physician. As her illness had progressed through the previous year, he had left the decision-making largely to Sara, her family, and the oncology team. Still, he had seen her and her husband regularly and listened to their concerns. That desperate morning, Morris was the one person Rich called before getting into the ambulance. He headed to the emergency room and met Sara and Rich when they arrived.

Morris said that the pneumonia might be treatable. But, he told Rich, "I'm worried this is it. I'm really worried about her." And he told him to let the family know that he said so.

Upstairs in her hospital room, Morris talked with Sara and Rich about the ways in which the cancer had been weakening her, making it hard for her body to fight off infection. Even if the antibiotics halted the infection, he said, he wanted them to remember that there was nothing that would stop the cancer.

Sara looked ghastly, Morris told me. "She was so short of breath. It was uncomfortable to watch. I still remember the attending"— the oncologist who admitted her for the pneumonia treatment. "He was actually kind of rattled about the whole case, and for him to be rattled is saying something."

After her parents arrived, Morris talked with them, too, and when they were finished Sara and her family agreed on a plan. The medical team would continue the antibiotics. But if things got worse they would not put her on a breathing machine. They also let him call the palliative-care team to visit. The team prescribed a small dose of morphine, which immediately eased her breathing. Her family saw how much her suffering diminished, and suddenly they didn't want any more suffering. The next morning they were the ones to hold back the medical team.

"They wanted to put a catheter in her, do this other stuff to her," her mother, Dawn, told me. "I said, '*No*. You aren't going to do anything to her.' I didn't care if she wet her bed. They wanted to do lab tests, blood-pressure measurements, finger sticks. I was very uninterested in their bookkeeping. I went over to see the head nurse and told them to stop."

In the previous three months, almost nothing we'd done to Sara —none of our chemotherapy and scans and tests and radiation—

had likely achieved anything except to make her worse. She may well have lived longer without any of it. At least she was spared at the very end.

That day Sara fell into unconsciousness as her body continued to fail. Through the next night, Rich recalled, "there was this awful groaning." There is no prettifying death. "Whether it was with inhaling or exhaling, I don't remember, but it was horrible, horrible, horrible to listen to."

Her father and her sister still thought that she might rally. But when the others had stepped out of the room, Rich knelt down weeping beside Sara and whispered in her ear. "It's okay to let go," he said. "You don't have to fight anymore. I will see you soon."

Later that morning, her breathing changed, slowing. At 9:45 A.M., Rich said, "Sara just kind of startled. She let a long breath out. Then she just stopped."

MALCOLM GLADWELL

The Treatment

FROM *The New Yorker*

IN THE WORLD of cancer research, there is something called a Kaplan-Meier curve, which tracks the health of patients in the trial of an experimental drug. In its simplest version, it consists of two lines. The first follows the patients in the "control arm," the second the patients in the "treatment arm." In most cases those two lines are virtually identical. That is the sad fact of cancer research: nine times out of ten, there is no difference in survival between those who were given the new drug and those who were not. But every now and again—after millions of dollars have been spent, and tens of thousands of pages of data collected, and patients followed, and toxicological issues examined, and safety issues resolved, and manufacturing processes fine-tuned—the patients in the treatment arm will live longer than the patients in the control arm, and the two lines on the Kaplan-Meier will start to diverge.

Seven years ago, for example, a team from Genentech presented the results of a colorectal-cancer drug trial at the annual meeting of the American Society of Clinical Oncology—a conference attended by virtually every major cancer researcher in the world. The lead Genentech researcher took the audience through one slide after another— *click, click, click*—laying out the design and scope of the study until he came to the crucial moment: the Kaplan-Meier. At that point, what he said became irrelevant. The members of the audience saw daylight between the two lines for a patient population in which that almost never happened, and they leaped to their feet and gave him an ovation. Every drug researcher in the world dreams of standing in front of thousands of people at

ASCO and clicking on a Kaplan-Meier like that. "It is why we are in this business," Safi Bahcall says. Once he thought that this dream would come true for him. It was in the late summer of 2006, and is among the greatest moments of his life.

Bahcall is the CEO of Synta Pharmaceuticals, a small biotechnology company. It occupies a one-story brick 1970s building outside Boston, just off Route 128, where many of the region's high-tech companies have congregated, and that summer Synta had two compounds in development. One was a cancer drug called elesclomol. The other was an immune modulator called apilimod. Experimental drugs must pass through three phases of testing before they can be considered for government approval. Phase 1 is a small trial to determine at what dose the drug can be taken safely. Phase 2 is a larger trial to figure out if it has therapeutic potential, and Phase 3 is a definitive trial to see if it actually works, usually in comparison with standard treatments. Elesclomol had progressed to Phase 2 for soft-tissue sarcomas and lung cancer and had come up short in both cases. A Phase 2 trial for metastatic melanoma—a deadly form of skin cancer—was also underway. But that was a long shot: nothing ever worked well for melanoma. In the previous thirty-five years, there had been something like seventy large-scale Phase 2 trials for metastatic-melanoma drugs, and if you plotted all the results on a single Kaplan-Meier there wouldn't be much more than a razor's edge of difference between any two of the lines.

That left apilimod. In animal studies and early clinical trials for autoimmune disorders, it seemed promising. But when Synta went to Phase 2 with a trial for psoriasis, the results were underwhelming. "It was ugly," Bahcall says. "We had lung cancer fail, sarcoma next, and then psoriasis. We had one more trial left, which was for Crohn's disease. I remember my biostats guy coming into my office, saying, 'I've got some good news and some bad news. The good news is that apilimod is safe. We have the data. No toxicity. The bad news is that it's not effective.' It was heartbreaking."

Bahcall is a boyish man in his early forties, with a round face and dark, curly hair. He was sitting at the dining-room table in his sparsely furnished apartment in Manhattan, overlooking the Hudson River. Behind him a bicycle was leaning against a bare wall, giving the room a post-college feel. Both his parents were astrophysicists, and he, too, was trained as a physicist, before leaving academia for the business world. He grew up in the realm of the abstract and the theoretical—with theorems and calculations and precise measurements. But drug development was different, and when he spoke about the failure of apilimod there was a slight catch in his voice.

Bahcall started to talk about one of the first patients ever treated with elesclomol: a twenty-four-year-old African American man. He'd had Kaposi's sarcoma; tumors covered his lower torso. He'd been at Beth Israel Deaconess Medical Center in Boston, and Bahcall had flown up to see him. On a Monday in January 2003, Bahcall sat by his bed and they talked. The patient was just out of college. He had an IV in his arm. You went to the hospital and you sat next to some kid whose only wish was not to die, and it was impossible not to get emotionally involved. In physics, failure was disappointing. In drug development, failure was *heartbreaking*. Elesclomol wasn't much help against Kaposi's sarcoma. And now apilimod didn't work for Crohn's. "I mean, we'd done charity work for the Crohn's & Colitis Foundation," Bahcall went on. "I have relatives and friends with Crohn's disease, personal experience with Crohn's disease. We had Crohn's patients come in and talk in meetings and tell their stories. We'd raised money for five years from investors. I felt terrible. Here we were with our lead drug, and it had failed. It was the end of the line."

That summer of 2006, in one painful meeting after another, Synta began to downsize. "It was a Wednesday," Bahcall said. "We were around a table, and we were talking about pruning the budget and how we're going to contain costs, one in a series of tough discussions, and I noticed my chief medical officer, Eric Jacobson, at the end of the table, kind of looking a little unusually perky for one of those kinds of discussions." After the meeting, Bahcall pulled Jacobson over: "Is something up?" Jacobson nodded. Half an hour before the meeting, he'd received some news. It was about the melanoma trial for elesclomol, the study everyone had given up on. "The consultant said she had never seen data this good," Jacobson told him.

Bahcall called back the management team for a special meeting. He gave the floor to Jacobson. "Eric was, like, 'Well, you know we've got this melanoma trial,'" Bahcall began, "and it took a moment to jog people's memories, because we'd all been so focused on

Crohn's disease and the psoriasis trials. And Eric said, 'Well, we got
the results. The drug worked! It was a positive trial!'" One person
slammed the table, stood up, and hollered. Others peppered Eric
with questions. "Eric said, 'Well, the group analyzing the data is trying to disprove it, and they can't disprove it.' And he said, 'The
consultant handed me the data on Wednesday morning, and she
said it was boinking good.' And everyone said, 'What?' Because Eric
is the sweetest guy, who never swears. A bad word cannot cross his
lips. Everyone started yelling, 'What? What? What did she say, Eric?
Eric! Eric! Say it! Say it!'"

Bahcall contacted Synta's board of directors. Two days later he
sent out a company-wide e-mail saying that there would be a meeting that afternoon. At four o'clock, all 130 employees trooped into
the building's lobby. Jacobson stood up. "So the lights go down,"
Bahcall continued. "Clinical guys, when they present data, tend to
do it in a very bottom-up way: this is the disease population, this is
the treatment, and this is the drug, and this is what was randomized, and this is the demographic, and this is the patient pool, and
this is who had toenail fungus, and this is who was Jewish. They go
on and on and on, and all anyone wants is, Show us the fucking
Kaplan-Meier! Finally he said, 'All right, now we can get to the efficacy.' It gets really silent in the room. He clicks the slide. The two
lines separate out beautifully—and a gasp goes out, across a hundred and thirty people. Eric starts to continue, and one person
goes like this"—Bahcall started clapping slowly—"and then a couple of people joined in, and then soon the whole room is just going like this—clap, clap, clap. There were tears. We all realized
that our lives had changed, the lives of patients had changed, the
way of treating the disease had changed. In that moment, everyone
realized that this little company of a hundred and thirty people
had a chance to win. We had a drug that worked, in a disease where
nothing worked. That was the single most moving five minutes of
all my years at Synta."

In the winter of 1955, a young doctor named Emil Freireich arrived at the National Cancer Institute in Bethesda, Maryland. He
had been drafted into the army and had been sent to fulfill his
military obligation in the public-health service. He went to see
Gordon Zubrod, then the clinical director for the NCI and later

one of the major figures in cancer research. "I said, 'I'm a hematologist,'" Freireich recalls. "He said, 'I've got a good idea for you. Cure leukemia.' It was a military assignment." From that assignment came the first great breakthrough in the war against cancer.

Freireich's focus was on the commonest form of childhood leukemia—acute lymphoblastic leukemia (ALL). The diagnosis was a death sentence. "The children would come in bleeding," Freireich says. "They'd have infections. They would be in pain. Median survival was about eight weeks, and everyone was dead within the year." At the time three drugs were known to be useful against ALL. One was methotrexate, which, the pediatric pathologist Sidney Farber had shown seven years earlier, could push the disease into remission. Corticosteroids and 6-mercaptopurine (6-MP) had since proved useful. But even though methotrexate and 6-MP could kill a lot of cancer cells, they couldn't kill them all, and those that survived would regroup and adjust and multiply and return with a vengeance. "These remissions were all temporary—two or three months," Freireich, who now directs the adult-leukemia research program at the M. D. Anderson Cancer Center in Houston, says. "The authorities in hematology didn't even want to use them in children. They felt it just prolonged the agony, made them suffer, and gave them side effects. That was the landscape."

In those years the medical world had made great strides against tuberculosis, and treating TB ran into the same problem as treating cancer: if doctors went after it with one drug, the bacteria eventually developed resistance. Their solution was to use multiple drugs simultaneously that worked in very different ways. Freireich wondered about applying that model to leukemia. Methotrexate worked by disrupting folic-acid uptake, which was crucial in the division of cells; 6-MP shut down the synthesis of purine, which was also critical in cell division. Putting the two together would be like hitting the cancer with a left hook and a right hook. Working with a group that eventually included Tom Frei, of the NCI, and James Holland, of the Roswell Park Cancer Institute in Buffalo, Freireich started treating ALL patients with methotrexate and 6-MP in combination, each at two-thirds its regular dose to keep side effects in check. The remissions grew more frequent. Freireich then added the steroid prednisone, which worked by a mechanism different from that of either 6-MP or methotrexate; he could give it at full dose and not worry about the side effects getting out of control. Now he had a left hook, a right hook, and an uppercut.

"So things are looking good," Freireich went on. "But still everyone dies. The remissions are short. And then out of the blue came the gift from heaven"—another drug, derived from periwinkle, that had been discovered by Irving Johnson, a researcher at Eli Lilly. "In order to get two milligrams of drug, it took something like two train-car loads of periwinkle," Freireich said. "It was expensive. But Johnson was persistent." Lilly offered the new drug to Freireich. "Johnson had done work in mice, and he showed me the results. I said, 'Gee whiz, I've got ten kids on the ward dying. I'll give it to them tomorrow.' So I went to Zubrod. He said, 'I don't think it's a good idea.' But I said, 'These kids are dying. What's the difference?' He said, 'Okay, I'll let you do a few children.' The response rate was fifty-five percent. The kids jumped out of bed." The drug was called vincristine, and by itself it was no wonder drug. Like the others, it worked only for a while. But the good news was that it had a unique mechanism of action—it interfered with cell division by binding to what is called the spindle protein—and its side effects were different from those of the other drugs. "So I sat down at my desk one day and I thought, Gee, if I can give 6-MP and meth at two-thirds dose and prednisone at full dose and vincristine has different limiting toxicities, I bet I can give a full dose of that, too. So I devised a trial where we would give all four in combination." The trial was called VAMP. It was a left hook, a right hook, an uppercut, and a jab, and the hope was that if you hit leukemia with that barrage it would never get up off the canvas.

The first patient treated under the experimental regimen was a young girl. Freireich started her off with a dose that turned out to be too high, and she almost died. She was put on antibiotics and a respirator. Freireich saw her eight times a day, sitting at her bedside. She pulled through the chemo-induced crisis, only to die later of an infection. But Freireich was learning. He tinkered with his protocol and started again, with Patient No. 2. Her name was Janice. She was fifteen, and her recovery was nothing short of miraculous. So was the recovery of the next patient and the next and the next, until nearly every child was in remission, without need of antibiotics or transfusions. In 1965 Frei and Freireich published one of the most famous articles in the history of oncology, "Progress and Perspective in the Chemotherapy of Acute Leukemia," in *Advances in Chemotherapy*. Almost three decades later, a perfectly healthy Janice graced the cover of the journal *Cancer Research*.

What happened with ALL was a formative experience for an entire generation of cancer fighters. VAMP proved that medicine didn't need a magic bullet—a superdrug that could stop all cancer in its tracks. A drug that worked a little bit could be combined with another that worked a little bit and another that worked a little bit, and, as long as all three worked in different ways and had different side effects, the combination could turn out to be spectacular. To be valuable, a cancer drug didn't have to be especially effective on its own; it just had to be novel in the way it acted. And from the beginning, this was what caused so much excitement about elesclomol.

Safi Bahcall's partner in the founding of Synta was a cell biologist at Harvard Medical School named Lan Bo Chen. Chen, who is in his mid-sixties, was born in Taiwan. He is a mischievous man with short-cropped straight black hair and various quirks—including a willingness to say whatever is on his mind, a skepticism about all things Japanese (the Japanese occupied Taiwan during the war, after all), and a keen interest in the marital prospects of his unattached coworkers. Bahcall, who is Jewish, describes him affectionately as "the best and worst parts of a Jewish father and the best and worst parts of a Jewish mother rolled into one." (Sample e-mail from Chen: "Safi is in Israel. Hope he finds wife.")

Drug hunters like Chen fall into one of two broad schools. The first school, that of "rational design," believes in starting with the disease and working backward—designing a customized solution based on the characteristics of the problem. Herceptin, one of the most important of the new generation of breast-cancer drugs, is a good example. It was based on genetic detective work showing that about a quarter of all breast cancers were caused by the overproduction of a protein called HER2. HER2 kept causing cells to divide and divide, and scientists set about designing a drug to turn HER2 off. The result is a drug that improved survival in 25 percent of patients with advanced breast cancer. (When Herceptin's Kaplan-Meier was shown at ASCO, there was stunned silence.) But working backward to a solution requires a precise understanding of the problem, and cancer remains so mysterious and complex that in most cases scientists don't have that precise understanding. Or they think they do, and then, after they turn off one mechanism, they discover that the tumor has other deadly tricks in reserve.

The other approach is to start with a drug candidate and then hunt for diseases that it might attack. This strategy, known as "mass screening," doesn't involve a theory. Instead, it involves a random search for matches between treatments and diseases. This was the school to which Chen belonged. In fact, he felt that the main problem with mass screening was that it wasn't mass enough. There were countless companies outside the drug business—from industrial research labs to photography giants like Kodak and Fujifilm—that had millions of chemicals sitting in their vaults. Yet most of these chemicals had never been tested to see if they had potential as drugs. Chen couldn't understand why. If the goal of drug discovery was novelty, shouldn't the hunt for new drugs go as far and wide as possible?

"In the early eighties, I looked into how Merck and Pfizer went about drug discovery," Chen recalls. "How many compounds are they using? Are they doing the best they can? And I come up with an incredible number. It turns out that mankind had, at this point, made tens of millions of compounds. But Pfizer was screening only six hundred thousand compounds, and Merck even fewer, about five hundred thousand. How could they screen for drugs and use only five hundred thousand, when mankind has already made so many more?"

An early financial backer of Chen's was Michael Milken, the junk-bond king of the 1980s, who, after being treated for prostate cancer, became a major cancer philanthropist. "I told Milken my story," Chen said, "and very quickly he said, 'I'm going to give you four million dollars. Do whatever you want.' Right away Milken thought of Russia. Someone had told him that the Russians had had, for a long time, thousands of chemists in one city making compounds, and none of those compounds had been disclosed." Chen's first purchase was a batch of 22,000 chemicals, gathered from all over Russia and Ukraine. They cost about ten dollars each and came in tiny glass vials. With his money from Milken, Chen then bought a $600,000 state-of-the-art drug-screening machine. It was a big, automated Rube Goldberg contraption that could test ninety-six compounds at a time and do a hundred batches a day. A robotic arm would deposit a few drops of each chemical onto a plate, followed by a clump of cancer cells and a touch of blue dye. The mixture was left to sit for a week and then reexamined. If the cells were still alive, they would show as blue. If the chemical killed the cancer cells, the fluid would be clear.

Chen's laboratory began by testing his compounds against prostate-cancer cells, since that was the disease Milken had. Later he screened dozens of other cancer cells as well. In the first round, his batch of chemicals killed everything in sight. But plenty of compounds, including pesticides and other sorts of industrial poisons, will kill cancer cells. The trouble is that they'll kill healthy cells as well. Chen was looking for something that was selective—that was more likely to kill malignant cells than normal cells. He was also interested in sensitivity—in a chemical's ability to kill at low concentrations. Chen reduced the amount of each chemical on the plate a thousandfold and tried again. Now just one chemical worked. He tried the same chemical on healthy cells. It left them alone. Chen lowered the dose another thousandfold. It still worked. The compound came from the National Taras Shevchenko University of Kiev. It was an odd little chemical, the laboratory equivalent of a jazz musician's riff. "It was pure chemist's joy," Chen said. "Homemade, random, and clearly made for no particular purpose. It was the only one that worked on everything we tried."

Mass screening wasn't as elegant or as efficient as rational drug design. But it provided a chance of stumbling across something by accident—something so novel and unexpected that no scientist would have dreamed it up. It provided for serendipity, and the history of drug discovery is full of stories of serendipity. Alexander Fleming was looking for something to fight bacteria but didn't think the answer would be provided by the mold that grew on a petri dish he accidentally left out on his bench. That's where penicillin came from. Pfizer was looking for a new heart treatment and realized that a drug candidate's unexpected side effect was more useful than its main effect. That's where Viagra came from. "The end of surprise would be the end of science," the historian Robert Friedel wrote in the 2001 essay "Serendipity Is No Accident." "To this extent, the scientist must constantly seek and hope for surprises." When Chen gathered chemical compounds from the farthest corners of the earth and tested them against one cancer-cell line after another, he was engineering surprise.

What he found was exactly what he'd hoped for when he started his hunt: something he could never have imagined on his own. When cancer cells came into contact with the chemical, they seemed to go into crisis mode: they acted as if they had been attacked with a blowtorch. The Ukrainian chemical, elesclomol, worked by gathering up copper from the bloodstream and bringing it into cells' mitochondria, sparking an electrochemical reaction. His focus was on the toxic, oxygen-based compounds in the cell called ROS, reactive oxygen species. Normal cells keep ROS in check. Many kinds of cancer cells, though, generate so much ROS that the cell's ability to keep functioning is stretched to the breaking point, and elesclomol cranked ROS up even further, to the point that the cancer cells went up in flames. Researchers had long known that heating up a cancer cell was a good way of killing it, and there had been plenty of interest over the years in achieving that effect with ROS. But the idea of using copper to set off an electrochemical reaction was so weird—and so unlike the way cancer drugs normally worked—that it's not an approach anyone would have tried by design. That was the serendipity. It took a bit of "chemist's joy," constructed for no particular reason by some bench scientists in Kiev, to show the way. Elesclomol was wondrously novel. "I fell in love," Chen said. "I can't explain it. I just did."

When Freireich went to Zubrod with his idea for VAMP, Zubrod could easily have said no. Drug protocols are typically tested in advance for safety in animal models. This one wasn't. Freireich freely admits that the whole idea of putting together poisonous drugs in such dosages was "insane," and, of course, the first patient in the trial had nearly been killed by the toxic regimen. If she had died from it, the whole trial could have been derailed.

The ALL success story provided a hopeful road map for a generation of cancer fighters. But it also came with a warning: those who pursued the unexpected had to live with unexpected consequences. This was not the elegance of rational drug design, where scientists perfect their strategy in the laboratory before moving into the clinic. Working from the treatment to the disease was an exercise in uncertainty and trial and error.

If you're trying to put together a combination of three or four drugs out of an available pool of dozens, how do you choose which to start with? The number of permutations is vast. And, once you've settled on a combination, how do you administer it? A child gets sick. You treat her. She goes into remission, and then she relapses. VAMP established that the best way to induce remission was to treat the child aggressively when she first showed up with leukemia. But do you treat during the remission as well, or only when the child relapses? And, if you treat during remission, do you treat as aggressively as you did during remission induction, or at a lower level? Do you use the same drugs in induction as you do in remission and as you do in relapse? How do you give the drugs, sequentially or in combination? At what dose? And how frequently—every day, or do you want to give the child's body a few days to recover between bouts of chemo?

Oncologists compared daily 6-MP plus daily methotrexate with daily 6-MP plus methotrexate every four days. They compared methotrexate followed by 6-MP, 6-MP followed by methotrexate, and both together. They compared prednisone followed by full doses of 6-MP, methotrexate, and a new drug, cyclophosphamide (CTX), with prednisone followed by half doses of 6-MP, methotrexate, and CTX. It was endless: vincristine plus prednisone and then methotrexate every four days or vincristine plus prednisone and then methotrexate daily? They tried new drugs and different combinations. They tweaked and refined and gradually pushed the cure rate from 40 percent to 85 percent. At St. Jude Children's Research Hospital in Memphis—which became a major center of ALL research—no fewer than sixteen clinical trials, enrolling 3,011 children, have been conducted in the past forty-eight years.

And this was just childhood leukemia. Beginning in the 1970s, Lawrence Einhorn, at Indiana University, pushed cure rates for testicular cancer above 80 percent with a regimen called BEP: three to four rounds of bleomycin, etoposide, and cisplatin. In the 1970s Vincent T. DeVita, at the NCI, came up with MOPP for advanced Hodgkin's disease: mustargen, oncovin, procarbazine, and prednisone. DeVita went on to develop a combination therapy for breast cancer called CMF—cyclophosphamide, methotrexate, and 5-fluorouracil. Each combination was a variation on the combination that came before it, tailored to its target through a series of iterations. The often-asked question "When will we find a cure for cancer?" implies that there is some kind of master code behind the disease waiting to be cracked. But perhaps there isn't a master code. Perhaps there is only what can be uncovered, one step at a time, through trial and error.

When elesclomol emerged from the laboratory, then, all that was known about it was that it did something novel to cancer cells in the laboratory. Nobody had any idea what its best target was. So Synta gave elesclomol to an oncologist at Beth Israel in Boston, who began randomly testing it out on his patients in combination with paclitaxel, a standard chemotherapy drug. The addition of elesclomol seemed to shrink the tumor of someone with melanoma. A patient whose advanced ovarian cancer had failed multiple rounds of previous treatment had some response. There was dramatic activity against Kaposi's sarcoma. They could have gone on with Phase 1s indefinitely, of course. Chen wanted to combine elesclomol with radiation therapy, and another group at Synta would later lobby hard to study elesclomol's effects on acute myeloid leukemia (AML), the commonest form of adult leukemia. But they had to draw the line somewhere. Phase 2 would be lung cancer, soft-tissue sarcomas, and melanoma.

Now Synta had its targets. But with this round of testing came an even more difficult question. What's the best way to conduct a test of a drug you barely understand? To complicate matters further, melanoma, the disease that seemed to be the best of the three options, is among the most complicated of all cancers. Sometimes it confines itself to the surface of the skin. Sometimes it invades every organ in the body. Some kinds of melanoma have a mutation involving a gene called BRAF; others don't. Some late-stage melanoma tumors pump out high levels of an enzyme called LDH. Sometimes they pump out only low levels of LDH, and patients with low-LDH tumors lived so much longer that it was as if they had a different disease. Two patients could appear to have identical diagnoses, and then one would be dead in six months and the other would be fine. Tumors sometimes mysteriously disappeared. How did you conduct a drug trial with a disease like this?

It was entirely possible that elesclomol would work in low-LDH patients and not in high-LDH patients, or in high-LDH patients and not in low-LDH ones. It might work well against the melanoma that confined itself to the skin and not against the kind that invaded the liver and other secondary organs; it might work in the early stages of metastasis and not in the later stages. Then there was the prior-treatment question. Because tumors quickly become resistant to drugs, new treatments sometimes work better on "naive" patients—those who haven't been treated with other forms of chemotherapy. So elesclomol might work on chemo-naive patients and not on prior-chemo patients. And in any of these situations, elesclomol might work better or worse depending on which other drug or drugs it was combined with. There was no end to the possible combinations of patient populations and drugs that Synta could have explored.

At the same time, Synta had to make sure that whatever trial it ran was as big as possible. With a disease as variable as melanoma, there was always the risk in a small study that what you thought was a positive result was really a matter of spontaneous remissions, and that a negative result was just the bad luck of having patients with an unusually recalcitrant form of the disease. John Kirkwood, a melanoma specialist at the University of Pittsburgh, had done the math: in order to guard against some lucky or unlucky artifact, the treatment arm of a Phase 2 trial should have at least seventy patients.

Synta was faced with a dilemma. Given melanoma's variability, the company would ideally have done half a dozen or more versions of its Phase 2 trial: low-LDH, high-LDH, early-stage, late-stage, prior-chemo, chemo-naive, multidrug, single-drug. There was no way, though, that they could afford to do that many trials with seventy patients in each treatment arm. The American biotech industry is made up of lots of companies like Synta, because small startups are believed to be more innovative and adventurous than big pharmaceutical houses. But not even big firms can do multiple Phase 2 trials on a single disease—not when trials cost more than $100,000 per patient and not when, in pursuit of serendipity, they are simultaneously testing that same experimental drug on two or three other kinds of cancer. So Synta compromised. The company settled on one melanoma trial: fifty-three patients were given elesclomol plus paclitaxel, and twenty-eight, in the control group, were given paclitaxel alone, representing every sort of LDH level, stage of disease, and prior-treatment status. That's a long way from half a dozen trials of seventy each.

Synta then went to Phase 3: 651 chemo-naive patients, drawn from 150 hospitals in fifteen countries. The trial was dubbed SYM-METRY. It was funded by the pharmaceutical giant Glaxo Smith Kline. Glaxo agreed to underwrite the cost of the next round of clinical trials and—should the drug be approved by the Food and Drug Administration—to split the revenues with Synta.

But was this the perfect trial? Not really. In the Phase 2 trial, eles-clomol had been mixed with an organic solvent called Cremo-phore and then spun around in a sonicator, which is like a mini washing machine. Elesclomol, which is rock-hard in its crystalline form, needed to be completely dissolved if it was going to work as a drug. For SYMMETRY, though, sonicators couldn't be used. "Many countries said that it would be difficult, and some hospitals even said, 'We don't allow sonication in the preparation room,'" Chen explained. "We got all kinds of unbelievable feedback. In the end we came up with something that, after mixing, you use your hand to shake it." Would hand shaking be a problem? No one knew.

Then a Synta chemist, Mitsunori Ono, figured out how to make a water-soluble version of elesclomol. When the head of Synta's chemistry team presented the results, he "sang a Japanese drinking song," Chen said, permitting himself a small smile at the eccen-tricities of the Japanese. "He was very happy." It was a great ac-complishment. The water-soluble version could be given in higher doses. Should they stop SYMMETRY and start again with elesclo-mol 2.0? They couldn't. A new trial would cost many millions of dollars more and set the whole effort back two or three years. So they went ahead with a drug that didn't dissolve easily against a dif-ficult target, with an assortment of patients who may or may not have been ideal—and crossed their fingers.

SYMMETRY began in late 2007. It was a double-blind, random-ized trial. No one had any idea who was getting elesclomol and who wasn't, and no one would have any idea how well the patients on elesclomol were doing until the trial data were unblinded. Day-to-day management of the study was shared with a third-party con-tractor. The trial itself was supervised by an outside group, known as a data-monitoring committee. "We send them all the data in some database format, and they plug that into their software pack-age, and then they type in the code and press 'Enter,'" Bahcall said. "And then this line"—he pointed at the Kaplan-Meier in front of him—"will, hopefully, separate into two lines. They will find out in thirty seconds. It's, literally, those guys press a button and for the next five years, ten years, the life of the drug, that's really the only bit of evidence that matters." It was January 2009, and the last of the 651 patients were scheduled to be enrolled in the trial in the next few weeks. According to protocol, when the results began to come in, the data-monitoring committee would call Jacobson, and Jacobson would call Bahcall. "ASCO starts May 29," Bahcall said. "If we get our data by early May, we could present at ASCO this year."

In the course of the SYMMETRY trial, Bahcall's dining-room-table talks grew more reflective. He drew Kaplan-Meiers on the back of napkins. He talked about the twists and turns that other biotech companies had encountered on the road to the marketplace. He told wry stories about Lan Bo Chen, the Jewish mother and Jewish father rolled into one—and, over and over, he brought up the name of Judah Folkman. Folkman died in 2008, and he was a legend. He was the father of angiogenesis—a wholly new way of attacking cancer tumors. Avastin, the drug that everyone cheered at ASCO seven years ago, was the result of Folkman's work.

Folkman's great breakthrough had come while he was working with mouse melanoma cells at the National Naval Medical Center: when the tumors couldn't set up a network of blood vessels to feed themselves, they would stop growing. Folkman realized that the body must have its own system for promoting and halting blood-vessel formation, and that if he could find a substance that prevented vessels from being formed he would have a potentially powerful cancer drug. One of the researchers in Folkman's laboratory, Michael O'Reilly, found what seemed to be a potent inhibitor: angiostatin. O'Reilly then assembled a group of mice with an aggressive lung cancer and treated half with a saline solution and half with angiostatin. In the book *Dr. Folkman's War* (2001), Robert Cooke describes the climactic moment when the results of the experiment came in:

> With a horde of excited researchers jampacked into a small laboratory room, Folkman euthanized all fifteen mice, then began handing them one by one to O'Reilly to dissect. O'Reilly took the first mouse, made

an incision in its chest, and removed the lung. The organ was overwhelmed by cancer. Folkman checked a notebook to see which group
the mouse had been in. It was one of those that had gotten only saline.
O'Reilly cut into the next mouse and removed its lung. It was perfect.
What treatment had it gotten? The notebook revealed it was angiostatin.

It wasn't Folkman's triumph that Bahcall kept coming back to,
however. It was his struggle. Folkman's great insight at the Naval
Medical Center occurred in 1960. O'Reilly's breakthrough experiment occurred in 1994. In the intervening years, Folkman's work
was dismissed and attacked and confronted with every obstacle.

At times Bahcall tried to convince himself that elesclomol's path
might be different. Synta had those exciting Phase 2 results and
the endorsement of the Glaxo deal. "For the results not to be real,
you'd have to believe that it was just a statistical fluke that the patients who got drugs are getting better," Bahcall said, in one of
those dining-room-table moments. "You'd have to believe that the
fact that there were more responses in the treatment group was
also a statistical fluke, along with the fact that we've seen these
signs of activity in Phase 1, and the fact that the underlying biology
strongly says that we have an extremely active anticancer agent."

But then he would remember Folkman. Angiostatin and a companion agent also identified by Folkman's laboratory, endostatin,
were licensed by a biotech company called EntreMed. And EntreMed never made a dime off either drug. The two drugs failed to
show any clinical effects in both Phase 1 and Phase 2. Avastin was a
completely different anti-angiogenesis agent, discovered and developed by another team entirely and brought to market a decade
after O'Reilly's experiment. What's more, Avastin's colorectal-cancer trial — the one that received a standing ovation at ASCO — was
the drug's second round. A previous Phase 3 trial for breast cancer
had been a crushing failure. Even Folkman's beautifully elaborated
theory about angiogenesis may not fully explain the way Avastin
works. In addition to cutting off the flow of blood vessels to the tumor, Avastin seems also to work by repairing some of the blood
vessels feeding the tumor, so that the drugs administered in combination with Avastin can get to the tumor more efficiently.

Bahcall followed the fortunes of other biotech companies the
way a teenage boy follows baseball statistics, and he knew that noth-

ing ever went smoothly. He could list, one by one, all the break-through drugs that had failed their first Phase 3 or had failed multiple Phase 2s or that turned out not to work the way they were supposed to work. In the world of serendipity and of trial and error, failure was a condition of discovery, because, when something was new and worked in ways that no one quite understood, every bit of knowledge had to be learned, one experiment at a time. You ended up with VAMP, which worked, but only after you compared daily 6-MP and daily methotrexate with daily 6-MP and methotrexate every four days, and so on, through a great many iterations, none of which worked very well at all. You had results that looked "boinking good," but only after a trial with a hundred compromises.

Chen had the same combination of realism and idealism that Bahcall did. He was the in-house skeptic at Synta. He was the one who worried the most about the hand shaking of the drugs in the SYMMETRY trial. He had never been comfortable with the big push behind melanoma. "Everyone at Dana-Farber"—the cancer hospital at Harvard—"told me, 'Don't touch melanoma,'" Chen said. "'It is so hard. Maybe you save it as the last, after you have already treated and tried everything else.'" The scientists at Synta were getting better and better at understanding just what it was that elesclomol did when it confronted a cancer cell. But he knew that there was always a gap between what could be learned in the laboratory and what happened in the clinic. "We just don't know what happens in vivo," he said. "That's why drug development is still so hard and so expensive, because the human body is such a black box. We are totally shooting in the dark." He shrugged. "You have to have good science, sure. But once you shoot the drug in humans you go home and pray."

Chen was sitting in the room at Synta where Eric Jacobson had revealed the "boinking good" news about elesclomol's Phase 2 melanoma study. Down the hall was a huge walk-in freezer, filled with thousands of chemicals from the Russian haul. In another room was the Rube Goldberg drug-screening machine, bought with Milken's money. Chen began to talk about elesclomol's earliest days, when he was still scavenging through the libraries of chemical companies for leads and Bahcall was still an ex-physicist looking to start a biotech company. "I could not convince anyone that elesclomol had potential," Chen went on. "Everyone around me tried to stop it, including my research partner, who is a Nobel laureate. He just hated it." At one point Chen was working with Fujifilm. The people there hated elesclomol. He worked for a while for the Japanese chemical company Shionogi. The Japanese hated it. "But you know who I found who believed in it?" Chen's eyes lit up: "Safi!"

Last year, on February 25, Bahcall and Chen were at a Synta board meeting in midtown Manhattan. It was five-thirty in the afternoon. As the meeting was breaking up, Bahcall got a call on his cell phone. "I have to take this," he said to Chen. He ducked into a nearby conference room, and Chen waited for him with the company's chairman, Keith Gollust. Fifteen minutes passed, then twenty. "I tell Keith it must be the data-monitoring committee," Chen recalls. "He says, 'No way. Too soon. How could the DMC have any news just yet?' I said, 'It has to be.' So he stays with me and we wait. Another twenty minutes. Finally Safi comes out, and I looked at him and I knew. He didn't have to say anything. It was the color of his face."

The call had been from Eric Jacobson. He had just come back from Florida, where he had met with the DMC on the SYMMETRY trial. The results of the trial had been unblinded. Jacobson had spent the last several days going over the data, trying to answer every question and double-check every conclusion. "I have some really bad news," he told Bahcall. The trial would have to be halted: more people were dying in the treatment arm than in the control arm. "It took me about a half hour to come out of primary shock," Bahcall said. "I didn't go home. I just grabbed my bag, got into a cab, went straight to LaGuardia, took the next flight to Logan, drove straight to the office. The chief medical officer, the clinical guys, statistical guys, operational team were all there, and we essentially spent the rest of the night, until about one or two in the morning, reviewing the data." It looked as if patients with high-LDH tumors were the problem: elesclomol seemed to fail them completely. It was heartbreaking. Glaxo, Bahcall knew, was certain to pull out of the deal. There would have to be many layoffs.

The next day Bahcall called a meeting of the management team. They met in the Synta conference room. "Eric has some news,"

Bahcall said. Jacobson stood up and began. But before he got very far he had to stop, because he was overcome with emotion, and soon everyone else in the room was, too.

On December 7, 2009, Synta released the following statement:

> Synta Pharmaceuticals Corp. (NASDAQ: SNTA), a biopharmaceutical company focused on discovering, developing, and commercializing small molecule drugs to treat severe medical conditions, today announced the results of a study evaluating the activity of elesclomol against acute myeloid leukemia (AML) cell lines and primary leukemic blast cells from AML patients, presented at the Annual Meeting of the American Society of Hematology (ASH) in New Orleans . . .
>
> "The experiments conducted at the University of Toronto showed elesclomol was highly active against AML cell lines and primary blast cells from AML patients at concentrations substantially lower than those already achieved in cancer patients in clinical trials," said Vojo Vukovic, M.D., Ph.D., Senior Vice President and Chief Medical Officer, Synta. "Of particular interest were the ex vivo studies of primary AML blast cells from patients recently treated at Toronto, where all 10 samples of leukemic cells responded to exposure to elesclomol. These results provide a strong rationale for further exploring the potential of elesclomol in AML, a disease with high medical need and limited options for patients."

"I will bet anything I have, with anybody, that this will be a drug one day," Chen said. It was January. The early AML results had just come in. Glaxo was a memory. "Now, maybe we are crazy, we are romantic. But this kind of characteristic you have to have if you want to be a drug hunter. You have to be optimistic, you have to have supreme confidence, because the odds are so incredibly against you. I am a scientist. I just hope that I would be so romantic that I become deluded enough to keep hoping."

ANDREW GRANT

Cosmic Blueprint of Life

FROM *Discover*

IN THE LATEST scientific version of Genesis, life begins, paradoxically, with an act of destruction. After 10 billion years of guzzling the hydrogen in its core, a sun-size star runs out of nuclear fuel and becomes unstable. It goes through a series of convulsions and expels a shell of searing-hot atoms—including hydrogen, carbon, and oxygen. The star fizzles into an inert cinder, and its atoms drift off, seemingly lost in the interstellar gloom.

But next the story takes a surprise turn, from destruction to construction. Some of those rogue atoms float into a nearby gas cloud and stick to fine grains of dust there. Even at a frigid −440 degrees Fahrenheit, the atoms bump and crash into each other, merging to form simple molecules. Over millions of years, one relatively dense region of the cloud begins to collapse in on itself. An infant star takes shape at the center. In the surrounding areas, temperatures rise, molecules evaporate from their icy dust grains, and a new round of more intricate chemical reactions begins.

Then comes the most wondrous part of the whole tale. Those reactions weave the simple atoms of hydrogen, carbon, and oxygen into complex organic molecules. Such carbon-bearing compounds are the raw material for life—and they seem to emerge spontaneously, inexorably, in the enormous stretches between the stars. "The abundance of organics and their role in getting life started may make a big, big difference between a giant universe with a lot of life and one with very little," says Scott Sandford of NASA's Ames Research Center in Moffett Field, California, who studies organic molecules from space.

The notion that the underlying chemistry of life could have begun in the far reaches of space, long before our planet even existed, used to be controversial, even comical. No longer. Recent observations show that nebulas throughout our galaxy are bursting with prebiotic molecules. Laboratory simulations demonstrate how intricate molecular reactions can occur efficiently even under exceedingly cold, dry, near-vacuum conditions. Most persuasively, we know for sure that organic chemicals from space could have landed on Earth in the past—because they are doing so right now. Detailed analysis of a meteorite that landed in Australia reveals that it is chock-full of prebiotic molecules.

Similar meteorites and comets would have blanketed Earth with organic chemicals from the time it was born about 4.5 billion years ago until the era when life appeared, a few hundred million years later. Maybe this is how Earth became a living world. Maybe the same thing has happened in many other places as well. "The processes that made these materials and dumped them on our planet are universal. They should happen anywhere you make stars and planets," Sandford says.

The first persuasive hints of life's possible cosmic ancestry came in 1953, courtesy of a renowned experiment devised by chemists Stanley Miller and Harold Urey. From studies of ancient rocks, geologists had a rough sense of our planet's original chemical composition. Biologists, meanwhile, had uncovered the amazingly complex organic molecules that allow living cells to survive. Miller and Urey wanted to see if pure chemistry could help explain how the former transformed into the latter.

The two researchers prepared a closed system of glass flasks and tubes and injected a gaseous mixture of methane, ammonia, hydrogen, and water—four basic compounds thought to be abundant in Earth's primitive atmosphere. Then Miller and Urey applied an electric current to simulate the energy unleashed by lightning strikes. Within a week their concoction had produced several intriguing prebiotic compounds. Many scientists interpreted this as hard experimental evidence that the building blocks of life could have emerged on Earth from nonbiological reactions.

In many ways, though, the experiment supported the opposite view. Even the simplest life forms incorporate two amazingly complex types of organic molecules: proteins and nucleic acids. Proteins perform the basic tasks of metabolism. Nucleic acids (specifically RNA and DNA) encode genetic information and pass it along from one generation to the next. Although the Miller-Urey experiment produced amino acids, the fundamental units of proteins, it never came close to manufacturing nucleobases, the molecular building blocks of DNA and RNA. Furthermore, it is likely that Miller and Urey erred by simulating Earth's early atmosphere with gases containing hydrogen, which reacts easily, as opposed to carbon dioxide, a gas that is far less reactive but was probably far more plentiful at the time. "Interesting chemicals could not have been made as easily as the experiment made it seem," says the astrobiologist Douglas Whittet of Rensselaer Polytechnic Institute in upstate New York.

If life could not so easily have begun on Earth, a few voices argued, perhaps it originated from beyond. The most notable advocate of that hypothesis was the influential British cosmologist Fred Hoyle, who coined the term Big Bang. His 1957 science-fiction novel, *The Black Cloud,* envisioned a living, intelligent dust cloud in space; it foreshadowed his later support of panspermia, the theory that life evolves in space and spreads throughout the universe. Starting in the 1960s, Hoyle wrote a series of academic papers describing how bacterial cells could make their way from interstellar dust grains to comets and eventually down to planets like Earth.

Most of Hoyle's peers considered his ideas borderline delusional. Back then almost nobody thought prebiotic molecules, let alone entire microbes, could survive the harsh vacuum of space. "Everyone assumed space was too cold and too low-density to form molecules," says the National Radio Astronomy Observatory (NRAO) astrochemist Anthony Remijan, a leading expert in interstellar chemistry. "That assumption became 'fact' without any evidence behind it at all."

One of Remijan's mentors, the astronomer Lew Snyder, then at NRAO, dared to disagree. He did not share Hoyle's vision of bacteria hitching rides across the galaxy, but he thought that interesting molecules might subsist in the alleged desert of interstellar space. Snyder had a strategy for finding them, too. He knew that many chemical compounds are dipolar—they have a positively charged side and a negatively charged one—and that charged particles in motion release energy. If molecules were freely floating as gases, Snyder realized, some of them should spin like batons and create a faint radio-wave signal. Even better, each type of molecule should have its own unique energy signature: it should broadcast at a specific set of frequencies that could be detected and identified by astronomers using radio telescopes on Earth.

Starting in the mid-1960s, Snyder applied for observing time on the main radio telescopes, to no avail. The scientists in charge of the observatories agreed with the consensus view that space could not support complex chemistry. In December 1968 Snyder traveled to Austin, Texas, for a meeting of the American Astronomical Society, where he and a colleague, David Buhl, hoped to change some minds. At the end of their talk, the famed physicist Charles Townes (who won a Nobel Prize for his work in the development of the laser) stood up and announced that he had found ammonia molecules near the center of the Milky Way using the radio telescope at the University of California, Berkeley. "Suddenly the people at NRAO decided we weren't crazy anymore," Snyder says, "and asked us for a list of molecules we wanted to look for."

Early in 1969 Snyder and Buhl set up shop at NRAO's Green Bank Telescope in West Virginia and chose their first target: formaldehyde, an organic molecule made up of two hydrogen atoms and an oxygen atom tethered to an atom of carbon. Sure enough, when they pointed the 140-foot radio dish at a massive cloud of gas and dust near the center of the Milky Way, there was a distinctive dip in the radio signal at 4.8 gigahertz—the music of formaldehyde. The same signal appeared in cloud after cloud. After waiting for more than a year to get observing time, Snyder needed just a few nights at the telescope to demonstrate that complex organic molecules, formaldehyde in particular, permeate the galaxy. He soon found hydrogen cyanide (88.6 gigahertz) in the Orion nebula and isocyanic acid (87.9 gigahertz) in a cloud called Sagittarius B2. "After that, we could have gotten telescope time to do anything," Snyder says. "We could have looked for interstellar flu germs."

Within a few years, Snyder and other radio astronomers had identified dozens of organic molecules, including formic acid (which causes the sting in ant bites) and methanol (a simple alcohol). Although none of these molecules reached the complexity of

Miller and Urey's amino acids, some of them can form proteins and other biologically important compounds when mixed together in water on Earth. Contrary to all expectation, interstellar clouds proved to be very friendly environments for breeding complex chemistry. Now a whole new discipline—astrochemistry—began to emerge, and its emboldened practitioners set out to learn more about what is cooking in those colorful nebulas.

Despite the large and growing catalog of space chemicals coming from the radio observatories, the astronomer J. Mayo Greenberg of the University of Leiden in the Netherlands suspected that his colleagues were missing a vital piece of the puzzle. The radio astronomers were searching for free-floating gas molecules in space, but nebulas also contain dust, microscopic grains of carbon and silicon. What would happen, Greenberg wondered, if interstellar gas molecules like formaldehyde collided with frigid grains of dust? They would freeze there instantly, he surmised, creating another kind of environment in which chemical reactions, driven by starlight, could take place. At temperatures just a few degrees above absolute zero, the molecules would still vibrate. These vibrating molecules—just like the rotating dipolar ones Snyder observed—could absorb and emit radiation. The frozen chemicals Greenberg was postulating would show up not in radio, however, but at infrared wavelengths. Starting in the 1970s, Greenberg was vindicated by a team of astronomers at the University of California, San Diego. They pointed a variety of infrared telescopes at interstellar dust clouds and discovered dips at specific frequencies corresponding to molecules including methanol, ammonia, and water ice.

Now that Greenberg knew interstellar space harbored frozen molecules as well as gaseous ones, he wanted to know how these chemicals interact under such extraordinary conditions. Theory alone could not provide the answer; this question called for some hands-on experiments. So in 1976 Greenberg hired Louis Allamandola, a recent Berkeley PhD graduate in low-temperature chemistry, to re-create the kinds of reactions that might take place on microscopic icy grains thousands of light-years away.

Allamandola's solution was to create an apparatus that could replicate the exotic cold depths of space—in essence, an extraterrestrial version of the Miller-Urey experiment. With his colleague

Fred Baas, he installed equipment to chill a shoebox-size chamber to within several degrees of absolute zero and depressurize it to a near vacuum. Then he set up a plasma lamp to fire beams of ultraviolet light at the chamber, much like the radiation present in planet- and star-forming regions of dust clouds. Finally, in true Miller-Urey fashion, he threw in a gaseous mixture of simple molecules, mimicking what was then known about the composition of interstellar clouds, and watched the results.

Allamandola's simulations, carried out first at Leiden and now at NASA's Ames Research Center, revealed not only that some chemical reactions really do occur at extremely low temperatures, but also that these reactions produce other reactive chemicals, thereby providing the spark for more molecular hookups. Ultraviolet radiation spices things up as well: it heats the grains and breaks up some of the molecules into reactive fragments, which in turn bond with other fragments to form new kinds of molecules.

Once again, nature proved extremely adept at brewing complex molecules. In current versions of Allamandola's experiment, the resulting icy mixtures contain dozens of prebiotic molecules, among them the same amino acids that Miller and Urey found. In fact, Allamandola's nebula-in-a-box has yielded an even richer chemical palette. He has manufactured intricate molecular rings containing carbon, nitrogen, and hydrogen; fatty-acid-like molecules that look and behave like the membranes protecting living cells; and nucleic acids or nucleotides, the primary components of RNA and DNA.

Creating molecules in the lab does not prove that the same molecules exist on dust grains in distant nebulas, but so far Allamandola's technique has an impressive track record. By 1990 he had published a list of simple compounds his group at Ames had created in simulations. By 2000 radio astronomers had found almost all of them in various dust clouds throughout our galaxy, suggesting that the interplay between ice and gas may be one of the most important mechanisms for synthesizing the precursors of life.

Still, Allamandola's research could not explain how compounds moved from the far reaches of space to the surface of Earth, where life actually took hold. Addressing this question meant bridging the gap between diffuse interstellar clouds and the condensed objects that ultimately emerge from them. When dense regions of a cloud collapse, the massive inner part becomes a star while the rest forms a swirling disk of gas and dust that may give rise to planets. (We now know that many, perhaps most, stars produce such planetary systems.) As large planets come together, the process involves such heat and pressure that all traces of preexisting organic matter are destroyed. Not all material in the disk gets treated so brutally, however. Some of it remains nearly intact in comets and asteroids, smaller conglomerations of ice and rock. When bits of these objects struck Earth as meteorites, they could have delivered organic molecules back onto its surface.

Convincing evidence that meteorites could be rich sources of organic molecules came in 1969, when a 200-pound meteorite hurtled to the ground in Murchison, Australia. Analysis indicates that the rock contains millions of organic compounds, including amino acids that could not have come from terrestrial contamination. Two years ago Zita Martins from Leiden showed that the meteorite contains nucleobases. David Deamer of the University of California, Santa Cruz, even found fatty-acid-like molecules similar to those Allamandola created in the lab. Other meteorites—including Murray, which landed in Kentucky in 1950, and Allende, which made landfall in Mexico in 1969—have been shown to contain similar organic compounds.

Meteorites carrying the same complex chemicals have been striking Earth since it formed 4.5 billion years ago. "The things we see landing on Earth now are probably representative of what was landing on us back during Earth's infancy," says NASA's Sandford, who has traveled the world searching for samples from on high. In 1984 he found a rock from Mars, and in 1989 a piece of the moon, all right here on Earth. During a six-week tour of Antarctica, he slept in a tent under the midnight sun and rode a snowmobile to ice fields littered with meteorites by day.

Gradually, painfully, through some four decades of effort, Sandford and the other scientists have teased out different strands of the story of prebiotic chemistry. Carbon, hydrogen, oxygen, and other atoms knock about in nebulas, sometimes freely and sometimes bound up with ice and dust. They arrange themselves into elaborate molecular structures. Meteorites abound with organic compounds, which rain down on any nearby planets.

Helping to weave all those strands into a single, elegant narrative is an Emory University astrochemist with a providential name: Susanna Widicus Weaver. Through a series of models and experiments, she has demonstrated that ultraviolet radiation can break chemical bonds and split molecules into highly reactive fragments called radicals. It is difficult for radicals to do much at −440 degrees F, but when the temperature warms even slightly (as when a star begins to form), the radicals merge to form larger molecules. "You can take methanol [CH_3OH], break it apart, and make several types of radicals, and then those can all find each other," Weaver says. "In just two or three steps on the grain surface, you can go from a simple mixture to something a lot more complex, like methyl formate [$HCOOCH_3$]." In a major 2008 paper, Weaver predicted an abundance of such radicals in dust clouds. A thorough search of interstellar ice grains by infrared astronomers should determine whether radicals indeed play a primary role in constructing prebiotic molecules. If they do, astrochemists in the lab could see what other complex combinations result from these radicals and then search for those molecules in space.

Weaver's models also demonstrate that once the temperature in the dust cloud reaches about −280 degrees F, most of the molecules evaporate from the ice on dust grains and enter a gas phase, allowing them to react a lot more quickly and to form complex molecules. Molecular players might include acetone (the stuff in nail polish remover), methyl formate, and ethylene glycol (antifreeze), she notes. That explains why radio astronomers have found more complex molecules in the warmer, more active star-birthing regions of dust clouds than in the colder, darker areas.

But then the story becomes less clear. Radio astronomers have yet to identify anything as complex as an amino acid, so astrochemists do not know exactly how complex these gaseous molecules can get. We know that meteorites contain amino acids and even nucleobases, but not whether they scooped up those molecules from dust clouds or created them later, on their interplanetary course. "We really don't know where the chemistry in the dust cloud stops and where the chemistry in meteorites starts up," Weaver says. She notes that the answer has tremendous implications for one of science's most fundamental questions: how common is life throughout the universe?

If meteorites create most of the direct chemical precursors of life, our solar system might be an unusual case. According to Weaver, the size of our sun, the region of the galaxy in which it formed, even how long it took for the planets to form—all these characteristics are different in other star systems and may influence the chemical inventory available to any Earthlike planets orbiting there. But if dust clouds can manufacture these molecules on their own, then life is probably prevalent throughout the universe. "No matter which dust cloud you look at, things look very similar chemically," Allamandola says.

A new generation of sensitive, high-resolution telescopes will help resolve the debate by probing both dust clouds and the protoplanetary disks from which asteroids, comets, and planets form. Remijan and his colleagues are salivating over the scientific potential of the Atacama Large Millimeter/submillimeter Array (ALMA) in Chile, a network of sixty-six radio dishes that will provide unprecedented resolution and sensitivity when it becomes fully operational in late 2012. And two space-based infrared observatories—the European Space Agency's Herschel Space Observatory, which has been gathering data since 2009, and NASA's James Webb Space Telescope, scheduled to launch in 2018—will allow astronomers to search for the infrared signatures of the specific icy radicals that Weaver's model predicts.

As much as astronomers love new instruments to play with, nothing beats seeing extraterrestrial chemistry in action. We cannot yet send probes to the Orion Nebula, but we can look for clues closer to home. Europa, a large satellite of Jupiter, is covered with a thick shell of ice crisscrossed with long brownish or pinkish fractures. Saturn's much smaller moon, Enceladus, features a network of icy volcanoes spewing ammonia, formaldehyde, and other organic molecules. These water-rich moons might have replicated the organic chemistry found within interstellar dust clouds. One clue involves the changing colors of the moons. "Some of the same processes that take place in deep space may occur in these icy bodies in the supercold outer regions of the solar system," Allamandola says.

Taking the big view, Remijan marvels at all he and his colleagues have achieved. Not so long ago, deep space seemed static and dull; now it looks like the possible breeding ground for a blueprint of life that might be shared all across the universe. Yet the greatest enigma remains untouched: How did a collection of organic molecules, whatever their origin, make the leap to life on Earth?

"The overall goal is to take this chemical inventory, mix it all together, and form a self-replicating molecule like RNA," Remijan says. He has personally helped detect eleven interstellar molecules with the Green Bank Telescope in West Virginia but recognizes that a full list of the organic chemicals out there is only a start. What scientists really need is a snapshot of the chemistry that was happening on Earth 4 billion years ago. That might actually be possible.

Titan, another of Saturn's moons, has a thick, methane-tinged atmosphere that is reminiscent of early Earth's. It even has pools of hydrocarbons on its surface, the only known bodies of liquid on any world other than our own. But Titan's −290 degree F surface temperature means that all of its chemistry moves in slow motion. Studies of Titan could therefore provide a peek at how complex prebiotic molecules came together on Earth, and potentially on many other worlds too.

In a new study, researchers from the United States and France conducted a new Miller-Urey–style experiment that mixed the organic molecules found in Titan's atmosphere. They ended up with all of the nucleobases that make up RNA and DNA. The study suggests the beginning of a new synthesis that reframes the old questions in a deeper and more meaningful way. It may not be a question of whether life's chemistry began in space or in meteorites or on the surface of a planet (or moon). All three environments may very well have lent a hand.

"When life was forming on Earth, it probably used a large variety of sources, some from the planet and some from the sky," Sandford says. "Life doesn't care about the 'Made in . . .' label on the molecules."

STEPHEN HAWKING AND
LEONARD MLODINOW

The (Elusive) Theory of Everything

FROM *Scientific American*

A FEW YEARS AGO the City Council of Monza, Italy, barred pet owners from keeping goldfish in curved fishbowls. The sponsors of the measure explained that it is cruel to keep a fish in a bowl because the curved sides give the fish a distorted view of reality. Aside from the measure's significance to the poor goldfish, the story raises an interesting philosophical question: How do we know that the reality we perceive is true? The goldfish is seeing a version of reality that is different from ours, but can we be sure it is any less real? For all we know, we too may spend our entire lives staring out at the world through a distorting lens.

In physics the question is not academic. Indeed, physicists and cosmologists are finding themselves in a similar predicament to the goldfish's. For decades we have strived to come up with an ultimate theory of everything—one complete and consistent set of fundamental laws of nature that explain every aspect of reality. It now appears that this quest may yield not a single theory but a family of interconnected theories, each describing its own version of reality, as if it viewed the universe through its own fishbowl.

This notion may be difficult for many people, including some working scientists, to accept. Most people believe that there is an objective reality out there and that our senses and our science directly convey information about the material world. Classical science is based on the belief that an external world exists whose properties are definite and independent of the observer who perceives them. In philosophy that belief is called realism.

Those who remember Timothy Leary and the 1960s, however, know of another possibility: one's concept of reality can depend on the mind of the perceiver. That viewpoint, with various subtle differences, goes by names such as antirealism, instrumentalism, or idealism. According to those doctrines, the world we know is constructed by the human mind employing sensory data as its raw material and is shaped by the interpretive structure of our brains. This viewpoint may be hard to accept, but it is not difficult to understand. There is no way to remove the observer—us—from our perception of the world.

The way physics has been going, realism is becoming difficult to defend. In classical physics—the physics of Newton, which so accurately describes our everyday experience—the interpretation of terms such as *object* and *position* is for the most part in harmony with our commonsense, "realistic" understanding of those concepts. As measuring devices, however, we are crude instruments. Physicists have found that everyday objects and the light we see them by are made from objects—such as electrons and photons—that we do not perceive directly. These objects are governed not by classical physics but by the laws of quantum theory.

The reality of quantum theory is a radical departure from that of classical physics. In the framework of quantum theory, particles have neither definite positions nor definite velocities unless and until an observer measures those quantities. In some cases, individual objects do not even have an independent existence but rather exist only as part of an ensemble of many. Quantum physics also has important implications for our concept of the past. In classical physics the past is assumed to exist as a definite series of events, but according to quantum physics, the past, like the future, is indefinite and exists only as a spectrum of possibilities. Even the universe as a whole has no single past or history. So quantum physics implies a different reality than that of classical physics—even though the latter is consistent with our intuition and still serves us well when we design things such as buildings and bridges.

These examples bring us to a conclusion that provides an important framework with which to interpret modern science. In our view, there is no picture- or theory-independent concept of reality. Instead we adopt a view that we call model-dependent realism: the idea that a physical theory or world picture is a model (generally of a mathematical nature) and a set of rules that connect the elements of the model to observations. According to model-dependent realism, it is pointless to ask whether a model is real, only whether it agrees with observation. If two models agree with observation, neither one can be considered more real than the other. A person can use whichever model is more convenient in the situation under consideration.

Do Not Attempt to Adjust the Picture

The idea of alternative realities is a mainstay of today's popular culture. For example, in the science-fiction film *The Matrix* the human race is unknowingly living in a simulated virtual reality created by intelligent computers to keep them pacified and content while the computers suck their bioelectrical energy (whatever that is). How do we know we are not just computer-generated characters living in a Matrix-like world? If we lived in a synthetic, imaginary world, events would not necessarily have any logic or consistency or obey any laws. The aliens in control might find it more interesting or amusing to see our reactions, for example, if everyone in the world suddenly decided that chocolate was repulsive or that war was not an option, but that has never happened. If the aliens did enforce consistent laws, we would have no way to tell that another reality stood behind the simulated one. It is easy to call the world the aliens live in the "real" one and the computer-generated world a false one. But if—like us—the beings in the simulated world could not gaze into their universe from the outside, they would have no reason to doubt their own pictures of reality.

The goldfish are in a similar situation. Their view is not the same as ours from outside their curved bowl, but they could still formulate scientific laws governing the motion of the objects they observe on the outside. For instance, because light bends as it travels from air to water, a freely moving object that we would observe to move in a straight line would be observed by the goldfish to move along a curved path. The goldfish could formulate scientific laws from their distorted frame of reference that would always hold true and that would enable them to make predictions about the future motion of objects outside the bowl. Their laws would be more complicated than the laws in our frame, but simplicity is a matter of taste. If the goldfish formulated such a theory, we would have to admit the goldfish's view *as* a valid picture of reality.

A famous real-world example of different pictures of reality is the contrast between Ptolemy's Earth-centered model of the cosmos and Copernicus's sun-centered model. Although it is not uncommon for people to say that Copernicus proved Ptolemy wrong, that is not true. As in the case of our view versus that of the goldfish, one can use either picture as a model of the universe, because we can explain our observations of the heavens by assuming either Earth or the sun to be at rest. Despite its role in philosophical debates over the nature of our universe, the real advantage of the Copernican system is that the equations of motion are much simpler in the frame of reference in which the sun is at rest.

Model-dependent realism applies not only to scientific models but also to the conscious and subconscious mental models we all create to interpret and understand the everyday world. For example, the human brain processes crude data from the optic nerve, combining input from both eyes, enhancing the resolution, and filling in gaps such as the one in the retina's blind spot. Moreover, it creates the impression of three-dimensional space from the retina's two-dimensional data. When you see a chair, you have merely used the light scattered by the chair to build a mental image or model of the chair. The brain is so good at model-building that if people are fitted with glasses that turn the images in their eyes upside down, their brains change the model so that they again see things the right way up—hopefully before they try to sit down.

Glimpses of the Deep Theory

In the quest to discover the ultimate laws of physics, no approach has raised higher hopes—or more controversy—than string theory. String theory was first proposed in the 1970s as an attempt to unify all the forces of nature into one coherent framework and, in particular, to bring the force of gravity into the domain of quantum physics. By the early 1990s, however, physicists discovered that string theory suffers from an awkward issue: there are five different string theories. For those advocating that string theory was the unique theory of everything, this was quite an embarrassment. In the mid-1990s researchers started discovering that these different theories—and yet another theory called supergravity—actually describe the same phenomena, giving them some hope that they would amount eventually to a unified theory. The theories are indeed related by what physicists call dualities, which are like mathematical dictionaries for translating concepts back and forth. But, alas, each theory is a good description of phenomena only under a certain range of conditions—for example, at low energies. None can describe every aspect of the universe.

String theorists are now convinced that the five different string theories are just different approximations to a more fundamental theory called M-theory. (No one seems to know what the "M" stands for. It may be "master," "miracle," or "mystery," or all three.) People are still trying to decipher the nature of M-theory, but it seems that the traditional expectation of a single theory of nature may be untenable and that to describe the universe we must employ different theories in different situations. Thus, M-theory is not a theory in the usual sense but a network of theories. It is a bit like a map. To faithfully represent the entire Earth on a flat surface, one has to use a collection of maps, each of which covers a limited region. The maps overlap one another, and where they do, they show the same landscape. Similarly, the different theories in the M-theory family may look very different, but they can all be regarded as versions of the same underlying theory, and they all predict the same phenomena where they overlap, but none works well in all situations.

Whenever we develop a model of the world and find it to be successful, we tend to attribute to the model the quality of reality or absolute truth. But M-theory, like the goldfish example, shows that the same physical situation can be modeled in different ways, each employing different fundamental elements and concepts. It might be that to describe the universe we have to employ different theories in different situations. Each theory may have its own version of reality, but according to model-dependent realism, that diversity is acceptable, and none of the versions can be said to be more real than any other. It is not the physicist's traditional expectation for a theory of nature, nor does it correspond to our everyday idea of reality. But it might be the way of the universe.

AMY IRVINE

Spectral Light

FROM *Orion*

IN THE DEAD of night the human brain is most capable of distillation—of boiling things down to basic black and white. *Smoke
means fire. Breaking glass signals intrusion.* From an evolutionary
standpoint, this kind of rudimentary thought process might be a
most valuable survival skill—the kind that allows a body to respond
to threats even in a state of half-sleep. My husband, Herb, is a lawyer, the kind of man who has been trained to think before he acts
—to examine all angles and consider complexities. But at three
A.M. on an uncharacteristically cold and moonless night in late
spring, even he is reduced. And through that reduction, he would
come to see how things that lurk too starkly, even at opposing ends
of the spectrum, can shift. As if fundamentals could be that supple.
As if values—like the presence of all colors in relation to the sheer
absence of them—could be so pliant. As if the natural order of
things—like the age-old relationship between predator and prey
—could flex into a new arrangement altogether.

The dogs would start it. Their frenzied barks, their teeth gnash-
ing against the glass of the back door, would draw my husband out
of bed and into his jeans in a single motion. In the mudroom he
would stumble through a sea of writhing canines, pull on his boots
with one hand and turn the knob with the other. Two aging Aus-
sies and a half-blind border collie mix would spill out into the dark
yard and charge toward the goat pen. They would make it halfway
before stopping dead in their tracks and high-tailing it back to the
porch. Herb would hear the screams then, the desperate cries for
help. He would fumble in the doorway for the porch light, two-
stepping with the returning dogs, and there his sleep-riddled mind

would already be drawing conclusions so swiftly it would feel, he would say later, like pure instinct.

And here I should point out that my husband, despite his profession, is a man who could have been born into the Paleolithic—the kind of guy who has built a life sustained by wildness more than any other element. After college Herb left Michigan for the West and never looked back. On the other side of the Continental Divide he found the kind of unfettered topography that he needed —for he's a man who is happiest when ambling over great stretches of soil or stone. He loves the basics, the way they ignite his senses: The procurement of food, shelter, warmth. The silky curves of women, skylines, rivers. Then there is his deeply held belief that he is a sort of Dr. Dolittle, and indeed, I have been witness to his extraordinary ability to communicate with animals. Domestic or untamed, creatures of all sorts seem to enter quickly into some kind of understanding with him.

It is this latter quality that explains why my husband's guns have never been loaded—despite the fact that we have made our home in one of the more wild parts of the West, where black bears, mountain lions, bobcats, and elk are as common as livestock. Where large tracts of untrammeled public land still eclipse both alfalfa fields and subdivisions of "ranchettes." Herb had stored in various places a .22 Smith & Wesson six-shooter, a 12-gauge shotgun, and three rifles in .22, .30-06, and 7mm magnum calibers—an inheritance from his grandfather, who had been an avid hunter in both the Great Lakes region and in Africa. All but the .22s had lain in their cases since his grandfather had died nearly eleven years prior —and those two firearms had only been used to shoot beer cans off fence posts on the occasional Sunday afternoon. Looking back, I think we both took a certain pride—and a smug one at that—in having no need for guns in what is largely a gun-toting community of roughneck ranchers, folks who let loose bullets daily on coyotes and prairie dogs.

So it is mind-boggling that Herb would conclude as he did on that night. Call it a natural impulse, or call it one of the ill effects of living in a culture steeped in sensational news and violent movies, but his mind instantly crafted the assumption that the hair-raising cries coming across the dark yard were of human origin. Somehow, he decided—in our critter-laden outback of a neighborhood that sits seven miles from a tiny, low-crime kind of town—that some

heinous, unspeakable assault was being committed by one deranged human upon another. And as he charged away from the
now-cowed dogs into the colorless void that lay beyond the porch
light's glare, his brain illuminated with one white, shining thought:
This is what the world has come to. Standing empty-handed in the ink-
well of night, he was ready to face squarely some malevolence in
his own species.

Herb turned, detouring away from the pen and into an adjoining shed, where he flipped on the light and took quick inventory
of several of his grandfather's firearms. He then knelt to rummage
for ammunition in a random collection of boxes. This took some
doing—my husband is not the most organized of men. And in the
process he failed to hear the intruder climb back over the impos-
ingly tall fence that contained the goats and circle around the
shed. It was only as he realized that the cartridges that matched
these particular firearms were elsewhere that he heard the pad-
ding approach behind him. He stood and turned. On the threshold, only four feet away, stood a three-hundred-pound black bear.

Our daughter, Ruby, and I were not there that night—and in
hindsight, as well as in the spirit of thinking so fundamentally
about things, I can't decide if that was a good or a bad thing. Would
our presence have changed in any way Herb's course of action, or
the bear's? Would the dogs have been more aggressive? And what
might I have done to alter the outcome? Through countless re-
plays of the situation, Herb and I would be reminded that vari-
ables come in many hues, and each one has the potential to change
the overall effect—the way Warhol's varied silk-screens of Marilyn
Monroe changed the essence of the subject simply by changing the
colors. Of course we also have to consider that we can only see
things through the lens we were born peering into—while other
species are able to perceive things entirely invisible to the human
eye.

The bear stalled on the threshold for a moment, and my husband stalled briefly too, before realizing he could not possibly summon a single word of conversation with what stood before him. For
the first time in his life, Herb was tongue-tied. When he finally
spoke (*Yo, dude,* unbelievably), the bear fixed his gaze on him and
took a step forward. Fortunately, Herb had been training to bench-
press 315 pounds in honor of his fortieth birthday, and so, rather
than continuing the conversation, he lunged at the half-open door

and heaved his body against it—effectively shoving the bear back outside. The bear stood there for a few minutes, then shuffled across the driveway and into the woods.

Not man, but beast. Of course. Herb's mind quickly reconfigured to what should have been his first impression all along: big animal with teeth and claws has found easy food in what had been a rather unforgiving emergence from the winter den—the late frosts having nipped springtime staples such as young forbs and grasses. Meanwhile, our daughter's pet goat, a white, bottle-fed Nubian named Dora the Explorer, was still screaming. And at last Herb recognized the wails as hers.

Curiously, the dogs remained on the porch, not uttering a sound. When Herb exited the shed and headed for the house, he made note of the quiet, for three dogs barking in unison has always been enough to keep wild animals at bay. He hadn't even made it ten feet from the shed when the bear reemerged from the woods. Herb scrambled back inside and waited. When he opened the door a second time, the bear stepped out again and came right at him. The two repeated this pas de deux over the course of an hour. It was sometime during those sixty minutes that the goat ceased her cries.

Perhaps the reason Herb failed to grasp the situation more quickly was the same reason our dogs were so quiet: they were stymied by the bear's unlikely behavior, its seemingly sheer fearlessness, its clearly predatory intent—for these are not attributes we see in the wildlife on our mesa. Here, seated between the 14,000-foot peaks of the San Juan Mountains and the red canyons and rivers of the Colorado Plateau, there is still a great deal of food and space for humans and animals alike. Unlike their cousins trying to eke out a wild lifestyle within the slim margins of the nation's national parks, and unlike those poor ursines who have had the misfortune of claiming turf near dense populations of humanity, our black bears have had the luxury of keeping almost exclusively to themselves.

There were explanations: Perhaps the bear was sick. Or maybe it was a juvenile orphaned before learning how to acquire food properly. But it's also possible that this bear was part of an escalating and global trend—for some biologists say animals everywhere appear to be changing in new and unsettling ways. One recent study concluded that human impacts are forcing animals to evolve at a pace three hundred times faster than they would naturally. And there is evidence to support that the traits affected are not only size and reproductive capacities but behavior too. David Baron, the author of *The Beast in the Garden: A Modern Parable of Man and Nature,* contends that some regional trends in cougar attacks are "upward and exponential." Boulder, Colorado, is his prime example —where an environmentally minded populace has sought to live closer to nature by building homes on the "edge," the transitional area between forest highlands and desert plains. A zone where deer feed. And where lions hunt. The equation has been disastrous for everyone. The deer overpopulated in backyards while the big cats began stalking, mauling—and sometimes killing—a relatively significant number of household pets and humans. Of course the offending cats were destroyed. But here's the kicker: whenever a guilty lion was removed from an area, others quickly moved in— only to exhibit the same new tactics.

Add to this global scenario of changing behavior the ill effects of climate change. Barren-ground grizzlies, for example, have faced a diminished supply of coastal-plain plants—a food source that is dwindling in a warming Arctic. Without this essential nutritional source, the bears have been forced to alter their foraging and feeding habits, and some have become malnourished in the process. One such grizzly recently killed and consumed two experienced Alaskan-bush backpackers who, in the opinion of the investigating officer, "did most everything right" in their efforts to deter bears from their camp.

Given this new context, it is probably inappropriate to say that such predator behavior is aberrant; rather, these animals are adapting fittingly to a drastically altered environment. There's irony here: the more humans coif the natural world to our liking, the more we push out into the last wild places for recreation and real estate, the more we are finding ourselves back on the food chain —as a menu item.

Herb and I are complicit in this twenty-first-century showdown between wildlife and people; the five-acre parcel we purchased for a home site had been a bull pasture until the rancher subdivided it to subsidize his retirement. The east half of the property is hemmed in by neighboring pastures of grass and alfalfa—a scene utterly bucolic, punctuated by the brays of livestock. The backside of the property is altogether different. A woodland of oak, piñon,

and juniper slopes down to a lush creek bottom lined with cottonwoods, wild iris, and tall grasses. The draw carved by the creek begins high on the forested uplands to our south and serves as a natural corridor for wild ungulates that move between desert lowlands
in winter and high mountain meadows in summer. The predators
—black bear, mountain lion, bobcat, and coyote—all follow suit.
Our 800-square-foot house sits smack dab on the dividing line of
these two worlds—and depending on which way you turn when
you walk out the door, the rules for how to behave, what to watch
for, can be very different.

Ruby and I return home the day after the bear's visit, and already
Herb's story has spread across the mesa like a runaway ditch fire.
Like so many small towns in the West, we're a mixed group here
—a blend of traditional rural folk and transplants who have fled
cities and suburbs alike. For weeks afterward, I am reminded of the
dichotomy between the two camps; each time I am asked to re-
count the story on Herb's behalf, I receive one of two pat responses.

The New West: *Did he try and talk to the bear? Did he project peaceful
energy?*

The Old West: *I hope he shot the son of a bitch.*

To each individual, I nod. *Yes, he did try to communicate. And yes, he
shot it.*

The double affirmative isn't duplicitous. In terms of philosophy
and aesthetic sensibilities, it is easy to side with the New West's romantic notions about preserving natural landscapes and living in
and among wildlife. But the old-timers have a point. They largely
blame the newcomers, who flock to places like Telluride (a resort town thirty-four miles to the east), where predominantly well-
heeled, left-leaning residents supported the Colorado Division of
Wildlife's termination of the spring bear hunt. The town's abundant PETA enthusiasts and Humane Society donors also applauded
the prohibition of hounds for the remaining autumn hunt—
deeming it a cruel example of unfair chase. In response, our en-
clave's more traditional, rural crowd could be heard grumbling,
Now them bears are thick as thieves . . .

It's true that resort towns in the West tend to have the biggest
bear problems. Garbage cans, greasy barbecues, and bowls of pet

food get left out by individuals who tend to be rather naive about wildlife. When the bear comes sniffing for easy extra calories (in late summer a black bear must consume 200,000 calories a day in order to survive winter hibernation), such folks snap pictures when they should be clanging pans or throwing rocks—actions which, with a healthy, unconditioned bear, are almost always enough to scare it off. For minor first offenses, wildlife officers will tranquilize a bear, punch an ID tag through its ear, and relocate the animal— to places with more open country, places like my neighborhood, where they become *our* problem (and perhaps this was the case with Herb's visitor). But in severe cases, or repeat offenses, the bears are put down with a big-caliber bullet—executions that are all on the taxpayer's dime.

Bear stories like Herb's linger on our tongues, in our imaginations, like erotica. They are titillating precisely because they are the closest that most of us come to igniting the ancient physiological and psychological tinder of the predator-prey relationship that lies dormant in the human body. (*Come get me, Mommy!* I see the glowing embers in my daughter's eyes, hear the quickening of her breath, when she asks me to chase her.) And perhaps this is why the cause to protect North America's predators is so feverish, and why it is equaled in pitch only by the efforts to exterminate them. Each side is glaring, garish even, in its shriek of righteousness— and so it is with bears the way it is with everything else: we respond from a black-and-white paradigm; the potent dualities of *us versus them* resound with a faint, prehistoric echo. Instead of man against weather, or man against beast, though, it's Republicans vs. Democrats, tree-huggers vs. wise-users, Buddhists vs. Bible thumpers. The appeal of such binary thinking is that we are able to name not only who we are but also what we are not. We draw the dividing line like a firebreak, and it holds back the advancing enemy while we retreat to safer ground.

But I am the descendant of rural ranchers on one side and artists, scholars, mountaineers, and businessmen on the other. As a daughter of the American West—both the old version and the new —what I have felt about my homeland could easily be characterized as a form of cultural schizophrenia, a psychic swing between my frontier-busting forebears and my Patagonia-clad, Sierra Club card–carrying contemporaries. For many years, I chose a side—

shoring up my persona by way of education (higher), occupation (as both a national park ranger and a paid public-lands advocate for the Southern Utah Wilderness Alliance), and recreation (bourgeois style, like rock climbing, river running, and skiing). And as I grew into this role, I grew apart from the other side of my family and their cowboy ways. I kept them at arm's length with a subtle (so I thought) sense of superiority. Feeling right served as a shield for my own mind—which felt as if it would shatter if I attempted a mental straddle between two worlds.

What I missed was this: hunting is a vital part of life for both sides of the family—whether it's hooking rainbow trout to grill for a streamside Mother's Day brunch, shooting antelope to complement the garden harvest feast at summer's end, or plucking pheasants to roast alongside the Thanksgiving turkey. For such events my mother's and my father's sides still sometimes come together—the men combining efforts to bring home animal flesh, the women uniting in the kitchen to cook it. The universal acts of procuring and preparing our sustenance have always served as our species' most common denominator—and among my kin, they have always made faint all other disparities.

By calling through the shed window, Herb finally got our most geriatric dog, Jack, to come off the porch just far enough to distract the bear while he made a break for the house. For an old guy, the Aussie put on quite a show, and Herb finally got the head start he needed. He and all three dogs just squeaked inside the house; when they turned to look back through the glass, they saw the bear's snout pressed up against it.

The creature pawed at the door, attempting entry. Herb dashed to the bedroom closet, grabbed from the top shelf the only gun kept in the house—the .22 revolver, complete with cartridges. Not that this particular gun could have done much harm—for a bear, a perfect shot would still be nothing more than a bee sting. But Herb was thinking more complexly by this point. There was no time to call for help from the neighbors, and the nearest law enforcement was, at best, twenty minutes away. Besides, he wanted to get to the goat. He was banking on the fact that if the impact of the shot didn't scare off the bear, its report would.

Herb beckoned the two younger dogs, and together the three of them sneaked out the back entrance and crept up on the bear, which was still on the front porch, facing off with Jack through the glass door. Herb got as close as he could, and as the bear turned in his direction, he fired a round at the animal's underbelly—the only place a low-caliber bullet could have any kind of impact. Before Herb could blink, the bear turned and disappeared into the dark of the woods, black devoured by black. Then Herb headed to the goat pen to retrieve Dora's flayed body.

It wasn't the best scenario; the bear was still alive. It could come back for its kill—or for Herb. Nevertheless, Herb was momentarily relieved. His strategy had been a serious gamble, for the animal could have turned on him just as quickly as it had fled into the woods. I can imagine my husband at that moment, contemplating all that was happening along with that which might have been: his jaw would have been set like a steel trap. And yet the encounter would have resulted in bright eyes, flushed skin, and a larger-than-life grin—indications that an ancient inner sense of vitality had been pricked.

Herb was adamant that we not tell Ruby, who was only three and a half at the time, what had happened to her goat. He begged me to speak euphemistically—to say that Dora got sick and *passed on*. But I knew our daughter would see that something was wrong. I thought it worse to lie to her, to undermine her intuitive perceptions by telling her that things were not what they seemed. Besides, to deny the bloody realities of animals eating animals—including our family's consumption of meat—would only distance my daughter from her budding relationship with the natural world. I needed to believe Ruby could handle the fact that something had tried to eat her goat. Just as I was banking on the fact that she would be able to face on her dinner plate the elk I hoped to shoot in the fall, the chickens we had raised and would butcher, and be able to eat both with reverence, gratitude, and delight.

And so I tell her.

She cries.

And after her initial outburst of grief, her pale, tearstained face blooms red with fury: "I hate that bear, Mommy. Bears are bad, bad, bad." I think then that maybe Herb had been right. Maybe this is too much for her. But it is too late to recant.

"We can be sad for Dora, sweetheart. And we can be sad for the bear too—because if he's still alive, he will probably be destroyed."

Yes, and yes again. I hold my breath and wait for some sort of resolution.

For the next two days, we find the bear's tracks, punctuated with small splats of blood, encircling our fence line. The local game warden surmises that the single small round Herb put in the animal will probably fester in the gut—eventually killing it. On the third day, the bear returns in the middle of the night and digs up Dora's body—which had been buried three feet deep beneath a big rock slab at the far edge of the property. And then we never see sign of him again.

For several weeks Ruby acts out the drama of goat and bear with her toys, and each night at bedtime asks me to repeat the story of what happened. One night she awakes in terror. Shaking, howling, she scrambles onto my lap and tells me she had dreamed that a bear was trying to kill her. I think of Carl Jung, who suggested that the image of the bear in the unconscious is a representation of one's own potency. To run from a bear in your dreams is to flee from your own potential. To turn and face such an animal is to reckon with the Other—not just its beloved aspects but also that, perhaps especially that, which is wild, ravenous, even terrifying— and with the parts of our own wildness that we fear more with each passing generation, with each species' extinction, with each acre of land razed.

I tell Ruby that if the bear comes again, she must stand her ground, ask it what it wants. I stroke her strawberry-blond curls as she falls back asleep. A few hours later, she wakes again, whimpering.

"Mommy, the bear came back, and when I asked him what he wanted, he said he was hungry. So I gave him a carrot."

Ruby is no longer terrified, but tentative. She falls back into sleep, and I am still sitting next to her when she starts to giggle. Then she sits straight up, her eyes shining in the pewter moon-shower falling through the window.

"Mommy, the bear came back, and this time he looked just like Winnie-the-Pooh!"

For a moment, I cringe at my daughter's reduction of a wild creature to a cartoon character. But then I see that she has, on a deep level, bent the bear into something she can manage—and in this way she has digested her conflict with the animal and its deeds. Afterward I notice in my daughter a deeper appreciation for the animals around her—she loves them more than ever. And yet: she now holds a realistic and healthy respect for those that have the potential to harm her.

I learn more slowly than my daughter. The day after her dream a neighboring rancher stops by to inquire if we've seen the bear around. He whistles at the claw marks on the shed's threshold and has a good laugh at Herb's small pistol. But, as the new long-haired attorney on the mesa, my husband scores points for having a gun at all—and a few more for being willing to use it. And when, in order to prove his adequacy, he pulls out his 7mm mag, the rifle his grandfather had used for killing Cape buffalo, he really gets a slap on the back. "Next time, son, you drop that bastard dead in his tracks."

Herb just shrugs and smiles. I, however, feel compelled to interject my belief that we don't want to kill interloping bears—that we merely want to keep them at bay. The rancher cocks his frayed ball cap and juts his grizzled chin at me.

"Notice how the bear that paid your husband a visit thought nothing of your three dogs? That's because you nature lovers thought you were doin' right for the bears by making it illegal to hunt 'em with hounds. Now you got bears strollin' right by dogs, into backyards and barnyards, with no fear at all. "

Facing my neighbor, I feel a powerful impulse to pull back. This is where civil convention dictates that I silently agree to disagree, that I make some remark about the weather. Later I can air my opposition among like-minded people who will fan my flames of indignation. Emboldened by their passionate agreement, I'll feel justified in penning letters to the editor, e-mails to the Division of Wildlife—any venue that is capable of presenting the issue in black and white, any venue that is impersonal enough to isolate my beliefs from my neighbor's.

And yet. In Aspen during a two-week period this past summer, a bear sauntered right through a fur salon, another broke into a house and attacked the owner, and another bit into a woman's thigh while she lay sleeping on her deck. During the same time frame, a bear broke into a steel enclosure down the road from our house, killing five Shetland sheep and maiming two others—only to return in broad daylight for more. Three days after the offending bear was trapped and removed, another one moved in and killed three additional sheep. And in the nearby tourist town of Ouray, at least two more bears ate an elderly woman who, every evening for years—despite harsh reprimands from state and local officials—had watched from a fenced-in porch as bears came into her yard to feed on the dog chow she set out for them. The coroner's report concluded that the woman had been dragged out of her makeshift observation cage and devoured by the very animals she fed.

Standing on my own bear-clawed threshold, I am caught in the spell of a familiar misanthropy, only this time I begin to sense how it stunts my understanding of the world. And suddenly I find myself willing to consider my neighbor's perspective, to extend an open-mindedness toward his knowledge and experience that I haven't even granted my own rural family members. It comes down to this: by retreating from that which we oppose, we render lifeless all opportunities for intimacy, and for community. To smile and step away is as fatal to possibility as is brandishing a finger of blame.

And so, after a long, awkward silence I offer my neighbor a seat on the porch and a cold beer. Then I lean forward. I seek luminosity—the deep bruise of blue that hung on the fence alongside the man's coyote hides, complemented by the soft rose of empathy that emanated as he knelt in my goat pen the summer before, showing me how to revive two kids half dead with scours. I was new to goat-keeping then. With my young animals, my neighbor was as tender and gentle as I've ever seen a man. And in eyeing these two tints of him at once, I find a newfound level of humility reflecting back.

"Tell me," I say, haltingly, "how you would restore the equilibrium."

"For starters," he says, "git yourselves some outside working dogs —no more welfare critters. Then load one of them bigger guns you got there, and for god's sake, keep it where you can use it."

F. Scott Fitzgerald wrote: "The test of a first-rate intelligence is the ability to hold two opposing ideas in mind at the same time and still retain the ability to function." But perhaps it isn't brains so much as courage—the courage to say yes, and yes again. At the very least, I am learning to bring into singular focus my double-edged essence. For if my preschool-age daughter can behold the whole of each animal, then surely the rest of us can embrace two seemingly opposing elements with every nuance, every context, every color in between.

We'll need a language of delicacy to articulate such complex thoughts and feelings—one that can carry us across the muddy mire of moral, spiritual, political, and environmental ambiguities. And if we wield our words with heartfelt compassion and respect, it just might be enough to repair the psychic fissures we have suffered in this age of sharp divisions.

Now I keep a loaded rifle within arm's reach. We have two new dogs that roam our fence line, day and night. And I find myself hoping that hounds will give chase during the next bear season. It's not a contradiction to say all this—and then to say I am still rooting for the bears, for their rightful place on our mesa, and across the remaining wildlands in the West. Indeed, as my family prepares for the rigors of the autumn elk hunt in the Colorado high country, I am reminded that it is no small thing to inhabit our place on the carnivorous continuum—a place where we not only consume animals but, in turn, we consent to the possibility of being consumed. This place, an edge of sorts, awakens us to our biological inheritance, and we become viscerally, sensually invested in our surroundings and their ability to sustain us.

These adjustments to my view of the world have not made me a more typical westerner; nor have I become a more conventional environmentalist. But if our model of advocacy, no matter what the cause, requires that we stridently defend our territory without leaning across the fence to consider, wholeheartedly, another view, if we cannot embrace the Other in both its delightful and repelling pigments, then the world has little chance to be spared. For this is what it means to forge meaningful conduits between our existence and every other bit of biota. Swallowing the spectrum whole is to devour the exquisite breadth of life. After all, diversity is the strength of a people. Of an ecosystem.

The hunter and the hunted. The Old West and the New. The wild and the tame. We must be lithe enough to stretch between.

ROWAN JACOBSEN

The Spill Seekers

FROM *Outside*

SHE NEEDED TO GET OUT. Wide of beam, forty-three feet long, and 11,000 pounds of lead in her keel, she'd been built with oceans in mind. Her name was *Dolphin's Waltz,* and she was sick of putzing around the shallows of Alabama's Mobile Bay, where she docked. Now stately breakers rolled across her bow, muddy waters giving way to the gray-green Gulf of Mexico. She was on the hunt.

We passed within kissing distance of shrieking rigs, hunkered down and drilling away like mosquitoes; dodged a swarm of shrimping boats that looked like giant waterborne grasshoppers, their spars deployed as they searched for oil slicks; and then made for Dauphin Island Pass and the wide open. A Coast Guard chopper slashed overhead, and a blimp hung in the southern sky like an alternate moon. We were the only pleasure boat around. A west wind snapped the jib taut as dolphins hot-dogged across our bow wave, exploding into the air. For being smack in the middle of America's biggest environmental disaster, it was pretty fucking nice.

You might say July 2010 was an odd time for a pleasure cruise in the Gulf of Mexico, and maybe a sailboat—slow, bulky, with the upwind quarter of the world off-limits—is an odd way to do any sort of journalism. I won't argue. I wasn't going to scoop the *Wall Street Journal,* but that was the point. Perhaps, as the press hordes stampeded past us in their helicopters and vans, chasing the latest oil sighting, they were missing the real story.

So we'd take it slow, on the gulf's schedule. I'd challenged our skipper, Josh Deupree, to a wind-powered tour of a particularly stunning corner. I'd heard rumors of cleanup operations gone awry—workers pelting each other with eggs from Louisiana rookeries—so we'd head west to investigate the situation on Petit Bois Island, known as the most pristine wilderness island in the gulf and vital habitat for more than 250 species of birds. From Petit Bois we'd cross Mississippi Sound to see the marshes and seagrass meadows of the coast, taking in as many marine ecosystems as we could in three days.

A sort of nagging consumer guilt, and the desire for seafaring in its purest form, was the motivation to try it without fossil fuels— once out in the gulf, at least. Sailboats are docked in slips like parking spaces. Exiting had involved cranking up, making several right-angle turns, negotiating a narrow harbor, passing under a bridge, then following a dredged shipping channel across Mobile Bay and out into the gulf.

As soon as we'd cleared the bridge and caught the shipping channel, we killed the engine and raised the sails. Close-hauled to the wind, *Dolphin's Waltz* heeled over and shot straight out of the channel toward Dauphin Island Pass. The smell of diesel faded, replaced by the slow oscillation of wind and wave. She was free.

Not that we were making any kind of statement. Just getting to the boat, our crew had of course burned gobs of petroleum products. I flew from Vermont to Newark, where I looked down upon a grid of refineries and tanks, then to Houston, same damn thing, and then to Mobile. Oil was everywhere and in everything, not just the gulf.

Like something out of Mordor, a refinery's orange methane flare blazed atop a black pillar on the shore of the bay. Fed by six rivers, Mobile Bay covers 413 square miles and spits out some 62,000 cubic feet of water every second, making it North America's fourth-largest estuary by flow. Its mouth was crawling with boats as we sailed out. Vessels of Opportunity—boats hired by BP to patrol for oil—darted across the channel. The program was the biggest gold rush on the coast. Even the smallest boats made $1,600 per day.

About 3,000 VOOs, as everyone called them, were operating out of Alabama, Mississippi, and Louisiana. They came in every size and shape: sportfishing behemoths, Boston Whalers, pontoon boats. I tried not to think about how much gas they were burning in the effort to save us all from oil. All flew the triangular VOO flag and seemed to be zipping about with minimal coordination.

As we passed within twenty yards of one diesel-burning rig, a man on the catwalk waved his arms and shouted. I thought maybe we were too close, but Jimbo Meador, one of our crew, said, "He wishes he was on this boat and not that goddamn rig." Virtually everyone I'd spoken with on the Gulf Coast had said, "Oh, you have to meet Jimbo Meador. He's the real-life Forrest Gump." Well, Jimbo's buddy Winston Groom did dedicate the book to him. And, yes, like Gump, Jimbo once ran a huge shrimping business out of Alabama. And, yes, for the accent, Tom Hanks studied Jimbo's speech. ("But I don't know why he bothered," says Jimbo. "He sounded like an idiot.") But there the similarity ends.

Born and raised on the Alabama shore, Jimbo is one of the best fly-fishermen on the gulf and, because he does it all by kayak and stand-up paddleboard, one of the fittest sixty-eight-year-olds you'll ever see. The ocean is his life. "If I get too far from salt water, I start to get nervous," he told me. With his deep tan and wavy light gray hair, he looks as if sand and sea foam crystallized into a man and strode out of the surf. If anyone cared about the health of this place, it was Jimbo.

Our other crew member, Bill Finch, was equally invested. A senior fellow with the Ocean Foundation, the fifty-one-year-old had been instrumental in creating one of the country's best oyster-restoration projects. A mile and a half of new oyster reef had been "planted" on Coffee Island, off the Alabama coast, in April—the beginning of Bill's Hundred Miles of Reefs vision, part of the Restore Coastal Alabama 100–1000 project for Mississippi Sound and Mobile Bay. It had seemed a heartening success until BP's Deepwater Horizon blew on April 20 and started spewing 205 million gallons of black goo.

I'd met Bill in early May when I went to document the oyster reefs before the oil arrived, and he told me something new and surprising at every turn in our conversation: Alabama has the greatest biodiversity of any state east of the Mississippi; the greatest diversity of freshwater fish and mussels; the greatest concentration of turtle species in the world; and so on. Sure, with such awful casinos, condos, and refineries, parts of the Gulf Coast do live up to the Redneck Riviera reputation. But few folks know about the other gulf, the natural miracle with arguably the finest beaches in the nation, the best diving, the best birding, the most productive wetlands, the thriving communities of shrimp and bluefin tuna and sperm whales. If we were to hold up the Gulf of Mexico in one hand and the Arctic National Wildlife Refuge in the other, then agree to protect one and exploit one, the decision would be a no-brainer. Bill had helped me see that.

He'd taken me to Grand Bay National Wildlife Refuge, a 10,000-acre crown jewel of biodiversity shared by Mississippi and Alabama, where we stood in the salt marshes, squinting into a tide pool at the turquoise streaks of male sailfin mollies displaying for females. "Look at that!" Bill had said. "Isn't it stunning?" With his gray beard, balding pate, and general demeanor, Bill struck me as the Lorax of the coast—an impression cemented when he suddenly bounded across the marsh through knee-deep muck to stop an airboat from plowing into the cordgrass.

Behind us, on the salt pan, willets piped their pennywhistle love songs. A rail had just hatched her chicks. At the edge of the marsh, acres of oysters spawned. It was, normally, the most hopeful time of year. And Grand Bay sure as hell looked normal. Miles of rush stretched to Mississippi, backed by a wall of some of the only tidal pine forest on earth. Waves rolled over vast seagrass beds. Beautiful.

But it had been hard to enjoy. Three miles offshore, a cannonade of tarballs was splattering Dauphin Island. Farther out, something dark and dead was gathering strength in the depths. When oil hits marshes, it chokes out the oxygen and nitrogen and the soil goes hypoxic. Then the marsh dies. Estuaries like Grand Bay are where at least 90 percent of the fish species in the gulf come to breed. If the estuaries die, the gulf is a goner.

We had to slalom rigs as we sailed, there were so many. East Coasters may stand on a beach and expect to see nothing but blue horizon, but Gulf Coasters enjoy nothing so sublime. Stand on any given shore in Alabama, Mississippi, Louisiana, or Texas and you can count as many as twenty oil rigs rising out of the haze on spindly legs, like the Martian invaders from *War of the Worlds*.

Most people seem to believe it's a horrible coincidence that BP's leak happened so close to the lower forty-eight's most bountiful fishing grounds, but the same factors that make the northern Gulf of Mexico so rich in oil also make it rich in life. For millions of years, the Mississippi River has been the gulf's great benefactor, pouring midwestern nutrients into its warm coastal waters and fueling extraordinary production of phytoplankton, the single-celled plants that are the foundation of the marine food web. That phytoplankton feeds some of the most fantastic stocks of oysters, shrimp, crabs, and fish on earth. What doesn't get eaten drifts to the bottom as it dies—"marine snow." As additional sediment buries it over eons, it gets pressure-cooked into that blackest of sauces, oil.

In the steamy Cretaceous, 100 million years ago, seas were much higher, and much of Texas and Louisiana lay beneath an expanded gulf. As global temperatures cooled and the gulf shrank, the easily obtained oil became accessible beneath the gulf states. That was the first oil boom. We burned through most of that in a few decades. That was the first oil bust, and by the 1980s it had turned East Texas and Louisiana into very ugly places. But in the 1990s, geologists discovered enormous oil reservoirs much deeper in the gulf than ever suspected. New technology provided the means to drill a mile underwater through three miles of bedrock, and oil prices went sky-high, making it all worthwhile.

More than 4,000 rigs pepper the gulf. You head their way to catch red snapper, which cluster beneath. The Deepwater Horizon blowout has focused attention on deep-water drilling, but there are still only a few dozen rigs capable of operating in water a mile deep. These massive platforms are the Parthenons of our time, soaring high-tech temples to the reigning god. But the majority of rigs are more like rinky-dink parish churches, mostly stuck to the seafloor close to shore. Many lie abandoned; others limp along sketchily, like the Mariner Energy rig that caught fire on September 2. The low-hanging fruit is gone.

Deep water is the source of most gulf drilling jobs. A few decades after the last of those reservoirs are drilled, they'll be tapped out. The gulf's petroleum era will end and the region's entire identity will change. To what, no one knows. A graveyard on the edge of a cesspool? It's not hard to imagine.

Our skipper, Josh Deupree, is one of the coast's top sailboat racers, yet he'd been trapped lately, hesitant to even take *Dolphin's Waltz* out of port for fear of mucking up her engine and hull with sticky oil. Luckily for us, though, the thirty-six-year-old couldn't resist the lure of an exploratory mission.

A few miles off the coast of Petit Bois Island, we finally got the boat dirty. Our first slick looked like a trail of brick-colored diarrhea running to the southern horizon. We punched through it and watched it disappear behind us. We hit a few more slicks as we approached the island. Sailing has been defined as hours of boredom interrupted by moments of terror. This was hours of beauty broken by moments of disgust.

A line of trees on the horizon: Petit Bois, a typical barrier island, a wandering carpet of dunes that gets piled just high enough on the inland side to support a few marshes and trees. These trees were mostly dead, casualties of Hurricane Katrina. Petit Bois is part of Gulf Islands National Seashore. In 1978 it was designated as wilderness, the Park Service's highest level of protection. Yet we could see blue shapes along the beach. Binoculars revealed them to be tents, each shading an assemblage of coolers, rakes, and plastic chairs.

The sun was oozing orangely into the sea as we anchored in the lee of the island. It appeared to be deserted, but the lagoons were fenced with bright yellow boom, and anchored farther away was a clot of crew boats, barges, and tenders. The barges were stacked with double-decker forty-foot steel boxes marked LIVING QUARTERS, shipping containers serving as windowless "flotels." So this was where Hazmat Nation was spending the evening. I wondered how many were squeezed into each tin.

Anchored near us was a gigantic three-story sportfishing boat flying a VOO flag. A couple of guys in camo fatigues lolled on the upper deck. The boat never budged in the sixteen hours we were anchored there. The VOO program is a boondoggle. The boats' ability to deal with any oil they find is very limited, as explained by a VOO captain who agreed to speak anonymously. "If we found anything," he said, "we'd call the shrimp boat assigned to us, they'd come, and we'd boom it off and suit up in Tyvek. You'd put a produce bag on the end of an aluminum pole and actually scoop oil up with those bags, then you'd put it in bigger plastic bags and drop it off. We got as much as we could, but it was almost pointless. You can't really clean that shit up."

The idea was to employ out-of-work fishermen, but my VOO mole—a sportfishing guide—explained that it hasn't worked that way. Despite the fact that his guide business had dried up after the

Deepwater Horizon blew on April 20, it had taken him until July 1 to get activated. "There were people not in the fishing business who formed corporations early on," he said. "One group had nine boats. Their friends' kids operated them. And you'd call BP and ask, 'When am I gonna get activated?' 'Well, we've got too many people.' A friend of mine was out there for two months in a twelve-foot johnboat. Two months! In a twelve-foot boat! There was no supervision. You'd just check in: 'See ya in twelve hours.' People were going out and fishing all day."

BP also launched the $500 million Gulf of Mexico Research Initiative, which might be coined the Scientists of Opportunity. I learned of some being offered $250 an hour for their research — on topics approved by BP, of course — but prohibited from discussing or publishing their findings for three years.

We slept on deck to beat the heat, slapping mosquitoes through the night—except Jimbo, who ignored them. At dawn he muttered "I got to marinate" and threw himself overboard. The water looked clear of oil, so I did, too, and immediately got stung by a jellyfish. These Jimbo tuned out as well. "I used to swim long distance," he said, pulling beautiful strokes around the boat. "I'd get stung constantly. Put yourself in a different mindset. They don't actually hurt."

Jimbo—who's a part owner of Dragonfly boats, the most enlightened fly-fishing craft ever conceived—had been fishing in the Bahamas with his old pal Jimmy Buffett when the Deepwater Horizon exploded. Jimbo and Jimmy hatched a plan, with the musician funding the design and construction by Dragonfly of two skiffs custom-made for wildlife rescue. The Shallow Water Attention Terminal, or SWAT, boats have whisper-soft trolling motors, a draft of just ten inches, mid-deck worktables, misting systems, canopies, Wi-Fi, video cameras, and "sea-mist green" hulls (so they merge with the waterline and don't spook birds). After the boats were built, however, Jimbo was informed by Fish and Wildlife that only trained specialists with federal permits have clearance to handle oiled birds; if anyone else tried to rescue one, they could be in violation of federal law under the Migratory Bird Treaty Act. Jimbo suggested donating a boat to the local Audubon Center. Nope, no permits. He would eventually have to donate it to a conservation nonprofit who could then lend it to Fish and Wildlife for rescue and research efforts. "Most frustrating thing I've ever experienced," he said.

Still stinging from the jellyfish, I kayaked to Petit Bois with Bill Finch. We dragged the kayak ashore and made our way through the dunes toward the gulf side. Blooming morning glory vines crawled over the sand and scrub. Bill nibbled on wild plants: sea rocket, a briny, mustardy green that tasted like Grey Poupon, and glasswort, which was crunchy and salty, the potato chip of the beach. Foraging is one of his passions. Not long before the Big Leak, he'd been out on the islands wandering over Native American middens—ancient heaps of oyster shells—grazing on glasswort and wolfberries, shucking oysters straight out of the water. "It's the way people must've eaten for ten thousand years," he said.

Birds flushed out of our path as we walked. The six barrier islands of Gulf Islands National Seashore are the first pit stop for many birds, as well as monarch butterflies, flying north from South America and the Yucatán in the spring. Emaciated fowl rain down on the islands, rest, refuel, and then scatter across America. In fall the islands are often the last staging area before the big flight over the gulf. The lonesome islands have a poignant feel. "Prepare yourself for enchantment!" the *Exploring Gulf Islands National Seashore* guidebook says of Petit Bois. "The feeling is primeval, as if you have been deposited on an oasis."

Things had changed. Stippling the high-tide line were tens of thousands of tar patties, and suddenly we weren't having a nice nature walk anymore. They looked like underbaked molasses ginger cookies and smelled like hot asphalt. Some were as big as Frisbees. A ghost crab was mining one, carrying clawfuls back home. On the beach, blue tents but nobody in sight. ATV tracks cut through the sand.

Our crewmates caught up to us as we trudged. A helicopter inspected us, and a flock of brown pelicans sailed past. "Hope you make it, boys," Jimbo called out. The temperature headed toward 100 degrees. The oil seemed to have an affinity for trash. Any piece of plastic was shellacked with it, as if it had some sort of molecular attraction to its own kind. But despite the horror on the beach, the water looked clear. Jimbo needed to marinate again. We swam. Jimbo bronzed.

I'm always struck by the energy of coasts; the friction of two worlds colliding draws so much life, like us, to hug the edges. We stood in the surf and watched clouds of mullet dart by. "Shit," said Jimbo, "one throw of my cast net and we'd have supper for a week." But all waters were closed to fishing. (We'd been eating tinned sardines from Portugal.) The dolphins had no such restrictions. They caught the breaks, surfed into eighteen inches of water, and scraped the sand as they scarfed up fish.

Farther down the beach, the water turned the color of iced tea. Crabs with aprons full of eggs skittered through it sideways. Suddenly a pack of Gators—heavy-duty, four-wheel-drive ATVs made by John Deere—came skidding around the end of the island. The cleanup crews were awake. The Gators sped past us toward the shade of the tents, and the mostly obese crews suited up in Tyvek and gloves and grabbed their tools and plastic trash bags. It looked punishingly hot in those suits. The long scoops used by the crews are slotted, so sand filters out as they dig up tarballs. It was like watching a bunch of little people clean the world's largest litter box, with older tarballs dry and easy to bag but recent arrivals messy and laborious. One of the ATVs veered toward us and slammed to a stop. "How'd y'all get here?" asked the driver in a thick Mississippi accent.

"Boat," said Josh.

"Hell, I know that," said the driver. "But how'd you get *here?*"

"Walked." We'd come about a mile.

The marvel of it all slowly sunk in. "You *walked?*" His passenger, a weary-looking man with a grizzled beard, narrowed his eyes: "Y'all better not be steppin' on any tarballs."

This was the crew boss (whose name I was never able to confirm). Things weren't going well for him. He had orders to keep all the Gators operating in a single track to avoid tearing up the dunes, but the rut was now so deep, they were starting to run aground and break down. "It's just killin' our Gators," he said. How were they supposed to get up and down the island without their Gators?

So far his crews had raked up sixty tons of oil, dumped it into petroleum-based garbage bags, and hauled them out to the barges. But Tropical Storm Alex had recently deposited a whole new layer of oil and sand on the beach, so they were essentially right back where they'd started. Worse, the new sand from Alex had covered huge mats of old oil, which they now had to dig through several inches of beach to find. He was trying to requisition some gas-powered leaf blowers to blow the new sand off the old.

Nothing was happening fast. The worker-safety guidelines were strict about heat. On a yellow-flag day—pretty much the best you can hope for there in the summertime—employees work forty minutes, then get a twenty-minute rest. We were into the red-flag days (temperatures in excess of 92 degrees), meaning twenty minutes on and forty minutes off. Should a *really* hot day rear its sweaty head, they'd bust out the cooling vests and A/C or work after dusk. The best that could reasonably be hoped for was about two hours of work per day out of any employee, and a fair amount of that seemed to be devoted to shuttling people to and from the potties stationed at one end of the six-mile island. "It's a logistical nightmare," sighed the boss. Recently, however, there'd been a breakthrough. "We got permission," he said, gesturing out to sea, "to take a pee in the ocean."

Bill looked a little dazed as we paddled back to the boat in silence, so I asked him about it later. "Petit Bois is so unconnected to the rest of the world," he said. "You can't hear anything but the tingle of the sand. To come over the dunes and see that regimental deployment—it kinda makes you stiffen up. And then the waves turning up tarballs instead of shells and pebbles. That long line of machines rolling toward you. It's been violated twice: first by the oil and then by the assault to save it. I don't know what the long-term impacts of the cleanup will be. The only thing I'm fairly certain of is that I'll be finding chunks of oil probably for as long as I'm able to keep going there. And I'll worry about it every time I see it. That just pisses me off."

Of course, the White House would tell Bill not to worry. In early August, not long after our visit, energy czar Carol Browner would go on the *Today* show and—in a gaffe redolent of Dubya's "mission accomplished" moment—say, "More than three-quarters of the oil is gone. It was captured, it was skimmed, it was burned, it was contained. Mother Nature did her part." First off, "the oil" here refers to the total leakage, but 17 percent of that was captured directly at the wellhead; it never even entered the water. An honest accounting requires considering only the crude that escaped. Displaying a sort of medieval understanding of biology, Browner also explained, "some of it will become very small microorganisms and disappear into the gulf."

Okay, Browner's no scientist, but she got her numbers from the Department of the Interior and the National Oceanic and Atmospheric Administration, and NOAA should know better—especially when you consider that its administrator is the respected marine ecologist Jane Lubchenco. Of the oil that actually entered the gulf, 9.6 percent had been either skimmed or burned, according to the feds; what Browner and the media seized upon was NOAA's misleading categorization of the other 90.4 percent, which was qualified as dispersed, dissolved, or residual.

The notion that "out of sight" (dispersed or dissolved in this case) equals "gone" has underpinned centuries of environmental abuse. With well over a million gallons of chemical dispersant dumped on the slicks, NOAA, it would seem, hoped to dispel most of the oil to the mythical land of Away and declare victory. But an independent Georgia Sea Grant report authored by five prominent marine scientists has shown that, as of early August, with natural degradation and evaporation taken into account, 70 to 79 percent of the oil remained in the gulf, much of it slimily blanketing the seafloor or hovering in vast plumes, its toxins potentially wrecking our food chain from the bottom up. And that's not even accounting for the oil known to be soaking into coastal wetlands.

Get used to the artful dodges.

I'd be lying if I said the Gulf Coast in the summer of 2010 was a miserable place. Most of it was stunning. But there was misery to be found, and over the course of several weeks, I found it. A few days after our sailing trip, I visited the Gulf Coast Research Lab in Ocean Springs, Mississippi, and rode in one of Jimbo's SWAT boats to a stretch of marsh grass so coated with oil that my fingers stuck to it. Sprawled across the grass was a long piece of absorbent boom, now black with oil, that had broken loose. Next to it was a big chunk of high-tech flotation foam, probably wreckage from the Deepwater Horizon. It had been reported to BP a week earlier.

At Grand Isle, Louisiana, the beach was closed for miles, blocked off with orange plastic construction screen. Signs on the road advertised ONSITE DISASTER RELIEF CATERING. The sand had been scrubbed, spun, and surf-washed so thoroughly by contractors that it sparkled. But at the far end of the island, at Grand Isle State Park, I found a young ranger named Leanne Sarco quietly Q-tipping oil off hermit crabs one at a time. She wore white and was oil-smeared from her shoes to her blond hair. Sarco led me along a path to the last wild beach on the island, a magnet for birds and other wildlife. Crude rippled in the tide pools. The sand was the color of coffee grounds. When you squeezed a handful, globs of black jelly oozed out. The state park had asked BP not to bring in its machinery for fear that the cleanup effort would be more damaging to the ecosystem than the oil. But it was not a pretty sight. She told me that the oil had seeped several feet below the surface and would be bubbling up for years.

The irony nobody wants to talk about is that all the lands in the Mississippi River Delta now hosting cleanup crews are dying anyway. The levees that unnaturally hold back the Mississippi present a perilous Catch-22, ostensibly protecting the land from flooding and storm surges yet helping to ensure that it will all fall into the gulf: square miles of Louisiana vanish every year. Even if we could somehow degrease the hundreds of miles of oily coastline without harming a single bird or blade of grass, the river delta would still most likely cease to exist sometime this century. It's sinking. New Orleans is on course to disappear beneath the waves, the American Atlantis.

That realization sucker-punched me the day I spent in Terrebonne Parish on boom patrol with Virgil Dardar, a fifty-two-year-old man of Cajun and Native American descent. Virgil, who goes by the nickname Kadoo, lives on Isle de Jean Charles, a three-mile spit of dry ground in the marshes thirty miles northwest of Grand Isle. For 170 years, his ancestors survived by fishing and hunting the marshes. They used to farm, too, until their soil became inundated with salt water. Hurricanes like Katrina, Rita, and Gustav have left the island a disaster zone of smashed homes, beached boats, and dead trees. Isle de Jean Charles is crumbling into the water in real time. Stand at the end, dangle your toes over the edge, and watch it go.

The paved road to Isle de Jean Charles cuts across two miles of open water that used to be marsh. Now, waves gnaw at it 24/7. It submerges completely every time a south wind accompanies the high tide, emerging a few hours later and a few millimeters smaller. Even at low tide, much of it is down to one jagged lane. A snaking line of orange cones leads cars along the safest path.

Kadoo has been an oysterman all his life. But the oil spill closed all the oyster grounds, possibly for years, so he'd hired on to patrol those same areas for oil or loose boom. He was surprisingly chipper about it. Good money, a fast skiff, and a hot breakfast and bag lunch every day. Best of all, he felt respected.

Indeed, as Kadoo drove me around with BP's Chatt Smith, I realized the company would be utterly lost without local watermen like Kadoo. The area is changing so fast that the maps are virtually useless. When we left the dock, we boated beneath power lines that had been on dry ground thirty years ago. They were sunk deep, their poles cocked at alarming angles, and the boat channel ducked right through them. Dead cypress and oak trees stuck out of the water, their skeletal trunks bleached by the sun. I asked Kadoo if he remembered them. "Oh, sure," he said, "them trees was alive fifteen years ago."

Bill Finch had explained to me how freakish the presence of a seawater-killed cypress community was: "You'll never see a salt marsh next to bald cypress unless something's gone wrong. In Louisiana, it's gone wrong. The marsh is retreating so fast that it's being slammed into these cypress, and they're dying." Ever seen a healthy cypress swamp? Moss-draped trunks stand in a few feet of fresh water, the air thick with honeysuckle and the thrum of buzzing cicadas. Fish stir the waters, and alligator eyes poke out of it. In the ferns, tree frogs trill. It's the original Dagobah, and you wouldn't be surprised if Yoda came tottering out of his mud hut to hail you as you passed. To know a place so supple and dense with life and then witness its transition to nothing but dead trunks and salt water must be profoundly demoralizing.

But that's what has happened in Terrebonne Parish. Southern Louisiana, which sits on nothing but Mississippi River mud up to several miles deep, has always sunk, but the sediment from the flooding river, as well as storm surge from the gulf, have always replenished it, slowly building up the land. Since the last ice age, which ended 10,000 years ago, the Mississippi has extended the marshes as much as fifty miles into the gulf. The river's levee system choked that off, and 10,000 miles of canals—dug through the marshes by an oil industry in search of new reservoirs—delivered two crushing blows to the erosion-halting marsh grasses, disrupting natural water-flow patterns, which left the grass drowned or dried out for extended periods, and bringing lethal salt water in.

As we navigated our way out of a liquid labyrinth, it was not lost on me that my guide through this land of the dead was actually named Virgil. We came to a spot on the map called Lake Tambour, then to Lake Barre, then Lake Felicity, but it was all water. Chatt explained that the chart was an older one and that each "lake" had once been enclosed among bayous. Long gone.

I asked Kadoo if the population of the island was going down. "It ain't comin' up," he said. "We lose a few more every hurricane."

"You know you're going to have to leave someday," Chatt said gently.

"When I die," said Kadoo. I asked what he planned to do when the BP job ended. "Stick with this spill-response stuff," he said. "I got my training. Always be a spill somewhere."

From space, southern Louisiana looks like a bunch of tattered clothes hanging in the gulf, and the raggedy ends simply disintegrate in the rinse cycle of each hurricane. This lost land is a literal gap in national security: the natural buffering of storm surges once enjoyed by New Orleans has been severely reduced over the past century. It's the canals dredged by the oil industry that give big storms the keys to the city, making all the difference between Katrina being a nuisance and being a catastrophe.

Many experts think that at some point this century, almost everything on the Gulf Coast south of I-10 may very well have been sliced clean off the United States. (Driving from Biloxi to Houston? Enjoy those gorgeous ocean views!) That is, unless there's an energized national effort to restore the gulf. We must resist the country's tendency to write off the whole area. Make no mistake: this is no eulogy. While you were being told that the Gulf of Mexico had been fatally damaged, I was sailing through a paradise of pompano, shorebirds, and the most abundant dolphin populations I've ever seen. The gulf is not just worth saving; we *must* save it—especially from itself.

While I was in Louisiana, there was an event at the Cajundome, in Lafayette, called the Rally for Economic Survival: 11,000 packed the place to hear the governor, the lieutenant governor, and, of all people, the executive director of the Louisiana Seafood Marketing and Promotion Board rail against the Obama administration for stealing their jobs by imposing a six-month moratorium on deepwater drilling.

"Enough is enough!" raged the lieutenant governor, Scott Angelle, in his thick Cajun accent. "Louisiana has a long and strong, distinguished history of fueling America, and we proudly do what few other states are willing to do . . . America is not yet ready to get all of its fuel from the birds and the bees and the flowers and the trees!"

True, but of the six billion to seven billion barrels of oil consumed by the United States each year, only about 10 percent comes from federal Gulf of Mexico waters; we get the same amount from both the Persian Gulf and Canada. Louisiana is no longer a significant source of crude, onshore or offshore. What it does supply is cheap labor and a pliant local government. In this, it's eerily reminiscent of Third World places ruined by oil. The BPs of the world would have you believe oil brings prosperity to the countries where it's discovered, but it brings misery so dependably that economists have a name for the phenomenon: the resource curse.

Ecuador, Venezuela, Iraq: bad things happen to countries "blessed" with oil. The Niger Delta is the Mississippi River Delta's separated-at-birth twin, offering the scariest cautionary tale of all. This tropical river delta held some of the greatest wetlands on earth, with abundant shellfish, crabs, and shrimp, the foundation of the economy and culture, but it also harbored vast oil reserves. In the past fifty years, Shell has grown preposterously wealthy off that oil, while Nigeria, with the tenth-largest oil reserves in the world, has become a postapocalyptic wasteland. Almost three times as much oil has spilled into the Niger Delta as was spilled by the Deepwater Horizon: 546 million gallons and counting. The creeks are black, and the crabs and shrimp are dead. There are always leaking, corroded wellheads and pipelines. Gangs of rebels and oil thieves roam the jungle. Flaring rigs fill the air with mercury, arsenic, and carcinogens. Disease is rampant. The government is cardboard.

Southern Louisiana is no Nigeria, but it's also no longer quite recognizable as the United States. The trailer homes on pilings, the dearth of education, the chronic disease, the fat parish chiefs —I know the Third World when I see it. Cajuns haven't grown rich on crude; Houston has. And when the oil runs out, there's nothing left to fall back on.

I bet Angelle would simply argue that oil is worth billions more than seafood. But that's only because we aren't sophisticated enough to put a value on all the multifarious "ecosystem services" the gulf provides: benefits of the natural world, resources and processes we all too often take for granted. If we were to add these things to the ledger—all that gulf seafood and the health savings from it, the hurricane protection and wildlife habitat in all those marshes, to name only a few—and apply the calculus of their self-perpetuating sustainability, the astronomical value would blow your mind. It leaves petroleum in the pit. We can get oil from a hundred different places. What we can't get elsewhere are the Gulf of Mexico's oyster reefs and wetlands. *Best on the planet.* How much are all those acres of disappearing land worth? What price the mental anxiety of a culture watching its homeland disintegrate? How much added value do you assign oyster reefs because they've never, ever blown up and killed anyone? It's only ignorance— an inability to tally all the gains and losses—that makes oil look good.

After Petit Bois, we'd sailed to the west end of Dauphin Island, anchoring near Katrina Cut, a rift in the island opened up by the hurricane. These barrier islands, which are a mere 3,000 years old— "not two generations of cypress," said Bill—get regularly knocked about by storms. Two rigs near our anchorage had beeped annoyingly through the night, and the next morning, when Jimbo and I looked to marinate, a strange black slick of congealed matter was drifting past the boat. We raised anchor and headed for Grand Bay.

If we were still hoping to find the old gulf, we met it sooner than expected. Looming beyond Grand Bay, a black bank of cloud sucked the color out of the western sky and rose imperiously into the troposphere, flicking snake tongues of lightning at the sea. The water turned a psychotic green. A mitt of wind came across the surface and swatted *Dolphin's Waltz* like a toy. "Better put the sugar to bed!" said Jimbo. We could see the Alabama shrimp fleet hauling ass for the shelter of Bayou la Batre, their hulls glowing white against the charcoal sky.

If Josh had said to me at that moment that the time had come to stop playing this eighteenth-century game, to crank up and play it safe, I would have bowed to his experience. But we instead came about and raised the biggest sail we had, a green spinnaker that billowed in front of us like a parachute and caught the storm's 30-knot gusts, headed toward Mobile Bay.

Our speed and direction had been decided for us: we were going home. I stood in the bow, handrail held tight, and let the cool rain needle my skin as waves hissed against the hull and splashed me. Horsepower had easily beaten wind power, leaving this part of Mississippi Sound deserted save for us. Frothing waves obscured the horizon, and I could see nothing but sails and water and cloud.

We sailed for hours on that course. The difference between powerboating and sailing is profound, but it's a feeling that sneaks up only after you've been at it for a while. It's a cool, crisp serenity caused, paradoxically, by a lack of control over your fate. You can't choose your weather. You can't bend the world to your will.

Entering Mobile Bay, we jibed to pick up the shipping channel, and the full brunt of the wind came across our deck. Josh's eyes flicked from the wind to the spinnaker to our course. "Harden up on that sheet a bit!" he called. Bill winched in the sheet, the spinnaker snapped into place, and *Dolphin's Waltz* heeled over and locked into a groove. She surged forward, 35,000 pounds of wood and fiberglass and lead cleaving the water into two foamy curls.

"She got a bone in her mouth now!" Jimbo shouted from the wheel.

The channel from Mobile Bay into the Mobile Yacht Club harbor is six and a half feet deep, more or less, quickly dropping to five if you go astray. *Dolphin's Waltz* draws six feet of water. Of course, when she's heeled over, flying across a strong wind, she draws a little less. Keep her at full speed, hard into the wind, and she can sneak through five or so feet of water. But when a boat with that much mass runs aground at that speed, very bad things happen.

The bridge loomed. "Jimbo," Josh said, "you just keep her pointed exactly at that top span." I asked Josh when he wanted to drop the sails. He waggled his head: "It's all about the braggin' rights, dawg."

I stationed myself in the bowsprit in classic "king of the world" position, the bay sluicing beneath, the bridge towering above. We had twenty feet of clearance above our mast and about ten on either side of the channel. "We comin' in hot!" yelled Josh.

Suddenly, a powerboat came out of the harbor straight at us. "Boat!" I shouted, gripping the handrail even tighter.

"We got right of way!" Josh yelled back. "We got rights over everybody right now!"

Sure enough, the powerboat scrambled sideways like a pedestrian dodging a runaway truck, and we blew into the harbor, past the dock and a handful of frozen onlookers. One of them snapped out of it, pointed, and shouted, "Josh, that's the coolest thing you have *ever* done!" Josh grinned.

Then we doused the sails, fired up the diesel, and motored back into our parking spot.

CHRISTOPHER KETCHAM

New Dog in Town

FROM *Orion*

WILD COYOTES HAVE SETTLED in or around every major city in the United States, thriving as never before, and in New York they have taken to golf. I'm told that the New Yorker coyotes spend a good deal of time near the tenth hole on the Van Cortlandt Park Golf Course in the Bronx. They apparently like to watch the players tee off among the Canada geese. They hunt squirrels and rabbits and wild turkeys along the edge of the forest surrounding the course, where there are big old hardwoods and ivy that looks like it could strangle a man — good habitat in which to den, skulk, plan. Sometimes in summer the coyotes emerge from the steam of the woods to chew golf balls and spit them onto the grass in disgust.

They also frequent the eighth and the ninth and the twelfth holes, where golfers have found raccoons with broken necks, the cadavers mauled. At the tenth hole, a coyote ran alongside a golf cart last summer, keeping pace with the vehicle as the golfers shook their heads in wonder. "I stop the cart, he stops," one golfer who was there told me. "I start it up, he follows. I jump out, he jumps back. I sit down in the cart, he comes forward. We hit for a while — we're swinging, and he's watching." Here the golfer, an animated southerner named Chris, mimes the animal, following with his head the coyote-tracked ball's trajectory up and up, along the fairway, then its long arc down. It was pleasing to Chris that coyotes like golf.

Until recently, I couldn't quite believe that coyotes were established New Yorkers. Among neophyte naturalists it's an anomaly, a bizarrerie, something like a miracle. Coyotes, after all, are natives of the high plains and deserts two thousand miles to the west. But for anyone who takes the time to get to know coyotes, their coming to the city is a development as natural as water finding a way downhill. It is also a lesson in evolution that has gone largely unheralded. Not in pristine wilderness, but here, amid the splendor of garbage cans filthy with food, the golf carts crawling on the fairway like alien bugs, in a park full of rats and feral cats and dullard chipmunks and thin rabbits and used condoms and bums camping out and drunks pissing in the brush, a park ringed by arguably the most urbanized ingathering of *Homo sapiens* in America—here the coyote thrives. It seemed to me good news.

The coyote, unlike its closest cousin, the wolf, is a true American. The coyote's earliest relatives began evolving in the Southwest 10 million years ago, with *Canis latrans* arriving roughly at the dawn of the Pleistocene Epoch, when huge predators roamed the continent. I imagine the coyote in its prehistoric form as a thing small and weak and quiet, slinking in the shadows alongside the megafauna of American prehistory. The little dog had to deal with the appetites of cave lions, which weighed upward of six hundred pounds; the predations of the saber-toothed cat; the fury of the short-faced bear, which, at a height of fourteen feet and a weight of up to nineteen hundred pounds, was the largest bear that ever lived. It tried not to get stomped by the mastodon and the mammoth and the stag-moose and the elephant-sized ground sloth and the armored glyptodon, a turtle as big as a Volkswagen.

Then, beginning some 12,000 years ago, the coyote got a break. In one of the great extinction events of prehistory, North America's megafauna, these giants of the continent, disappeared. What precipitated the mass extinction is unknown and is today the stuff of much speculation. The cause might have been climate change —the retreat of the glaciers, the warming of the planet— or perhaps it was a change in weather combined with overkill from newly arrived human predators who crossed the Bering Strait, armed with the technology of spears that the megafauna were not adapted to fend off. The coyote, fighting for so long in this hard world of giants, was among the few prehistoric American mammals to survive in the new environment. Other sizable fauna soon filled the extinction vacuum. But they were foreigners. Like the spear-chuck-ing humans, the new mammals were descendants of Asia. They are the creatures that we know now as bison, elk, moose, bighorn sheep, grizzly bear—invasives that we generally find ghettoized in our national parks.

Another invasive species that crossed from Asia was the gray wolf. Fast-forward several thousand years, and the coyote and the wolf have become mortal enemies. They have fought for space over the millennia, with the wolf claiming most of the American continent because the wolf is bigger, more aggressive, works in packs, and operates well in dense forest. Enter the white man, whose technology and avarice allowed for sweeping control over established predators. By 1900 white settlers had decimated the wolf population, which threatened their livestock and their children. This was accomplished not simply by unleashing gunpowder; the white man felled forests everywhere he went, which opened up the terrain and left no hiding place for wolves. The coyote, on the other hand, thrived in open spaces. It was as adaptable as the wolf was not; it had been adapting to predation in America for 10 million years. So the coyote took over the wolf's niche as top dog.

In the wake of white settlement, the coyote was reviled for its success. That we could not appreciate the elasticity of the native dog was fitting irony for the European species of human, so terribly successful at invading the continent and adapting to it ourselves (or, rather, forcing the continent to adapt to our new and increasingly invasive presence). Along with the Indian and the bison, the coyote was—remains—the pest par excellence of the American West, to this day classified in the law books of many western states as "vermin" or "nuisance" species. Tens of millions of coyotes have been slaughtered in the United States since 1900; federal and state governments over the last two decades have killed an estimated two million of them. This figure doesn't incorporate the tens, perhaps hundreds, of thousands of coyotes each year hunted, baited, trapped, snared, and poisoned by livestock ranchers and sport hunters who kill coyotes for recreation in contests and bounty hunts. The attempt at control has cost in the range of billions of dollars—no firm number is known—and it has failed spectacularly. Biologists note that the success of the coyote amid this carnage is largely due to a survival mechanism that renders the species impervious to the gun and the trap: when large numbers of coyotes are killed in any single ecosystem, the coyotes that remain produce bigger litters. The animal compensates for slaughter, in other words, by becoming more numerous, more problematic. I can only imagine this as a kind of Darwinian laughter: *Kill more of us, and more of us will come. Perhaps to be killed. So that there will be more of us. Ha ha!*

I first heard the sound of the eastern coyote in the sprawling Catskill Mountain range, a hundred miles north of New York City. The creatures screamed and shouted and yipped; I thought I was hallucinating. That was eight years ago. Only once did I get close, in October 2002, when a pack in fog yelled in my ear on a mountaintop. No sighting of the creatures that cold night, nothing to lay my eyes on. Just the high keening song, the crackle and whisper of my feet and theirs on the forest floor. Thereafter I made it a point to walk in the forest and climb the mountains at night, to listen for them, find their sign, their scat, their kills. Eight years later I've found a lot of scat, many prints in mud, and no coyotes.

But the song—it hung in my head. I'd listened to coyotes in the American West, and the song in the East was different. In the red rock of the desert, it's lone and sorrowful and begins with a bark (*Canis latrans,* after all, means "barking dog"). The pack sometimes —only sometimes—answers the loner, the voices clear and vibrant, like a Greek chorus. In the East, vocalizations never seem to open with the loner. The song instead begins with a scream upon a scream, followed by screeches, squeaks, eeks, heeing and hawing and ululations, dystonal and weird, that the western cousin can't match. I could say I hear the gamelan music of Indonesia, the off-time rhythms of Turkic Bosnia, girls screaming rape, men losing testicles.

A few years ago, late on a summer night in 2004, I thought I heard a lone coyote singing in the Bronx. This was not long after the first reported New York sightings, which already had begun to increase in frequency. If coyotes were on the move in the city, marking terrain, naturally, I thought, there should be communication among them. What I heard, wandering Van Cortlandt Park on that summer night, was a short spindrift cry, like something heard underwater, distant and muffled and indistinct, and later I assumed it was the work of a dog pretending at wildness while slobbering over an owner come home. Or perhaps it didn't happen at all and I imagined it because I'd been reading too many reports about coyotes. In the moment, though, I believed what I wanted, and I started howling. And waited. And howled. There was the hush of the city, the hoodoo silence of the buildings that surround the park, and in the silence I could hear the murmurs of men and women speaking in a thousand ways out of tune with each other, and finally, when I howled one last time, someone leaned out a window and cried, "Shut the fuck up."

From California to Maine, there are more coyotes than at any time since records have been kept, their territorial expansion unprecedented in speed and scope. "The coyote is the most successful colonizing mammal in recent history," Justina Ray, an ecologist with the Wildlife Conservation Society, tells me. They pressed eastward across the Plains and the Midwest. They went south into Alabama, Georgia, Florida. By the 1920s they had arrived in New York State, where they advanced at a mind-boggling rate of 116 miles per month. By 1942 they were in Vermont; by 1944 they were in New Hampshire; and by 1958 they were in western Massachusetts. I recently found their tracks like a palimpsest in the tidal flats at Cape Cod National Seashore—the messages of their movements chasing crabs—and one homeowner on the Cape described a den in his wooded backyard, thirty feet from his deck, where the pups stared out from inside a tree trunk. I talked with a Connecticut woman who welcomed a wounded coyote into her car, thinking it was a bashed-up dog, and took it into her house only to discover the creature going wild in her living room as it matured—wanting to get out. By the 1970s, coyotes were arriving in the cold sea-country of the Canadian Atlantic Provinces, having become fully established in New Brunswick by 1975, reaching southwestern Nova Scotia by 1980, Prince Edward Island by 1983, and floating on sea ice across the Cabot Strait to the island of Newfoundland by 1987.

That the coyote has expanded his range does not surprise biologists. What does confound is the suggestion, hotly debated, that the coyotes now taking over the eastern United States in fact represent a new subspecies of wild dog on the continent, the *Canis latrans varietas*. The western coyote is a smaller creature than the eastern cousin. The westerner weighs in at perhaps thirty pounds, looking somewhat like a fat fox. The eastern coyote grows as big as sixty pounds at his heftiest. The tracks I found on Cape Cod and in the Catskill Mountains suggest a big dog indeed.

So whence the bigger muscles, the extra weight, the new song? Perhaps natural selection in the face of bigger game, or the higher snows and colder weather of places like Chicago and New York, sparked the coyote's physical flowering. Perhaps coyotes in their dominance arrived at a sexual detente with the last wolves in the East and began breeding with their old enemies, which added to the girth of the eastern coyote and also gave him his new voice. Perhaps the wolves, in this same pivotal moment, realized they were outnumbered and preserved, in a copulative leap of hopelessness, what little remained of their genetic pool. I like to think that all these factors commingled and were further complicated by the reality of dealing with the human ecosphere — the byways and hidden passages of the city, the dynamism of interaction with cars, highways, apartment buildings replete with comings and goings, the all-night bodegas, the light of streetlamps, the conniving of rats, the surfeit of accident and possibility. I like to think the eastern coyote's build, its behavior, and, not least, its song reflect this complexity.

So the coyote runs across schoolyards in Philadelphia; he hides under a taxi on Michigan Avenue in Chicago. He is in Atlanta, and in Los Angeles, and Miami, and Washington, D.C. He follows into the cities our paths, our roads, our railways, our bike and hiking trails. In Seattle, a coyote ran into an elevator in a skyscraper for a ride, and another ended up in the luggage compartment of a tram at the SeaTac Airport. In Boston, biologists who radio-collared a female coyote during 2004 reported that the dog traveled freely across the towns of Revere, Medford, Somerville, and Cambridge, at one point crossing into Boston proper via a railroad line at three A.M. before bedding down in a railyard north of the Charles River. The dog, nicknamed Fog, had "little more than shrubs for her to sleep in." Stanley Gehrt, a biologist at Ohio State University who recently spent six years tracking the coyote populations of Chicago, concluded that there were at least two thousand of them living in the Windy City, and they were growing in number. Urban coyotes, Gehrt found, live longer than their country cousins; their range per pack is more compact, much like urban humans; and they hunt more often at night, very much like urban humans. Gehrt also found that coyotes howl in answer to the sirens from firehouses—calling to the sounds of men. "Originally known as ghosts of the plains, coyotes have become ghosts of the cities," Gehrt writes. "Coyotes are watching and learning from us."

I got a beer from a bodega in the Bronx and sat on a bench and thought about the ghost dog. To the American Indians, coyote is Trickster, the magician among the animals, the shadow creature, the player on the edges of human encampments. "Along the edge I am traveling, in a sacred manner," goes an old Lakota song honoring the Trickster. Sacred manner? Wile E. Coyote chasing the Roadrunner comes more readily to mind. Chuck Jones, the animator, pegged the Trickster, in cartoon Latin, as *Eatibus anythingus*. Which is true: coyotes eat garbage, darkness, rats, air—they'd lap my beer if I let them.

The Trickster in myth appears as rotten-minded as Wile E., as creepy and ill reputed, as underhanded as Bronx rats. But in the system of native myth, unlike Wile E., unlike Bronx rats, the coyote's lying and conniving and cheating is in the last act a leap of creation, bearing the new out of things that are busted down and old and not working. The cartoon hints at this: Wile E. falling off cliffs, born again from the smashed puffball of bones at the canyon bottom to try for a meal once more.

In the various coyote myths, the Trickster makes things happen by sheer pushy will and wackiness. He is a shape-shifter, blown apart, come together again. Coyote is sometimes the creator of the world itself in his tumbledown accidental manner; sometimes he brings fire to the hominids who are freezing in the cold; sometimes he gets the smartest and most beautiful girls pregnant when dumbstruck men can't get it up to perpetuate the line; sometimes he makes sure that animals get anuses when the Creator, whoever that fool is, forgets to do so. The Trickster, in other words, is a teacher of possibilities, pointing humankind down new paths when the poor bummed-out hominids are stumped.

My own amateur coyote study in New York's Van Cortlandt Park last autumn went not so well. Day after day I made the long trip by subway north from Brooklyn into the Bronx, my hopes up, maps out, binoculars in the backpack, notepad ready, boots laced high, a flashlight with extra batteries in case I found the creatures after dark. I got lost in the Van Cortlandt woods, scrambling near the border with Westchester. I got paranoid about muggers (who, like coyotes, never seem to show up). I got covered in mud tramping in washes looking for tracks. I got poison ivy up my leg and into my crotch.

The golfers at the Van Cortlandt Golf Course snickered at my efforts. "Saw more of your little friends just the other day," they'd tell me. "Haven't found any yet?" They'd laugh. "Coyotes don't do interviews," they'd tell me. They suggested I take up golf.

I took to wandering at night where I thought coyotes might be making their way into the Bronx. I imagined them arriving in Van Cortlandt Park via the Putnam Trail, a soil vein pounded smooth as glass, where I walked and walked. Or perhaps they followed the Old Croton Aqueduct Trail, the outmoded passage for the Catskill reservoir water that keeps the city alive (the Croton Aqueduct has long been supplanted by more modern piping). From the Bronx, the passage south onto the island of Manhattan is more difficult. Perhaps they cross the Harlem River, swimming the water, or, more likely, they walk the bridges at night. It is only some five miles from the tip of Manhattan to Central Park, which is the place to be if you're a coyote in Manhattan. On April Fool's Day 1999, a coyote named Lucky Pierre led reporters, helicopters, photographers, cops, and tourists in a chase across Central Park before succumbing to a tranquilizer dart. Pierre got his name because for a time he holed up in a cave across from the luxe Pierre Hotel. In the winter of 2004, a coyote was seen bounding among the ice floes on frozen Rockaway Inlet in Queens, near the dunes of Breezy Point, twenty-five miles south of Central Park. The animal apparently had gotten across Manhattan, across the East River, either dog-paddling in the water or hiking one of the bridges to Brooklyn, and thence across that borough to the shores where Brooklyn meets Queens and the sea. Cops in boats tried to capture the creature, but he dove into the cold surf, swam to shore, and was gone.

A coyote made it across Brooklyn? Incredible. I sometimes find it hard to cross Brooklyn.

On Super Bowl Sunday 2006, on Manhattan's Upper West Side, a coyote was found smashed at the side of a highway. That same year a second coyote was captured in Central Park. In 2008 a coyote appeared on the Bronx campus of the Horace Mann School,

the *New York Sun* reporting that it was out for a "jog" with the students and "offered no resistance" when animal-control officers
scooped it up. On February 4, 2010, a coyote, described as "timid
and skittish," stopped in the middle of a frozen Central Park pond
long enough to be captured on film by a photographer. Three days
later a trio of coyotes appeared out of nowhere on the Columbia University campus in upper Manhattan, then disappeared as
quickly. On March 24 a coyote was reported inside the Holland
Tunnel, then it was sighted wandering Tribeca. "Manhattan's coyote population continued its inexorable push southward," concluded the *New York Times*. There will of course be many more of
them, and we will welcome the romance of the wild dogs, until we
don't. The creatures will have to be hunted and killed once they
hunt and kill one too many of our domesticated foot warmers (or,
worse, the little ones in our own domesticated breed). We like our
gardens in the East, we like our vines run amok, our tall trees in
the backyard, the deer grazing on bluegrass, the pretense of wildness; we like our animals at home pretending at atavistic habits,
but we don't want a carcass at the door in the morning. In other
words, let's have gardens but not nature. Herein lies the irony of
the coyote's arrival in the urban East. He does not represent wildness; he is an adaptee to the garden. Without our subjugating the
land to the ridiculous extent that we have, he wouldn't be walking
alongside us.

My guide to the parks of the Bronx was a fifty-six-year-old New York
City Parks Department wildlife biologist named Dave Kunstler, who
gives the impression that he prefers the conversation of nighthawks
and tree frogs. It was October, the days growing short, and we hiked
the woods until dusk, looking for coyote dens where he suspected
they might be, finding none. We searched under rocks, behind
boulders, in tree trunks. We ended up purloining a golf cart at the
Van Cortlandt Golf Course to hunt them on the fairway. Kunstler
didn't seem much interested in the quest. What he mostly talked
about were invasive plants. Kunstler saw invasives everywhere pushing out the natives: porcelain-berry and Asiatic bittersweet and
mugwort; the *Ailanthus* and the Norway maple among the tall trees;
and elsewhere, whole stretches of forest swallowed in kudzu.

Many of the plants Kunstler pointed out could be classified as
weeds. Coyotes, it seems to me, are also a kind of weed species.

And their success is indicative of a larger problem facing the human race, the problem of weeds relentlessly encroaching, their effect the strangulation and diminishment of complex ecosystems everywhere. Weed species, writes David Quammen, "reproduce quickly, disperse widely when given a chance, tolerate a fairly broad range of habitat, take hold in strange places, succeed especially in disturbed ecosystems, and resist eradication once they're established. They are scrappers, generalists, opportunists." The coyote, exactly. Also, brown rats, cockroaches, crows, kudzu, raccoons, the white-tailed deer, ragweed, Russian thistle, feral cats, feral dogs, squirrels, wild turkeys—all of them weed species, all exploding in number across the country, but especially along the suburban-urban gradient. In his essay "Planet of Weeds," Quammen singles out another spectacularly successful weed species, *Homo sapiens,* and notes that other weeds down the food chain tend to follow where human beings tread. The planet of weeds, as Quammen describes it, is an impoverished place in its abundance because it heralds the end of diversity. And it is likely our inexorable future.

Kunstler and I rode around on the cart like a pair of tin cans, driving away beautiful geese, hundreds of them fleeing on the fairway—weeds with wings. We rode and rode.

Eventually we stopped a young man who was tending to the carts. "Sure, I seen one just yesterday."

I threw up my hands.

"They're probably watching us right now." Kunstler shrugged.

Not finding a single coyote suddenly made me depressed.

Where is Trickster? We need him. The hominids are screwing up and don't have a plan to fix things. We've got global warming and rising seas and peak oil and fish dying off and deserts spreading. We've got a planet of weeds, and we seem utterly incapable of adapting to forestall disaster. The coyote survived the great Pleistocene extinction and may very well survive the present one, the planet's sixth great extinction, an event that has been greatly accelerated by the industrialization of *Homo sapiens* to the point that many thousands of species have disappeared in the last century alone. One wonders whether, before it is all through, *Homo sapiens* might be among the deceased. Perhaps the message that Trickster brings to us is this: *The more of us you see, the more impoverished the world will be.*

DAN KOEPPEL

Taking a Fall

FROM *Popular Mechanics*

6:59:00 A.M., 35,000 Feet

You have a late night and an early flight. Not long after takeoff, you drift to sleep. Suddenly you're wide awake. There's cold air rushing everywhere, and sound. Intense, horrible sound. *Where am I?* you think. *Where's the plane?*

You're six miles up. You're alone. You're falling.

Things are bad. But now's the time to focus on the good news. (Yes, it goes beyond surviving the destruction of your aircraft.) Although gravity is against you, another force is working in your favor: time. Believe it or not, you're better off up here than if you'd slipped from the balcony of your high-rise hotel room after one too many drinks last night.

Or at least you will be. Oxygen is scarce at these heights. By now, hypoxia is starting to set in. You'll be unconscious soon, and you'll cannonball at least a mile before waking up again. When that happens, remember what you are about to read. The ground, after all, is your next destination.

Granted, the odds of surviving a six-mile plummet are extraordinarily slim, but at this point you've got nothing to lose by understanding your situation. There are two ways to fall out of a plane. The first is to free-fall, or drop from the sky with absolutely no protection or means of slowing your descent. The second is to become a wreckage rider, a term coined by the Massachusetts-based amateur historian Jim Hamilton, who developed the Free Fall Research Page—an online database of nearly every imaginable human plummet. That classification means you have the advantage of being attached to a chunk of the plane. In 1972 a Serbian flight attendant, Vesna Vulovic, was traveling in a DC-9 over Czechoslovakia when it blew up. She fell 33,000 feet, wedged between her seat, a catering trolley, a section of aircraft, and the body of another crew member, landing on—then sliding down—a snowy incline before coming to a stop, severely injured but alive.

Surviving a plunge surrounded by a semiprotective cocoon of debris is more common than surviving a pure free-fall, according to Hamilton's statistics; thirty-one such confirmed or "plausible" incidents have occurred since the 1940s. Free-fallers constitute a much more exclusive club, with just thirteen confirmed or plausible incidents, including perennial Ripley's Believe It or Not superstar Alan Magee, blown from his B-17 on a 1943 mission over France. The New Jersey airman, more recently the subject of a *MythBusters* episode, fell 20,000 feet and crashed into a train station; he was subsequently captured by German troops, who were astonished at his survival.

Whether you're attached to crumpled fuselage or just plain falling, the concept you'll be most interested in is *terminal velocity*. As gravity pulls you toward the earth, you go faster. But like any moving object, you create drag—more as your speed increases. When downward force equals upward resistance, acceleration stops. You max out.

Depending on your size and weight, and factors such as air density, your speed at that moment will be about 120 mph—and you'll get there after a surprisingly brief bit of falling: just 1,500 feet, about the same height as Chicago's Sears (now Willis) Tower. Equal speed means you hit the ground with equal force. The difference is the clock. Body meets Windy City sidewalk in twelve seconds. From an airplane's cruising altitude, you'll have almost enough time to read this entire article.

7:00:20 A.M., 22,000 Feet

By now, you've descended into breathable air. You sputter into consciousness. At this altitude you've got roughly two minutes until impact. Your plan is simple. You will enter a Zen state and decide to live. You will understand, as Hamilton notes, "that it isn't the fall that kills you—it's the landing."

Keeping your wits about you, you take aim.

But at what? Magee's landing on the stone floor of that French train station was softened by the skylight he crashed through a moment earlier. Glass hurts, but it gives. So does grass. Haystacks and bushes have cushioned surprised-to-be-alive free-fallers. Trees aren't bad, though they tend to skewer. Snow? Absolutely. Swamps? With their mucky, plant-covered surface, even more awesome. Hamilton documents one case of a sky diver who, upon total parachute failure, was saved by bouncing off high-tension wires. Contrary to popular belief, water is an awful choice. Like concrete, liquid doesn't compress. Hitting the ocean is essentially the same as colliding with a sidewalk, Hamilton explains, except that pavement (perhaps unfortunately) won't "open up and swallow your shattered body."

With a target in mind, the next consideration is body position. To slow your descent, emulate a sky diver. Spread your arms and legs, present your chest to the ground, and arch your back and head upward. This adds friction and helps you maneuver. But don't relax. This is not your landing pose.

The question of how to achieve ground contact remains, regrettably, given your predicament, a subject of debate. A 1942 study in the journal *War Medicine* noted "distribution and compensation of pressure play large parts in the defeat of injury." Recommendation: wide-body impact. But a 1963 report by the Federal Aviation Agency argued that shifting into the classic sky diver's landing stance—feet together, heels up, flexed knees and hips—best increases survivability. The same study noted that training in wrestling and acrobatics would help people survive falls. Martial arts were deemed especially useful for hard-surface impacts: "A 'black belt' expert can reportedly crack solid wood with a single blow," the authors wrote, speculating that such skills might be transferable.

The ultimate learn-by-doing experience might be a lesson from the Japanese parachutist Yasuhiro Kubo, who holds the world record in the activity's banzai category. The sky diver tosses his chute from the plane and then jumps out after it, waiting as long as possible to retrieve it, put it on, and pull the ripcord. In 2000, Kubo

—starting from 9,842 feet—fell for fifty seconds before recovering his gear. A safer way to practice your technique would be at one of the wind-tunnel simulators found at about a dozen U.S. theme parks and malls. But neither will help with the toughest part: sticking the landing. For that you might consider—though it's not exactly advisable—a leap off the world's highest bridge, France's Millau Viaduct; its platform towers 891 feet over mostly spongy farmland.

Water landings—if you must—require quick decision-making. Studies of bridge-jump survivors indicate that a feet-first, knifelike entry (aka "the pencil") best optimizes your odds of resurfacing. The famed cliff divers of Acapulco, however, tend to assume a head-down position, with the fingers of each hand locked together, arms outstretched, protecting the head. Whichever you choose, first assume the free-fall position for as long as you can. Then, if a feet-first entry is inevitable, the most important piece of advice, for reasons both unmentionable and easily understood, is to *clench your butt.*

No matter the surface, definitely don't land on your head. In a 1977 "Study of Impact Tolerance Through Free-Fall Investigations," researchers at the Highway Safety Research Institute found that the major cause of death in falls—they examined drops from buildings, bridges, and the occasional elevator shaft (oops!) —was cranial contact. If you have to arrive top-down, sacrifice your good looks and land on your face rather than the back or top of your head. You might also consider flying with a pair of goggles in your pocket, Hamilton says, since you're likely to get watery eyes—impairing accuracy—on the way down.

7:02:19 A.M., 1,000 Feet

Given your starting altitude, you'll be just about ready to hit the ground as you reach this section of instruction (based on the average adult reading speed of 250 words per minute). The basics have been covered, so feel free to concentrate on the task at hand. But if you're so inclined, here's some supplemental information—though be warned that none of it will help you much at this point.

Statistically speaking, it's best to be a flight crew member, a child, or traveling in a military aircraft. Over the past four decades, there have been at least a dozen commercial airline crashes with just one survivor. Of those documented, four of the survivors were crew, like the flight attendant Vulovic, and seven were passengers under the age of eighteen. That includes Mohammed el-Fateh Osman, a two-year-old wreckage rider who lived through the crash of a Boeing jet in Sudan in 2003, and, more recently, fourteen-year-old Bahia Bakari, the sole survivor of last June's Yemenia Airways plunge off the Comoros Islands.

Crew survival may be related to better restraint systems, but there's no consensus on why children seem to pull through falls more often. The Federal Aviation Agency study notes that kids, especially those under the age of four, have more flexible skeletons, more relaxed muscle tonus, and a higher proportion of subcutaneous fat, which helps protect internal organs. Smaller people— whose heads are lower than the seat backs in front of them—are better shielded from debris in a plane that's coming apart. Lower body weight reduces terminal velocity, and reduced surface area decreases the chance of impalement upon landing.

7:02:25 A.M., 0 Feet

The ground. Like a Shaolin master, you are at peace and prepared. *Impact.* You're alive. What next? If you're lucky, you might find that your injuries are minor, stand up, and smoke a celebratory cigarette, as the British tail gunner Nicholas Alkemade did in 1944 after landing in snowy bushes following an 18,000-foot plummet. (If you're a smoker, you're *super extra lucky,* since you've technically gotten to indulge during the course of an airliner trip.) More likely, you'll have tough work ahead.

Follow the example of Juliane Koepcke. On Christmas Eve, 1971, the Lockheed Electra she was traveling in exploded over the Amazon. The next morning the seventeen-year-old German awoke on the jungle floor, strapped into her seat, surrounded by fallen holiday gifts. Injured and alone, she pushed the death of her mother, who'd been seated next to her on the plane, out of her mind. Instead, she remembered advice from her father, a biologist: to find civilization when lost in the jungle, follow water. Koepcke waded from tiny streams to larger ones. She passed crocodiles and poked the mud in front of her with a stick to scare away stingrays. She had lost one shoe in the fall and was wearing a ripped miniskirt. Her only food was a bag of candy, and she had nothing but dark, dirty water to drink. She ignored her broken collarbone and her wounds, infested with maggots.

On the tenth day, she rested on the bank of the Shebonya River. When she stood up again, she saw a canoe tethered to the shoreline. It took her hours to climb the embankment to a hut, where, the next day, a group of lumberjacks found her. The incident was seen as a miracle in Peru, and free-fall statistics seem to support those arguing for divine intervention: according to the Geneva-based Aircraft Crashes Record Office, 118,934 people died in 15,463 plane crashes between 1940 and 2008. Even when you add failed-chute sky divers, Hamilton's tally of confirmed or plausible lived-to-tell-about-it incidents is only 157, with 42 occurring at heights over 10,000 feet.

But Koepcke never saw survival as a matter of fate. She can still recall the first moments of her fall from the plane, as she spun through the air in her seat. That wasn't under her control, but what happened when she regained consciousness was. "I had been able to make the correct decision — to leave the scene of the crash," she says now. And because of experience at her parents' biological research station, she says, "I did not feel fear. I knew how to move in the forest and the river, in which I had to swim with dangerous animals like caimans and piranhas."

Or, by now, you're wide awake, and the aircraft's wheels have touched safely down on the tarmac. You understand that the odds of any kind of accident on a commercial flight are slimmer than slim and that you will likely never have to use this information. But as a courtesy to the next passenger, consider leaving your copy of this guide in the seat-back pocket.

JARON LANIER

The First Church of Robotics

FROM THE *New York Times*

THE NEWS OF THE DAY often includes an item about some development in artificial intelligence (AI): a machine that smiles, a program that can predict human tastes in mates or music, a robot that teaches foreign languages to children. This constant stream of stories suggests that machines are becoming smart and autonomous, a new form of life, and that we should think of them as fellow creatures instead of as tools. But such conclusions aren't just changing how we think about computers—they are reshaping the basic assumptions of our lives in misguided and ultimately damaging ways.

I myself have worked on projects like machine-vision algorithms that can detect human facial expressions in order to animate avatars or recognize individuals. Some would say these are examples of AI, but I would say it is research on a specific software problem that shouldn't be confused with the deeper issues of intelligence or the nature of personhood. Equally important, my philosophical position has not prevented me from making progress in my work. (This is not an insignificant distinction: someone who refused to believe in, say, general relativity would not be able to make a GPS navigation system.)

In fact, the nuts and bolts of AI research can often be more usefully interpreted without the concept of AI at all. For example, IBM scientists recently unveiled a "question answering" machine that is designed to play the TV quiz show *Jeopardy*. Suppose IBM had dispensed with the theatrics, declared it had done Google one better, and come up with a new phrase-based search engine. This framing of exactly the same technology would have gained IBM's team as much (deserved) recognition as the claim of an artificial intelligence but would also have educated the public about how such a technology might actually be used most effectively.

Another example is the way in which robot teachers are portrayed. For starters, these robots aren't all that sophisticated; miniature robotic devices used in endoscopic surgeries are infinitely more advanced, but they don't get the same attention because they aren't presented with the AI spin.

Furthermore, these robots are just a form of high-tech puppetry. The children are the ones making the transaction take place — having conversations and interacting with these machines, but essentially teaching themselves. This just shows that humans are social creatures, so if a machine is presented in a social way, people will adapt to it.

What bothers me most about this trend, however, is that by allowing artificial intelligence to reshape our concept of personhood, we are leaving ourselves open to the flip side: we think of people more and more as computers, just as we think of computers as people.

In one recent example, Clay Shirky, a professor at New York University's Interactive Telecommunications Program, has suggested that when people engage in seemingly trivial activities like "re-Tweeting," relaying on Twitter a short message from someone else, something nontrivial — real thought and creativity — takes place on a grand scale within a global brain. That is, people perform machinelike activity, copying and relaying information; the Internet, as a whole, is claimed to perform the creative thinking, the problem solving, the connection-making. This is a devaluation of human thought.

Consider too the act of scanning a book into digital form. The historian George Dyson has written that a Google engineer once said to him: "We are not scanning all those books to be read by people. We are scanning them to be read by an AI." While we have yet to see how Google's book scanning will play out, a machine-centric vision of the project might encourage software that treats books as grist for the mill, decontextualized snippets in one big database, rather than separate expressions from individual writers. In this approach, the contents of books would be atomized into bits of information to be aggregated, and the authors themselves, the feeling of their voices, their differing perspectives, would be lost.

What all this comes down to is that the very idea of artificial intelligence gives us the cover to avoid accountability by pretending that machines can take on more and more human responsibility. This holds for things that we don't even think of as artificial intelligence, like the recommendations made by Netflix and Pandora. Seeing movies and listening to music suggested to us by algorithms is relatively harmless, I suppose. But I hope that once in a while the users of those services resist the recommendations; our exposure to art shouldn't be hemmed in by an algorithm that we merely want to believe predicts our tastes accurately. These algorithms do not represent emotion or meaning, only statistics and correlations.

What makes this doubly confounding is that while Silicon Valley might sell artificial intelligence to consumers, our industry certainly wouldn't apply the same automated techniques to some of its own work. Choosing design features in a new smartphone, say, is considered too consequential a game. Engineers don't seem quite ready to believe in their smart algorithms enough to put them up against Apple's chief executive, Steve Jobs, or some other person with a real design sensibility.

But the rest of us, lulled by the concept of ever more intelligent AI's, are expected to trust algorithms to assess our aesthetic choices, the progress of a student, the credit risk of a homeowner or an institution. In doing so, we only end up misreading the capability of our machines and distorting our own capabilities as human beings. We must instead take responsibility for every task undertaken by a machine and double-check every conclusion offered by an algorithm, just as we always look both ways when crossing an intersection, even though the light has turned green.

When we think of computers as inert, passive tools instead of people, we are rewarded with a clearer, less ideological view of what is going on—with the machines and with ourselves. So, why, aside from the theatrical appeal to consumers and reporters, must engineering results so often be presented in a Frankensteinian light?

The answer is simply that computer scientists are human and are as terrified by the human condition as anyone else. We, the technical elite, seek some way of thinking that gives us an answer to death, for instance. This helps explain the allure of a place like the Singularity University. The influential Silicon Valley institution preaches a story that goes like this: one day in the not-so-distant future, the Internet will suddenly coalesce into a super-intelligent AI, infinitely smarter than any of us individually and all of us combined; it will become alive in the blink of an eye and take over the world before humans even realize what's happening.

Some think the newly sentient Internet would then choose to kill us; others think it would be generous and digitize us the way Google is digitizing old books, so that we can live forever as algorithms inside the global brain. Yes, this sounds like many different science-fiction movies. Yes, it sounds nutty when stated so bluntly. But these are ideas with tremendous currency in Silicon Valley; these are guiding principles, not just amusements, for many of the most influential technologists.

It should go without saying that we can't count on the appearance of a soul-detecting sensor that will verify that a person's consciousness has been virtualized and immortalized. There is certainly no such sensor with us today to confirm metaphysical ideas about people, or even to recognize the contents of the human brain. All thoughts about consciousness, souls, and the like are bound up equally in faith, which suggests something remarkable: what we are seeing is a new religion, expressed through an engineering culture.

What I would like to point out, though, is that a great deal of the confusion and rancor in the world today concerns tension at the boundary between religion and modernity—whether it's the distrust among Islamic or Christian fundamentalists of the scientific worldview or even the discomfort that often greets progress in fields like climate-change science or stem-cell research.

If technologists are creating their own ultramodern religion, and it is one in which people are told to wait politely as their very souls are made obsolete, we might expect further and worsening tensions. But if technology were presented without metaphysical baggage, is it possible that modernity would not make people as uncomfortable?

Technology is essentially a form of service. We work to make the world better. Our inventions can ease burdens, reduce poverty and suffering, and sometimes even bring new forms of beauty into the world. We can give people more options to act morally, because people with medicine, housing, and agriculture can more easily afford to be kind than those who are sick, cold, and starving.

But civility, human improvement, these are still choices. That's why scientists and engineers should present technology in ways that don't confound those choices.

We serve people best when we keep our religious ideas out of our work.

JON MOOALLEM

The Love That Dare Not Squawk Its Name

FROM THE *New York Times Magazine*

THE LAYSAN ALBATROSS is a downy seabird with a seven-foot wingspan and a notched, pale yellow beak. Every November a small colony of albatrosses assembles at a place called Kaena Point, overlooking the Pacific at the foot of a volcanic range on the northwestern tip of Oahu, Hawaii. Each bird has spent the past six months in solitude, ranging over open water as far north as Alaska, and has come back to the breeding ground to reunite with its mate. Albatrosses can live to be sixty or seventy years old and typically mate with the same bird every year, for life. Their "divorce rate," as biologists term it, is among the lowest of any bird.

When I visited Kaena Point in November, the first birds were just returning, and they spent a lot of their time gliding and jackknifing in the wind a few feet overhead or plopped like cushions in the sand. There are about 120 breeding albatrosses in the colony, and gradually each will arrive and feel out the crowd for the one other particular albatross it has been waiting to have sex with again. At any given moment in the days before Thanksgiving, some birds may be just turning up while others sit there killing time. It feels like an airport baggage-claim area.

Once together, pairs will copulate and collaboratively incubate a single egg for sixty-five days. They take shifts: one bird has to sit at the nest while the other flaps off to fish and eat for weeks at a time. Couples preen each other's feathers and engage in elaborate mating behaviors and displays. "Like when you're in a couple," Marlene Zuk, a biologist who has visited the colony, explained to me. "All those sickening things that couples do that gross out everyone else but the two people in the couple? . . . Birds have the same thing." I often saw pairs sitting belly to belly, arching their necks and nuzzling their heads together to form a kind of heart shape. Speaking on Oahu a few years ago as first lady, Laura Bush praised Laysan albatross couples for making lifelong commitments to one another. Lindsay C. Young, a biologist who studies the Kaena Point colony, told me: "They were supposed to be icons of monogamy: one male and one female. But I wouldn't assume that what you're looking at *is* a male and a female."

Young has been researching the albatrosses on Oahu since 2003; the colony was the focus of her doctoral dissertation at the University of Hawaii, Manoa, which she completed last spring. (She now works on conservation projects as a biologist for hire.) In the course of her doctoral work, Young and a colleague discovered, almost incidentally, that a third of the pairs at Kaena Point actually consisted of two female birds, not one male and one female. Laysan albatrosses are one of countless species in which the two sexes look basically identical. It turned out that many of the female-female pairs, at Kaena Point and at a colony that Young's colleague studied on Kauai, had been together for four, eight, or even nineteen years—as far back as the biologists' data went in some cases. The female-female pairs had been incubating eggs together, rearing chicks, and just generally passing under everybody's nose for what you might call "straight" couples.

Young would never use the phrase "straight couples." And she is adamantly against calling the other birds "lesbians" too. For one thing, the same-sex pairs appear to do everything male-female pairs do *except* have sex, and Young isn't really sure, or comfortable judging, whether that technically qualifies them as lesbians or not. Moreover, the whole question is meaningless to her; it has nothing to do with her research. "Lesbian," she told me, "is a human term," and Young—a diligent and cautious scientist, just beginning to make a name in her field—is devoted to using the most aseptic language possible and resisting any tinge of anthropomorphism. "The study is about albatross," she told me firmly. "The study is not about humans." Often she seemed to be mentally peer-reviewing her words before speaking.

A discovery like Young's can disorient a wildlife biologist in the most thrilling way—if he or she takes it seriously, which has traditionally not been the case. Various forms of same-sex sexual activity have been recorded in more than 450 different species of animals by now, from flamingos to bison to beetles to guppies to warthogs. A female koala might force another female against a tree and mount her, while throwing back her head and releasing what one scientist described as "exhalated belchlike sounds." Male Amazon River dolphins have been known to penetrate each other in the blowhole. Within most species, homosexual sex has been documented only sporadically, and there appear to be few cases of individual animals who engage in it exclusively. For more than a century, this kind of observation was usually tacked onto scientific papers as a curiosity, if it was reported at all, and not pursued as a legitimate research subject. Biologists tried to explain away what they'd seen or dismissed it as theoretically meaningless—an isolated glitch in an otherwise elegant Darwinian universe where every facet of an animal's behavior is geared toward reproducing. One primatologist speculated that the real reason two male orangutans were fellating each other was nutritional.

In recent years, though, more biologists have been looking objectively at same-sex sexuality in animals—approaching it as real science. For Young, the existence of so many female-female albatross pairs disproved assumptions that she didn't even realize she'd been making and, in the process, raised a chain of progressively more complicated questions. One of the prickliest, it seemed, was how a scientist is supposed to talk about any of this, given how eager the rest of us have been to twist the sex lives of animals into allegories of our own. "This colony is literally the largest proportion of—I don't know what the correct term is: 'homosexual animals'? —in the world," Young told me. "Which I'm sure some people think is a great thing, and others might think is not."

It was a guarded understatement. Two years ago, Young decided to write a short paper with two colleagues on the female-female albatross pairs. "We were pretty careful in the original article to plainly and simply report what we found," she said. "It's definitely a little bit of a tricky subject, and one you want to be gentle on." But the journal that published the paper, *Biology Letters*, sent out a press release a few days after the California Supreme Court legalized gay marriage. At six the next morning, a Fox News reporter called Young on her cell phone. The resulting story joined others, including one in the *New York Times,* and as the news ricocheted around the Internet, a stampede of online commenters alternately celebrated Young's findings as a clear call for equality or denigrated them as "pure propaganda and selective science at its dumbest" and "an effort to humanize animals or devolve humans to the level of animals or to further an agenda." Many pointed out that animals also rape or eat their young; was America going to tolerate that too, just because it's "natural"?

A Denver-based publication for gay parents welcomed any and all new readers from "the extensive lesbian albatross parent community." The conservative Oklahoma senator Tom Coburn highlighted Young's paper on his website under the heading "Your Tax Dollars at Work," even though her study of the female-female pairs was not actually federally financed. Stephen Colbert warned on Comedy Central that "albatresbians" were threatening American family values with their "Sappho-avian agenda." A gay rights advocate e-mailed Young, asking her to fly a rainbow flag above each female-female nest to identify them and show solidarity. Even now, the first thing everyone wants to know from Young—sometimes the only thing—is, what does the existence of these lesbian albatrosses say about us?

"I don't answer that question," she told me.

A female Laysan albatross is physically capable of laying only one egg per year—that's just how it's built. Nevertheless, since as early as 1919, biologists have periodically found nests of albatrosses (and similar species of birds) with two eggs inside or with a second egg just outside the nest, as if it had rolled out. (This will inevitably happen; there's simply not enough room in the nest for two eggs and one Laysan albatross.) Scientists have a term for the phenomenon of extra eggs in a nest: a "supernormal clutch." But in the case of the albatross, they never had a watertight explanation.

In the early 1960s, one ornithologist tried to put the whole cumbersome mystery to rest by asserting that some of those female birds must simply be able to lay multiple eggs. The claim was apparently based on sketchy data, but supernormal clutches were so rare that it was hard to rack up enough observations to disprove the hypothesis. Real progress was finally made in 1968, when Harvey Fisher, a dean of midcentury albatross science, reported on seven years of daily observations made at 3,440 different nests on the Midway atoll in the middle of the Pacific. Fisher concluded that "two eggs in a nest are an indication that two females used the nest, although at different times." He was describing "egg dumping," whereby, for example, an inexperienced female accidentally lays her egg in the wrong nest. From then on, egg dumping was a default explanation for supernormal clutches in albatrosses. After all, Fisher had also declared that "promiscuity, polygamy and polyandry are unknown in this species." Lesbianism apparently never occurred to anyone—even enough to be cursorily dismissed. As Brenda Zaun recently told me, "It never dawned on anyone to sex the birds."

Zaun, a biologist with the U.S. Fish and Wildlife Service, was studying a Laysan colony on Kauai forty years after Fisher's publication. She realized that certain nests there seemed to wind up with two eggs in them year after year; the distribution of the supernormal clutches wasn't random, as it would presumably be if it were caused exclusively by egg dumping. On a hunch, Zaun pulled feathers from a sample of the breeding pairs associated with two-egg nests and sent them to Lindsay Young, asking her to draw DNA from the feathers and genetically determine the sexes of those birds in her lab. When the results showed that every bird was female, Young figured she'd messed up. So she did it again—and got the same result. Then she genetically sexed every bird at Kaena Point. "Where it wasn't totally clear, or I worried that maybe I mixed up the sample, I actually went back into the field and took new blood samples to do it again," Young told me. In the end she genetically sexed the birds in her lab four times, just to be sure. She found that 39 of the 125 nests at Kaena Point since 2004 belonged to female-female pairs, including more than 20 nests in which she'd never noticed a supernormal clutch. It seemed that certain females were somehow finding opportunities to quickly copulate with males but incubating their eggs—and doing everything else an albatross does while at the colony—with other females.

Young gave a talk about these findings at an international meeting of Pacific-seabird researchers. "There was a lot of murmuring in the room," she remembers. "Then, afterward, people were coming up to me and saying: 'We see supernormal clutches all the time. We assumed it was a male and a female.' And I'd say: 'Yeah? Well, you might want to look into that.'" Recently, journals have asked her to confidentially peer-review new papers about other species, describing similar discoveries. "I can't say which species," she explains, "but my guess is, in the next year, we're going to see a lot more examples of this."

It may seem surprising that scientists sometimes don't know the true sexes of the animals they spend their careers studying—that they can be tripped up in some *Tootsie*-like farce for so long. But it's easy to underestimate the pandemonium that they're struggling to interpret in the wild. Often biologists are forced to assign sexes to animals by watching what they do when they mate. When one albatross or boar or cricket rears up and mounts a second, it would seem to be advertising the genders of both. Unless, of course, that's not the situation at all.

"There is still an overall presumption of heterosexuality," the biologist Bruce Bagemihl told me. "Individuals, populations, or species are considered to be entirely heterosexual until proven otherwise." While this may sound like a reasonable starting point, Bagemihl calls it a "heterosexist bias" and has shown it to be a significant roadblock to understanding the diversity of what animals actually do. In 1999 Bagemihl published *Biological Exuberance,* a book that pulled together a colossal amount of previous piecemeal research and showed how biologists' biases had marginalized animal homosexuality for the last 150 years—sometimes innocently enough, sometimes in an eruption of anthropomorphic disgust. Courtship behaviors between two animals of the same sex were persistently described in the literature as "mock" or "pseudo" courtship—or just "practice." Homosexual sex between ostriches was interpreted by one scientist as "a nuisance" that "goes on and on." One man, studying Mazarine Blue butterflies in Morocco in 1987, regretted having to report "the lurid details of declining moral standards and of horrific sexual offenses" which are "all too often packed" into national newspapers. And a bighorn-sheep biologist confessed in his memoir, "I still cringe at the memory of seeing old D-ram mount S-ram repeatedly." To think, he wrote, "of those magnificent beasts as 'queers'—oh, God!"

"What Bagemihl's book really did," the Canadian primatologist and evolutionary psychologist Paul Vasey says, "is raise people's awareness around the fact that this occurs in quote-unquote nature —in animals. And that it can be studied in a serious, scholarly way." But studying it seriously means resolving a conundrum. At the heart of evolutionary biology, since Darwin, has been the idea that any genetic traits and behaviors that outfit an animal with an advantage—that help the animal make lots of offspring—will remain in a species, while ones that don't will vanish. In short, evolution gradually optimizes every animal toward a single goal: passing on its genes. The Yale ornithologist Richard Prum told me: "Our field is a lot like economics: we have a core of theory, like free-market theory, where we have the invisible hand of the market creating order—all commodities attain exactly the price they're worth. Homosexuality is a tough case, because it appears to violate that central tenet, that all of sexual behavior is about reproduction. The question is, why would anyone invest in sexual behavior that isn't reproductive?"—much less a behavior that looks to be starkly counterproductive. Moreover, if animals carrying the genes associated with it are less likely to reproduce, how has that behavior managed to stick around?

Given this big umbrella of theory, the very existence of homosexual behavior in animals can feel a little like impenetrable nonsense, something a researcher could spend years banging his or her head against the wall deliberating. The difficulty of that challenge, more than any implicit or explicit homophobia, may be why past biologists skirted the subject.

In the last decade, however, Paul Vasey and others have begun developing new hypotheses based on actual, prolonged observation of different animals, deciphering the ways given homosexual behaviors may have evolved and the evolutionary role they might play within the context of individual species. Different ideas are emerging about how these behaviors could fit within that traditional Darwinian framework, including seeing them as conferring reproductive advantages in roundabout ways. Male dung flies, for example, appear to mount other males to tire them out, knocking them out of competition for available females. Researchers speculate that young male bottlenose dolphins mount one another simply to establish trust and form bonds—but those bonds actually turn out to be critical to reproduction, since when males mature, they work in groups to cooperatively gain access to females.

These ideas generally aim to explain only particular behaviors in a particular species. So far, the only real conclusion this relatively small body of literature seems to point to, collectively, is a kind of deflating meta-conclusion: a single explanation of homosexual behavior in animals may not be possible, because thinking of "homosexual behavior in animals" as a single scientific subject might not make much sense. "Biologists want to build these unified theories to explain everything they see," Vasey told me. So do journalists, he added—all people, really. "But none of this lends itself to a linear story. My take on it is that homosexual behavior is not a uniform phenomenon. Having one unifying body of theory that explains why it's happening in all these different species might be a chimera."

The point of heterosexual sex, Vasey said, no matter what kind of animal is doing it, is primarily reproduction. But that shouldn't trick us into thinking that homosexual behavior has some equivalent organizing purpose—that the two are tidy opposites. "All this homosexual behavior isn't tied together by that sort of primary function," Vasey said. Even what the same-sex animals are doing varies tremendously from species to species. But we're quick to conceive of that great range of activities in the way it most handily tracks to our anthropomorphic point of view: put crassly, all those different animals just seem to be doing gay sex stuff with one another. As the biologist Marlene Zuk explains, we are hard-wired to read all animal behavior as "some version of the way people do things" and animals as "blurred, imperfect copies of humans."

When I visited Zuk at her lab at the University of California at Riverside last December, an online video clip of an octopus carrying a coconut shell around the seafloor, and periodically hiding under it, was starting to go viral. For a few days, people everywhere were flipping out about how intelligent and wily this octopus was. Not Zuk, though. "Oh, spare me," she said. To us, Zuk explained, that octopus's behavior reads as proof that "octopuses are at one with humans" because it just happens to look like something we do —how a toddler plays peekaboo under a blanket, say, or a bandit ducks into an alleyway dumpster to avoid the cops. But the octopus doesn't know that. Nor is it doing something so uncommon in the animal world. Zuk explained that caddis-fly larvae collect rocks and loom them together into intricate shelters. "But for some reason we don't think that's cool," she said, "because the caddis-fly larvae don't have big eyes like us."

Something similar may be happening with what we perceive to be homosexual sex in an array of animal species: we may be grouping together a big grab bag of behaviors based on only a superficial similarity. Within the logic of each species, or group of species, many of these behaviors appear to have their own causes and consequences—their own evolutionary meanings, so to speak. The Stanford biologist Joan Roughgarden told me to think of all these animals as "multitasking" with their private parts.

It's also possible that some homosexual behaviors don't provide a conventional evolutionary advantage; but neither do they upend everything we know about biology. For the last fifteen years, for example, Paul Vasey has been studying Japanese macaques, a species of two-and-a-half-foot-tall, pink-faced monkey. He has looked almost exclusively at why female macaques mount one another during the mating season. Vasey now says he is on to the answer: "It isn't functional," he told me; the behavior has no discernible purpose, adaptationally speaking. Instead, it's a byproduct of a behavior that does, and the supposedly streamlining force of evolution just never flushed that byproduct from the gene pool. Female macaques regularly mount males too, Vasey explained, probably to focus their attention and reinforce their bond as mates. The females are physically capable of mounting any gender of macaque. They've just never developed an instinct to limit themselves to one. "Evolution doesn't create perfect adaptations," Vasey said. As Zuk put it, "There's a lot of slop in the system—which," she was sure to add, "is not the same as saying homosexuality is a mistake."

About two dozen birds were knocking around when Lindsay Young and I arrived at Kaena Point one afternoon. Young dished about a few of them—"Her mate didn't show up last year"; "God, this one's annoying"—as they waddled by. Laysan albatrosses are not nearly as graceful on land as they are in the air; even they seem surprised by the size of their feet. (Later that week, at a nearby resort, I would recognize their gait while watching an out-of-shape snorkeler toddle back to his beach towel in rented flippers.) "I'm just writing down who's here," Young said, reading the numbers on the birds' leg bands and marking them on her clipboard. After trying and failing to get a clear view of one bird's leg with binoculars, she finally just walked to within a few feet of the animal and leaned over to look.

This is the luxury of studying Laysan albatrosses. Having evolved with no natural predators, the birds have no fight-or-flight instinct —you can basically go right up to one and grab it. In fact, Young did just this a short while later, slinking up to a male on all fours, sweeping it in by its flank, and, in one expert motion, straitjacketing the wings under one arm and clamping the beak shut in her other hand. Then she walked over and handed the thing to me; she needed to take an expensive tracking device off the bird's ankle. "Sorry, but it's like watching a thousand-dollar bill fly around," she said. She took some pliers from her backpack to twist off the anklet and, as I stood bear-hugging the albatross, she added: "They have a nice smell. It's a little musty."

Young and Marlene Zuk are now applying for a ten-year National Science Foundation grant to continue studying the female albatross pairs. One of the first questions they want to answer is how these birds are winding up with fertilized eggs. Typically, albatrosses fend off birds who aren't their mates. So Young has been trying to determine if males who arrive back at the colony before their own partners do are forcing themselves on these females or whether these females are somehow "soliciting" the males for sex. She was staking out Kaena Point on a daily basis, trying to watch these illicit copulations unfold for herself. This was Young's third year; so far, she'd managed to see it happen only twice.

Young and I ambled around for half an hour, maybe more. Then she pointed and, in a monotone, said, "So, that's a female-female pair." We crouched and watched the two birds, numbers 169 and 983. They sat under a spindly, native Hawaiian naio bush. They made *baa* sounds at each other. After a while, Young and I got up.

Another hour passed. (Usually, Young brings along a camping chair.) Occasionally, albatrosses danced in groups of two or three, raising their necks, groaning like vibrating cell phones, clacking their beaks or stomping. But most of the time, they didn't do much at all. "I've spent a lot of my career watching animals not have sex," Zuk later told me.

Homosexual activity is often observed in animal populations with a shortage of one sex—in the wild but more frequently at zoos. Some biologists anthropomorphically call this "the prisoner effect." That's basically the situation at Kaena Point: there are fewer male albatrosses than females (although not every male albatross has a mate). Because it takes two albatrosses to incubate an egg, switching on and off at the nest, a female that can't find a male (or maybe, Young says, who can't find "a good enough male") has no chance of producing a chick and passing on her genes. Quickly mating with an otherwise committed male, then pairing with another single female to incubate the egg, is a way to raise those odds.

Still, pairing off with another female creates its own problems: nearly every female lays an egg in November whether she has managed to get it fertilized or not, and the small, craterlike nests that albatross pairs build in the dirt can accommodate only one egg and one bird. So Young was also trying to figure out how a female-female pair decides which of its two eggs to incubate and which to chuck out of the nest—if the birds are deciding at all and not just knocking one egg out accidentally. From a strict Darwinian perspective, Young told me, "it doesn't pay for one bird to incubate the other's egg unless her partner is going to let her egg be incubated the following year." But presumably, neither female bird knows whether an egg is hers or the other bird's, much less whether it's fertilized or not. A Laysan albatross just knows to sit on whatever's under it. "They'll incubate anything—I have a photo of one incubating a volleyball," Young said.

And these were only preambles to more questions. With the male of an albatross pair replaced by another female, every step of the species' normal, well-honed process for fledging a chick seemed suddenly to present a fresh dilemma. Ultimately, either the rules of albatrossdom were breaking down and the lesbian couples were booting up some alternate suite of behaviors, governed by its own set of rules, or else science had never thoroughly understood the rules of albatrossdom to begin with. And that's the whole point for Young: it's the complexity and apparent flexibility of the species that fascinates her—the puzzle those female-female pairs create at Kaena Point just by existing. She's not trying to explain homosexual behavior. She's trying to explain the albatross. And that's why the rest of the world's politicized reaction to her work caught her by surprise.

Many people who contacted Young after the publication of her first albatross paper assumed she was a lesbian. She is not. Young's husband, a biological consultant, was actually an author of the paper, along with Brenda Zaun (who is also not gay, for what it's worth). Young found the assumption offensive—not because she was being mistaken for a gay person but because she was being mistaken for a bad scientist; these people seemed to presume that her research was compromised by a personal agenda. Still, some of the biologists doing the most incisive work on animal homosexuality are in fact gay. Several people I spoke to told me that their own sexual identities helped spur or maintain their interest in the topic; Bruce Bagemihl argued that gay and lesbian people are "often better equipped to detect heterosexist bias when investigating the subject simply because we encounter it so frequently in our everyday lives." With a laugh, Paul Vasey told me, "People automatically assume I'm gay." He is gay, he added, but that fact didn't seem to detract from his amusement.

In retrospect, the big, sloshing stew of anthropomorphic analyses that Young's paper provoked in the culture couldn't have been less surprising. For whatever reason, we're prone to seeing animals— especially animals that appear to be gay—as reflections, models, and foils of ourselves; we're extraordinarily, and sometimes irrationally, invested in them.

Only a few months before I visited Kaena Point, two penguins at the San Francisco Zoo became the latest in a tradition of captive same-sex penguin couples making global headlines. After six years together—in which the two birds even fostered a son, named Chuck Norris—the penguins split up when one of the males ran off with a female named Linda. The zoo's penguin keeper, Anthony Brown, told me he received angry e-mail, accusing him of separating the pair for political reasons. "Penguins make their own decisions here at the San Francisco Zoo," Brown assured me. And while he stressed that there is no scientific way of determining if animals are "gay," because the word connotes a sexual orientation, not just a behavior, he also noted that, this being the San Francisco Zoo, "there's definitely a lot of opinion here, internally, that we give in and call the penguins gay." Another male-male penguin couple who fostered a chick at the Central Park Zoo was subsequently immortalized in 2005 in the illustrated children's book

And Tango Makes Three. According to the American Library Association, there have been more requests for libraries to ban *And Tango Makes Three* every year than any other book in the country, for three years running.

What animals do—what's perceived to be "natural"—seems to carry a strange moral potency: it's out there, irrefutably, as either a validation or a denunciation of our own behavior, depending on how you happen to feel about homosexuality and about nature. During the Victorian era, observations of same-sex behavior in swans and insects were held up as evidence against the morality of homosexuality in humans, since at the dawn of industrialism and Darwinism, people were invested in seeing themselves as more civilized than the "lower animals." Robert Mugabe and the Nazis have employed the same reasoning, as did the 1970s antigay crusader Anita Bryant, who, Bruce Bagemihl notes, claimed in an interview that "even barnyard animals don't do what homosexuals do" and was unmoved when the interviewer pointed out what actually happens in barnyards. On the other hand, an Australian drag queen known as Dr. Gertrude Glossip has used Bagemihl's book to create a celebratory, interpretive gay-animal tour of the Adelaide zoo, marketed to gay and lesbian tourists. The book was also cited in a brief filed in the 2003 Supreme Court case that overturned a Texas state ban on sodomy and in a legislative debate on the floor of the British Parliament.

James Esseks, the director of the Lesbian, Gay, Bisexual and Transgender Project at the American Civil Liberties Union, told me he has never incorporated facts about animal behavior into a legal argument about the rights of human beings. It's totally beside the point, he said; people should not be discriminated against regardless of what animals do. (In her book *Sexual Selections,* Marlene Zuk writes, "People need to be able to make decisions about their lives without worrying about keeping up with the bonobos.") That being said, Esseks told me, polls show that Americans are more likely to discriminate against gays and lesbians if they think homosexuality is "a choice." "It shouldn't be the basis of a moral judgment," he said. But sometimes it is, and gay animals are compelling evidence that being gay isn't a choice at all. In fact, Esseks remembers reading a brief mention of animal homosexual behavior during an anthropology class in college in the mid-1980s. "And as a closeted guy, it made a difference to me," he told me. He remembers thinking: "Oh, hey, this is quote-unquote natural. This is normal. This is part of the normal spectrum of humanity—or life."

But later in our conversation, Esseks paused and stayed silent for a while. He was thinking like a lawyer again now, and found a hole in that line of reasoning. "I guess some of these animals could actually be quote-unquote making a choice," he said. How could we, as humans, ever know? "Huh," he said. "I'm just stopping to think that through. I'm not quite sure what to do with that." Esseks had stumbled right back into what he originally identified as the underlying problem. Those wanting to discriminate against gays and lesbians may have roped the rest of us into an argument over what's "natural" just by asserting for so long that homosexuality is not. But affixing any importance to the question of whether something is natural or unnatural is a red herring; it's impossible to pin down what those words mean even in a purely scientific context. (Zuk notes that animals don't drive cars or watch movies, and no one calls those activities "unnatural.") In the end there's just no coherent debate there to have. Animal research demonstrating the supposed "naturalness" of homosexuality has typically been embraced by gay rights activists and has put their opponents on the defensive. At the same time, research interpreted—or, maybe more often, misinterpreted—to be close to pinpointing that naturalness in a specific "gay gene" can make people on both sides anxious in a totally different way.

In 2007, for instance, the neurobiologist David Featherstone, at the University of Illinois at Chicago, and several colleagues, while searching for new drug treatments for Lou Gehrig's disease, happened upon a discovery: a specific protein mutation in the brain of male fruit flies made the flies try to have sex with other males. What the mutation did, more specifically, was tweak the fruit flies' sense of smell, making them attracted to male pheromones—mounting other males was the end result. To Featherstone, how fruit flies smell doesn't seem to have anything to do with human sexuality. "We didn't think about the societal implications—we're just a bunch of dorky biologists," he told me recently. Still, after publishing a paper describing this mutation, he received a flood of phone calls and e-mail messages presuming that he could, and would, translate this new knowledge into a way of changing people's sexual orientations. One e-mail message compared him with

Dr. Josef Mengele, noting "the direct line that leads from studies like this to compulsory eradication of gay sexuality . . . whether [by] burnings at the stake or injections with chemical suppressants. You," the writer added, "just placed a log on the pyre." (Earlier that year PETA and the former tennis star Martina Navratilova, among others, were waging similar attacks on a scientific study of gay sheep, presuming it was a precursor to developing a "treatment" for shutting off homosexuality in human fetuses.)

Still, many people who contacted Featherstone were actually grateful—for the same baseless prospect. Some confessed struggling with feelings for members of the same sex and explained to him, very disarmingly, the anguish they'd been living with and the hope his fruit-fly study finally offered them. There were poignant phone calls from parents, concerned about their gay children. "I felt bad in a way," Featherstone told me. It was hard not to be moved, and he would try to explain the implications, or lack thereof, of his research politely. "But there's also this liberal, modern side of me that's like: 'Take it easy, lady. Let your son be your son.'"

Not long ago, more than two years after the publication of the fruit-fly paper, a woman wrote to Featherstone about her college-aged daughter. The daughter couldn't shake an attraction to other girls but honestly felt she'd never be able to bring herself to accept it either. She was now contemplating suicide. "She feels that she is losing herself," the mother wrote, "that sweet, innocent light that is within her." Like many who reached out to Featherstone, the woman and her daughter seemed to take for granted that homosexuality was inborn—natural. Otherwise the situation wouldn't feel so torturously unfair. The mother begged Featherstone to rethink his unwillingness to turn his fruit-fly research into a treatment. "We all deserve a choice," she wrote.

Grasping for parallels with animals can create emotional truths, though it usually results in slushy logic. It's naive to slap conclusions about a given species directly onto humans.

But it's disingenuous to ignore the possibility of any connection. "A lot of zoologists are suspicious, I think, of applying the same evolutionary principles to humans that they apply to animals," Paul Vasey, the Japanese-macaque researcher, told me. There's an understandable tendency among some scientists to play down those links to stave off ideological misreading and controversy. "But broadly speaking, research on animals can inform research on humans," Vasey says. What we learn about one species can expand or reorient our approach to others; a well-supported finding about one animal's behavior can generate new hypotheses worth testing in another. "My research on Japanese macaques might influence how someone conducts their research on octopus or their research on moose. Or their research on humans," he said. In fact, it has influenced Vasey's own research on humans.

Since 2003, in addition to his investigation of female-female macaque sex, Vasey has also been studying a particular group of men in Samoa. "Westerners would consider them the equivalent of gay guys, I guess," he told me—they're attracted exclusively to other men. But they're not considered gay in Samoa. Instead, these men make up a third gender in Samoan culture, not men or women, called *fa'afafine*. (Vasey warned me that mislabeling the fa'afafine "gay" or "homosexual" in this article would jeopardize his ability to work with them in the future: while there's no stigma attached to being fa'afafine in Samoan culture, homosexuality is seen as different and often repugnant, even by some fa'afafine.)

In a paper published earlier this year, Vasey and one of his graduate students at the University of Lethbridge, Doug P. VanderLaan, report that fa'afafine are markedly more willing to help raise their nieces and nephews than typical Samoan uncles: they're more willing to baby-sit, help pay school and medical expenses, and so on. Furthermore, this heightened altruism and affection is focused only on the fa'afafine's nieces and nephews. They don't just love kids in general. They are a kind of superuncle. This offers support for a hypothesis that has been toyed around with speculatively since the 1970s, when E. O. Wilson raised it: if a key perspective of evolutionary biology urges us to understand homosexuality in any species as a beneficial adaptation—if the point of life is to pass on one's genes—then maybe the role of gay individuals is to somehow help their family members generate more offspring. Those family members will, after all, share a lot of the same genes.

Vasey and VanderLaan have also shown that mothers of fa'afafine have more kids than other Samoan women. And this fact supports a separate, existing hypothesis: maybe there's a collection of genes that, when expressed in a male, make him gay, but when expressed in a woman, make her more fertile. Like Wilson's theory, this idea was also meant to explain how homosexuality is maintained in a species and not pushed out by the invisible hand of Darwinian evolution. But unlike Wilson's hypothesis, it doesn't try to find a sneaky way to explain homosexuality as an evolutionary adaptation; instead, it imagines homosexuality as a byproduct of an adaptation. It's not too different from how Vasey explains why his female macaques insistently mount one another.

"What we're finding in Samoa now," Vasey told me, "is that it's not an either-or." Neither of the two hypotheses on its own can neatly explain the existence, or evolutionary contribution, of fa'afafine. "But when you put the two together," he said, "the situation becomes a whole lot more nuanced." It's significant that Vasey began his work in Samoa only after he'd gotten to the crux of the macaque situation. "The Japanese macaques," he told me, "in terms of my personal development, they raised my awareness of the possibility that homosexual behavior might not be an adaptation. I was more likely to put the two hypotheses together because I was just more sensitive, I guess, to the reality that the world . . . is organized so that adaptations and byproducts of adaptations coexist and hinge and impinge on each other. Humans are just another species."

Vasey and VanderLaan's work in Samoa doesn't come close to settling theoretical questions about homosexuality. But unlike many biologists I spoke to, Vasey still seemed at ease discussing the speculative and even philosophical ties between animal and human sexuality. He's not concerned with how foolishly or maliciously his work might be misread. "If somebody wanted to make something out of it, they could," Vasey told me, "but they'd just look like some kind of misinformed hillbilly."

Thus far, interpretations of his latest paper on the fa'afafine have been wildly contradictory but all equally overconfident. "New Gay Study Will Make Anti-Gay Activists Cry Uncle," one blog headline read. Another claimed, "Darwinian Fundamentalists Desperate to Rationalize Homosexuality," and cleared the way for a commenter to somehow bemoan Vasey's findings as "justification" for gay men "to sexually abuse their nephews."

"There's two mating right there," Lindsay Young called out.

They were right below her, ten yards away on a flat, vegetated ridge. It was late afternoon. One albatross lay on its stomach, wobbling with its wings pulled back—the way penguins slide over ice —while a second stood upright behind it, fat rippling down its telescoping neck, as it pumped its pelvis. "That looks pretty standard," Young said.

The birds carried on for a while. Then the male shivered and retracted. The female came to her feet and walked off. Young read the female's leg band with her binoculars. "You just hit the jackpot," she told me. The bird was part of a female-female pair. The male had another mate.

Young started scribbling notes, and we sat there rapidly rehashing the details. The sex didn't seem forced at all. In a rape, Young said—which, for all the talk of albatross monogamy, is not uncommon in the species—a male will pin a female's neck to the ground, or back her into a bush to tangle her up. (One study observed four different gangs of males forcing themselves on a single female, which lost an eye in the process.) But these two birds hardly seemed in a rush. Young made more notes. Then, with the male bird frozen right where he'd been left, the female slapped her rubbery feet on the ground, caught an updraft, and disappeared over the ocean.

The next morning Young still seemed to be assuring herself that her interpretation of what we'd seen was reasonable. "We didn't see how it started, but how it ended looked . . ."—she searched for a precise, nonanthropomorphic phrase. She couldn't really find one, and let out a self-effacing laugh. "Mutually beneficial?" she said. "I don't know!"

Dave Leonard, a friend of Young's, was tagging along. Leonard —tall, lanky, and tan, with a ponytail and a few days of scruff—is an ornithologist but works a desk job now for a state wildlife agency and seemed to be enjoying a morning outside. He brandished a gigantic telephoto lens in all directions and had trouble recovering after realizing he'd forgotten to pack his binoculars. Leonard knows his birds, but he was here as a bird lover, not a bird researcher, and wasn't overly concerned with scientific detachment. When Young pointed out a male albatross whinnying at every female that passed overhead, Leonard shook his head and joked, "I feel your pain, dude."

Eventually Young spotted a female from one of the female-female pairs calling to a male about fifteen feet away. The female was standing right where the male and his partner usually build their nest. Her head was straight up in the air, and she clapped her beak animatedly. In Young's experience, it was rare for a bird to call so determinedly to another that's not her partner; this would definitely count as "solicitation," she said, if the two birds wound up copulating. "Pull up a rock," she told me and Leonard.

We sat on the ground expectantly for a while. Eventually the male albatross took a few steps toward the calling female. Then it stopped and looked around. It was comical, given the circumstances.

"'Will anyone see me if I cheat?'" Young said. "I'm not sure if he's taking her up on it or just going, 'Why are you in my spot?'" She was doing the bird's interior monologue, narrating for one blameless, anthropomorphic moment.

The male stopped again and tucked his beak into the feathers behind his neck. Then he turned around and retreated. The taut sexual anticipation—at least as felt by us three humans—seemed to let up. "Well, his partner should be very proud of the self-control," Young said. Then she said, "I know when to cut my losses," gathered up her backpack and clipboard full of hard-earned data, and trudged off to watch some other birds.

More than 4,000 miles across the Pacific, at a place called Taiaroa Head in southeastern New Zealand, two female Royal albatrosses (a related species) were building their nest. Later that winter those two birds would become one of only a few known female-female pairs to successfully fledge a chick at Taiaroa Head in more than sixty years of continuous observation of the colony. (Two years before, the same two birds had engaged in a threesome, presiding over a single nest with the help of one male—just another "alternative mating strategy" albatrosses sometimes engage in, it turns out.)

The tourism board of Dunedin, a gay-friendly region of New Zealand, held a publicity-grabbing contest to name the "lesbian albatross" couple's chick. For months, as the paired females incubated their egg, a press officer at Tourism Dunedin issued releases, and news organizations around the world, from England to India, ran with the story. The PR woman also tried to interest me in a story about a flightless kakapo bird in the region named Sirocco who'd recently made a memorable appearance on the BBC—"He actually started to shag the presenter, Mark Carwardine!" she wrote to me—and "has avid followers on Facebook and Twitter!"

A biologist working with the albatrosses at Taiaroa Head, Lyndon Perriman, seemed to bristle at the idea of naming any albatrosses—"They are wild birds," he wrote to me in an e-mail message. He noted that the female-female pair made for an inconvenient tourist attraction because their nest was not visible from any of the public viewing areas. It seemed fitting: people's ideas about the couple were riveting enough; it wasn't necessary to see the actual birds. The chick hatched on February 1. Tourism Dunedin named it Lola. The shortlist also included Rainbow, Lady Gagabatross, and Ellen.

GEORGE MUSSER

Could Time End?

FROM *Scientific American*

IN OUR EXPERIENCE, nothing ever really ends. When we die, our bodies decay and the material in them returns to the earth and the air, allowing for the creation of new life. We live on in what comes after. But will that always be the case? Might there come a point sometime in the future when there is no "after"? Depressingly, modern physics suggests the answer is yes. Time itself could end. All activity would cease, and there would be no renewal or recovery. The end of time would be the end of endings.

This grisly prospect was an unanticipated prediction of Einstein's general theory of relativity, which provides our modern understanding of gravity. Before that theory, most physicists and philosophers thought time was a universal drumbeat, a steady rhythm that the cosmos marches to, never varying, wavering, or stopping. Einstein showed that the universe is more like a big polyrhythmic jam session. Time can slow down, or stretch out, or let it rip. When we feel the force of gravity, we are feeling time's rhythmic improvisation; falling objects are drawn to places where time passes more slowly. Time not only affects what matter does but also responds to what matter is doing, like drummers and dancers firing one another up into a rhythmic frenzy. When things get out of hand, though, time can go up in smoke like an overexcited drummer who spontaneously combusts.

The moments when that happens are known as singularities. The term actually refers to any boundary of time, be it beginning or end. The best known is the big bang, the instant 13.7 billion years ago when our universe—and, with it, time—burst into existence and began expanding. If the universe ever stops expanding and starts contracting again, it will go into something like the big bang in reverse — the big crunch — and bring time crashing to a halt.

Time needn't perish everywhere. Relativity says it expires inside black holes while carrying on in the universe at large. Black holes have a well-deserved reputation for destructiveness, but they are even worse than you might think. If you fell into one, you would not only be torn to shreds, but your remains would eventually hit a singularity at the center of the hole, and your timeline would end. No new life would emerge from your ashes; your molecules would not get recycled. Like a character reaching the last page of a novel, you would not suffer mere death but existential apocalypse.

It took physicists decades to accept that relativity theory would predict something so unsettling as death without rebirth. To this day they aren't quite sure what to make of it. Singularities are arguably the leading reason that physicists seek to create a unified theory of physics, which would merge Einstein's brainchild with quantum mechanics to create a quantum theory of gravity. They do so partly in the hope that they might explain singularities away. But you need to be careful what you wish for. Time's end is hard to imagine, but time's not ending may be equally paradoxical.

Edges of Time

Well before Albert Einstein came along, philosophers through the ages had debated whether time could be mortal. Immanuel Kant considered the issue to be an "antinomy" — something you could argue both ways, leaving you not knowing what to think.

My father-in-law found himself on one horn of this dilemma when he showed up at an airport one evening only to find that his flight had long since departed. The people at the check-in counter chided him, saying he should have known that the scheduled departure time of "twelve A.M." meant the first thing in the morning. Yet my father-in-law's confusion was understandable. Officially there is no such time as "twelve A.M." Midnight is neither ante meridiem nor post meridiem. It is both the end of one day and start of the next. In twenty-four-hour time notation, it is both 2400 and 0000.

Aristotle appealed to a similar principle when he argued that time can have neither beginning nor end. Every moment is both the end of an era and the start of something new; every event is both the outcome of something and the cause of something else. So how could time possibly end? What would prevent the last event in history from leading to another? Indeed, how would you even define the end of time when the very concept of "end" presupposes time? "It is not logically possible for time to have an end," asserts the University of Oxford philosopher Richard Swinburne. But if time cannot end, then the universe must be infinitely long-lived, and all the riddles posed by the notion of infinity come rushing in. Philosophers have thought it absurd that infinity could be anything but a mathematical idealization.

The triumph of the big bang theory and the discovery of black holes seemed to settle the question. The universe is shot through with singularities and could suffer a distressing variety of temporal cataclysms; even if it evades the big crunch, it might get done in by the big rip, the big freeze, or the big brake. But then ask what singularities (big or otherwise) actually are, and the answer is no longer so clear. "The physics of singularities is up for grabs," says Lawrence Sklar of the University of Michigan at Ann Arbor, a leading philosopher of physics.

The very theory that begat these monsters suggests they cannot really exist. At the big bang singularity, for example, relativity theory says that the precursors of every single galaxy we see were squashed into a single mathematical point—not just a tiny pinprick but a true point of zero size. Likewise, in a black hole, every single particle of a hapless astronaut gets compacted into an infinitesimal point. In both cases, calculating the density means dividing by zero volume, yielding infinity. Other types of singularities do not involve infinite density but an infinite something else.

Although modern physicists do not feel quite the same aversion to infinity that Aristotle and Kant did, they still take it as a sign they have pushed a theory too far. For example, consider the standard theory of ray optics taught in middle school. It beautifully explains eyeglass prescriptions and funhouse mirrors. But it also predicts that a lens focuses light from a distant source to a single mathematical point, producing a spot of infinite intensity. In reality, light gets focused not to a point but to a bull's-eye pattern. Its intensity may be high but is always finite. Ray optics errs because light is not really a ray but a wave.

In a similar vein, nearly all physicists presume that cosmic singularities actually have a finite, if high, density. Relativity theory errs because it fails to capture some important aspect of gravity or matter that comes into play near singularities and keeps the density under control. "Most people would say that they signal that the theory is breaking down there," says physicist James B. Hartle of the University of California, Santa Barbara.

To figure out what goes on will take a more encompassing theory, a quantum theory of gravity. Physicists are still working on such a theory, but they figure that it will incorporate the central insight of quantum mechanics: that matter, like light, has wavelike properties. These properties should smear the putative singularity into a small wad, rather than a point, and thereby banish the divide-by-zero error. If so, time may not, in fact, end.

Physicists argue it both ways. Some think time does end. The trouble with this option is that the known laws of physics operate within time and describe how things move and evolve. Time's end points are off the reservation; they would have to be governed not just by a new law of physics but by a new type of law of physics, one that eschews temporal concepts such as motion and change in favor of timeless ones such as geometric elegance. In one proposal in 2007, Brett McInnes of the National University of Singapore drew on ideas from the leading candidate for a quantum theory of gravity—string theory. He suggested that the primordial wad of a universe had the shape of a torus; because of mathematical theorems concerning tori, it had to be perfectly uniform and smooth. At the big crunch or a black-hole singularity, however, the universe could have any shape whatsoever, and the same mathematical reasoning need not apply; the universe would in general be extremely raggedy. Such a geometric law of physics differs from the usual dynamical laws in a crucial sense: it is not symmetrical in time. The end wouldn't just be the beginning played backward.

Other quantum gravity researchers think that time stretches on forever, with neither beginning nor end. In their view the big bang was simply a dramatic transition in the eternal life of the universe. Perhaps the prebangian universe started to undergo a big crunch and turned around when the density got too high—a big bounce. Artifacts of this prehistory may even have made it through to the present day. By similar reasoning, the singular wad at the heart of a black hole would boil and burble like a miniaturized star. If you fell into a black hole, you would die a painful death, but at least your timeline would not end. Your particles would plop into the wad and leave a distinct imprint on it, one that future generations might see in the feeble glow of light the hole gives off.

By supposing that time marches on, proponents of this approach avoid the need to speculate about a new type of law of physics. Yet they, too, run into trouble. For instance, the universe gets steadily more disordered with time; if it has been around forever, why is it not in total disarray by now? As for a black hole, how would the light bearing your imprint possibly manage to escape the hole's gravitational clutches?

The bottom line is that physicists struggle with antinomy no less than philosophers have. The late John Archibald Wheeler, a pioneer of quantum gravity, wrote, "Einstein's equation says 'this is the end' and physics says 'there is no end.'" Faced with this dilemma, some people throw up their hands and conclude that science can never resolve whether time ends. For them the boundaries of time are also the boundaries of reason and empirical observation. But others think the puzzle just requires some fresh thinking. "It is not outside the scope of physics," says physicist Gary Horowitz of the University of California, Santa Barbara. "Quantum gravity should be able to provide a definite answer."

How Time Slips Away

The HAL 9000 may have been a computer, but he was probably the most human character in *2001: A Space Odyssey*—expressive, resourceful, a bundle not just of wires but also of contradictions. Even his death was evocative of human death. It was not an event but a process. As Dave slowly pulled out his circuit boards, HAL lost his mental faculties one by one and described how it felt. He articulated the experience of regression in a way that people who die are often unable to. Human life is a complex feat of organization, the most complex known to science, and its emergence or submergence passes through the twilight between life and not life. Modern medicine shines a lantern into that twilight, as doctors save premature babies who once would have been lost and bring back people who have passed what was once a point of no return.

As physicists and philosophers struggle to grasp the end of time, many see parallels with the end of life. Just as life emerges out of lifeless molecules that organize themselves, time might emerge from some timeless stuff that brings itself to order. A temporal world is a highly structured one. Time tells us when events occur, for how long and in what order. Perhaps this structure was not imposed from the outside but arose from within. What can be made can be unmade. When the structure crumbles, time ends.

By this thinking, time's demise is no more paradoxical than the disintegration of any other complex system. One by one, time loses its features and passes through the twilight from existence to non-existence.

The first to go might be its unidirectionality—its "arrow" pointing from past to future. Physicists have recognized since the mid-nineteenth century that the arrow is a property not of time per se but of matter. Time is inherently bidirectional; the arrow we perceive is simply the natural degeneration of matter from order to chaos, a syndrome that anyone who lives with pets or young children will recognize. (The original orderliness might owe itself to the geometric principles that McInnes conjectured.) If this trend keeps up, the universe will approach a state of equilibrium, or "heat death," in which it cannot possibly get any messier. Individual particles will continue to reshuffle themselves, but the universe as a whole will cease to change, any surviving clocks will jiggle in both directions, and the future will become indistinguishable from the past. A few physicists have speculated that the arrow might reverse, so that the universe sets about tidying itself up, but for mortal creatures whose very existence depends on a forward arrow of time, such a reversal would mark an end to time as surely as heat death would.

Losing Track of Time

More recent research suggests that the arrow is not the only feature that time might lose as it suffers death by attrition. Another could be the concept of duration. Time as we know it comes in amounts: seconds, days, years. If it didn't, we could tell that events occurred in chronological order but couldn't tell how long they lasted. That scenario is what the University of Oxford physicist Roger Penrose presents in a book published in 2010, *Cycles of Time: An Extraordinary New View of the Universe.*

Throughout his career, Penrose really seems to have had it in for time. He and the University of Cambridge physicist Stephen Hawking showed in the 1960s that singularities do not arise only in special settings but should be everywhere. He has also argued that matter falling into a black hole has no afterlife and that time has no place in a truly fundamental theory of physics.

In his latest assault, Penrose begins with a basic observation about the very early universe. It was like a box of Legos that had just been dumped out on the floor and not yet assembled—a mishmash of quarks, electrons, and other elementary particles. From them, structures such as atoms, molecules, stars, and galaxies had to piece themselves together step by step. The first step was the creation of protons and neutrons, which consist of three quarks apiece and are about a femtometer (10^{-15} meter) across. They came together about 10 microseconds after the big bang (or big bounce, or whatever it was).

Before then, there were no structures at all—nothing was made up of pieces that were bound together. So there was nothing that could act as a clock. The oscillations of a clock rely on a well-defined reference such as the length of a pendulum, the distance between two mirrors, or the size of atomic orbitals. No such reference yet existed. Clumps of particles might have come together temporarily, but they could not tell time, because they had no fixed size. Individual quarks and electrons could not serve as a reference, because they have no size, either. No matter how closely particle physicists zoom in on one, all they see is a point. The only sizelike attribute these particles have is their so-called Compton wavelength, which sets the scale of quantum effects and is inversely proportional to mass. And they lacked even this rudimentary scale prior to a time of about 10 picoseconds after the big bang, when the process that endowed them with mass had not yet occurred.

"There's no sort of clock," Penrose says. "Things don't know how to keep track of time." Without anything capable of marking out regular time intervals, either an attosecond or a femtosecond could pass, and it made no difference to particles in the primordial soup.

Penrose proposes that this situation describes not only the distant past but also the distant future. Long after all the stars wink out, the universe will be a grim stew of black holes and loose particles; then even the black holes will decay away and leave only the particles. Most of those particles will be massless ones such as photons, and again clocks will become impossible to build. In alternative futures where the universe gets snuffed out by, say, a big crunch, clocks don't fare too well, either.

You might suppose that duration will continue to make sense in the abstract, even if nothing could measure it. But researchers question whether a quantity that cannot be measured even in principle really exists. To them the inability to build a clock is a sign that time itself has been stripped of one of its defining features. "If time is what is measured on a clock and there are no clocks, then there is no time," says the philosopher of physics Henrik Zinkernagel of the University of Granada in Spain, who also has studied the disappearance of time in the early universe.

Despite its elegance, Penrose's scenario does have its weak points. Not all the particles in the far future will be massless; at least some electrons will survive, and you should be able to build a clock out of them. Penrose speculates that the electrons will somehow go on a diet and shed their mass, but he admits he is on shaky ground. "That's one of the more uncomfortable things about this theory," he says. Also, if the early universe had no sense of scale, how was it able to expand, thin out, and cool down?

If Penrose is on to something, however, it has a remarkable implication. Although the densely packed early universe and ever emptying far future seem like polar opposites, they are equally bereft of clocks and other measures of scale. "The big bang is very similar to the remote future," Penrose says. He boldly surmises that they are actually the same stage of a grand cosmic cycle. When time ends, it will loop back around to a new big bang. Penrose, a man who has spent his career arguing that singularities mark the end of time, may have found a way to keep it going. The slayer of time has become its savior.

Time Stands Still

Even if duration becomes meaningless and the femtoseconds and attoseconds blur into one another, time isn't dead quite yet. It still dictates that events unfold in a sequence of cause and effect that is the same for all observers. In this respect, time is different from space. Two events that are adjacent within time—when I type on my keyboard, letters appear on my screen—are inextricably linked. But two objects that are adjacent within space—a keyboard and a Post-it note—might have nothing to do with each other. Spatial relations simply do not have the same inevitability that temporal ones do.

But under certain conditions, time could lose even this basic ordering function and become just another dimension of space. The idea goes back to the 1980s, when Hawking and Hartle sought to explain the big bang as the moment when time and space became differentiated. In 2007 Marc Mars of the University of Salamanca in Spain and José M. M. Senovilla and Raül Vera of the University of the Basque Country applied a similar idea not to time's beginning but to its end.

They were inspired by string theory and its conjecture that our four-dimensional universe—three dimensions of space, one of time—might be a membrane, or simply a "brane," floating in a higher-dimensional space like a leaf in the wind. We are trapped on the brane like a caterpillar clinging to the leaf. Ordinarily, we are free to roam around our 4-D prison. But if the brane is blown around fiercely enough, all we can do is hold on for dear life; we can no longer move. Specifically, we would have to go faster than the speed of light to make any headway moving along the brane, and we cannot do that. All processes involve some type of movement, so they all grind to a halt.

Seen from the outside, the timelines formed by successive moments in our lives do not end but merely get bent so that they are lines through space instead. The brane would still be 4-D, but all four dimensions would be space. Mars says that objects "are forced by the brane to move at speeds closer and closer to the speed of light, until eventually the trajectories tilt so much that they are in fact superluminal and there is no time. The key point is that they may be perfectly unaware that this is happening to them."

Because all our clocks would slow down and stop, too, we would have no way to tell that time was morphing into space. All we would see is that objects such as galaxies seemed to be speeding up. Eerily, that is exactly what astronomers really do see and usually attribute to some unknown kind of "dark energy." Could the acceleration instead be the swan song of time?

Your Time Is Up

By this late stage, it might appear that time has faded to nothingness. But a shadow of time still lingers. Even if you cannot define duration or causal relations, you can still label events by the time they occurred and lay them out on a timeline. Several groups of string theorists have recently made progress on how time might be stripped of this last remaining feature. Emil J. Martinec and Savdeep S. Sethi of the University of Chicago and Daniel Robbins of Texas A&M University, as well as Horowitz, Eva Silverstein of Stanford University, and Albion Lawrence of Brandeis University, among others, have studied what happens to time at black-hole singularities using one of the most powerful ideas of string theory, known as the holographic principle.

A hologram is a special type of image that evokes a sense of depth. Though flat, the hologram is patterned to make it look as though a solid object is floating in front of you in 3-D space. The holographic principle holds that our entire universe is like a holographic projection. A complex system of interacting quantum particles can evoke a sense of depth—that is to say, a spatial dimension that does not exist in the original system.

But the converse is not true. Not every image is a hologram; it must be patterned in just the right way. If you scratch a hologram, you spoil the illusion. Likewise, not every particle system gives rise to a universe like ours; the system must be patterned just so. If the system initially lacks the necessary regularities and then develops them, the spatial dimension pops into existence. If the system reverts to disorder, the dimension disappears whence it came.

Imagine, then, the collapse of a star to a black hole. The star looks 3-D to us but corresponds to a pattern in some 2-D particle system. As its gravity intensifies, the corresponding planar system jiggles with increasing fervor. When a singularity forms, order breaks down completely. The process is analogous to the melting of an ice cube: the water molecules go from a regular crystalline arrangement to the disordered jumble of a liquid. So the third dimension literally melts away.

As it goes, so does time. If you fall into a black hole, the time on your watch depends on your distance from the center of the hole, which is defined within the melting spatial dimension. As that dimension disintegrates, your watch starts to spin uncontrollably, and it becomes impossible to say that events occur at specific times or objects reside in specific places. "The conventional geometric notion of space-time has ended," Martinec says.

What that means in practice is that space and time no longer give structure to the world. If you try to measure objects' positions, you find that they appear to reside in more than one place. Spatial separation means nothing to them; they jump from one place to another without crossing the intervening distance. In fact, that is how the imprint of a hapless astronaut who passes the black hole's point of no return, its event horizon, can get back out. "If space and time do not exist near a singularity, the event horizon is no longer well defined," Horowitz says.

In other words, string theory does not just smear out the putative singularity, replacing the errant point with something more palatable while leaving the rest of the universe much the same. Instead it reveals a broader breakdown of the concepts of space and time, the effects of which persist far from the singularity itself. To be sure, the theory still requires a primal notion of time in the particle system. Scientists are still trying to develop a notion of dynamics that does not presuppose time at all. Until then, time clings stubbornly to life. It is so deeply ingrained in physics that scientists have yet to imagine its final and total disappearance.

Science comprehends the incomprehensible by breaking it down, by showing that a daunting journey is nothing more than a succession of small steps. So it is with the end of time. And in thinking about time, we come to a better appreciation of our own place in the universe as mortal creatures. The features that time will progressively lose are prerequisites of our existence. We need time to be unidirectional for us to develop and evolve; we need a notion of duration and scale to be able to form complex structures; we need causal ordering for processes to be able to unfold; we need spatial separation so that our bodies can create a little pocket of order in the world. As these qualities melt away, so does our ability to survive. The end of time may be something we can imagine, but no one will ever experience it directly, any more than we can be conscious at the moment of our own death.

As our distant descendants approach time's end, they will need to struggle for survival in an increasingly hostile universe, and their exertions will only hasten the inevitable. After all, we are not passive victims of time's demise; we are perpetrators. As we live, we convert energy to waste heat and contribute to the degeneration of the universe. Time must die that we may live.

JILL SISSON QUINN

Sign Here If You Exist

FROM *Ecotone*

THE FEMALE GIANT ICHNEUMON WASP flies, impressively for her near-eight-inch length, with the light buoyancy of cottonwood fluff, seemingly without direction, simply aloft. Despite her remarkable size, she is not bulky. Her three-part body makes up only about three inches of her total length, and is disproportionately slender; her thorax is connected to her abdomen by a Victorian-thin waist. Most of her maximum eight-inch span consists of an ovipositor half that length, which extends from the tip of her abdomen and trails behind her like a thread loose from a pant hem. Fully extended, she can be nearly as long as your *Peterson's Field Guide to Insects*.

Her overall appearance of fragility—the corseted middle, the filamentous tail—portrays in flight a façade of drifting. But both of the times I have seen a giant ichneumon wasp she was on a mission, in search of something very specific: a single species among the 1,017,018 described species of insects in the world (91,000 in the United States, 18,000 in Wisconsin, where I observed my second giant ichneumon). To comprehend this statistic, there are many things one needs to know: the definition of an insect, Linnaean taxonomy, the function of zero, the imaginary borders of states and countries. The female ichneumon wasp knows none of this. Yet it can locate a larva of the pigeon horntail—a type of wood wasp whose living body will nourish her developing young—hidden two inches deep in the wood of a dead tree in the middle of a forest.

Charles Darwin himself, it turns out, studied the ichneumon wasp. He mentions it specifically in an 1860 letter to the biologist

Asa Gray, a proponent of the idea that nature reveals God's benevolence. Darwin, on the other hand, swayed no doubt by the rather macabre details of this parasitic insect's life, writes: "I cannot persuade myself that a beneficent and omnipotent God would have designedly created the Ichneumonidae with the express intention of their feeding within the living bodies of Caterpillars"—and then, as if to reach the layman, he adds, "Or that a cat should play with mice." The tabby that curls in your lap and licks your temple, after all, has likely batted a live mouse between its paws until its brain swelled and burst. And the larvae of the giant ichneumon wasp eat, from the inside out and over the course of an entire season, the living bodies of the larvae of a fellow insect.

Like Darwin, I think I have put to rest my belief in a beneficent and omnipotent God—in any God, really. Contrary to what I once believed, it is easy to let go of God, whose essence has never been more than ethereal anyway, expanding like an escaping gas into the corners of whatever church you happened to attend, into the breath of whatever frightened, gracious, or insomnious prayer you found yourself emitting. But it is much more difficult to truly put to rest the belief in an afterlife, the kind where you might get to visit with all your dead friends and relatives. It will not be easy to let go of your deceased mother, who stands in her kitchen slicing potatoes and roast, who hacks ice from the sidewalk with shovels; she is marrow and bone, a kernel of morals, values, and lessons compacted like some astronomical amount of matter into tablespoons, one with sugar for your cereal, another for your fever, with a crushed aspirin and orange juice. You love her. You mark time and space by her: she is someone you are always either near to or very far from.

Can people live without the comfort of a creator? I think so. But relinquishing God—the Christian God, at least—does not leave everything else intact. A lack of the divine probably means that when you die what you consider your essence will cease to exist. You will no longer be able to commune with the people you love. Choosing to live without the assurance of an afterlife, therefore, feels like a kind of suicide, or murder.

Most parasites do not kill their hosts. You—your living, breathing self—are evidence of this, as you host an array of parasitic mi crobes. Only about 10 percent of the hundred trillion cells in your body are really your own; the rest are bacteria, fungi, and other "bugs." The majority of these microbes are mutualistic, meaning that both you and the microbe benefit from your relationship. A whopping 3.3 pounds of bacteria, representing five hundred separate species, live inside your intestines. You provide them with a suitable environment—the right moisture, temperature, and pH —and feed them the carbohydrates that you take in. They shoot you a solid supply of vitamins K and B_{12}, and other nutrients. But some microbes, like the fungi *Trichophyton* and *Epidermophyton*, which might take up residence beneath your toenail as you shower at the gym, are parasitic—they benefit from you, but you are harmed in some way by them. In the case of these two fungi, you would experience itching, burning, and dry skin. But you've probably never heard of anyone dying from athlete's foot, because it has never happened. Successful parasites—parasites that want to stay alive and reproduce—in general do not kill their hosts.

The giant ichneumon wasp is one of a few parasites that break this rule. Actually, it is not a parasite at all; it is more correctly called a parasitoid because its parasitism results in the death of the host. This is not to say the ichneumon wasp is not successful. It can afford to kill its host because its host has a very fast reproduction rate. If we did not have the ichneumon wasp, we also might not be living in wooden houses, because the wood-boring insects that these wasps parasitize would probably have killed all the trees. The wasp might look formidable, but in terms of its ecological role, it is a friend to humans.

This is what it does: A new giant ichneumon wasp hatches from its egg in a dark, paneled crib deep inside a dead or dying tree where the pregnant female placed it. Nearby, or sometimes directly beneath the egg just deposited, lies an unsuspecting horntail larva that has been chewing its cylindrical channels in the wood for sometimes two years. The wasp baby latches on to the exterior of the caterpillar and feeds on its fat and unvital organs until both are ready to metamorphose into adults. Then, when the host has chewed the pair nearly to the surface of the tree, and the giant ichneumon wasp larva, which cannot chew wood, has a clear exit, the ichneumon kills and consumes its host. The wasp metamorphoses, possibly over the course of an entire winter, then emerges. Often before the newly metamorphosed females have even passed through their exit holes, they will mate with one of the plethora of males that have alighted on the bark for just this purpose. It's a kind of ichneumon *quinceañera*, a spontaneous debutante ball.

The problem of where I would go after I died began with simple arithmetic. In our family there were five—my mother, my father, my two older sisters, and me. Yet the world never seemed to divide by fives or threes as easily as it did by twos: I stood *between* the double sinks my sisters occupied when we brushed our teeth; the chair where I sat for breakfast, lunch, and dinner was pulled up to our oval table just for meals, positioned at a point not opposite anyone, and then pushed away when we were done—it didn't even match our first dining room set; I sat in the middle of the back seat of the car, while my sisters each got a window; and when we bought a dozen donuts, the last two always had to be divided, somehow, into five equal pieces—or three, which was no easier, if my parents were dieting. At some point in my childhood, for some unknown reason —I have asked them, and they still can't say why—my parents bought four burial plots. I couldn't make any sense of this. I worried. Where would the last one of us who died—probably me—be laid to rest? All I could foresee was my parents and sisters lined up neatly next to one another for eternity. All I could do was fear my impending, everlasting physical absence from the people I loved the most. Now that my sisters and I have married and they have had children and I have moved away, I realize the accounting error was not in buying too few but in buying too many: there will likely be two empty plots next to my parents. I've become accustomed to physical distance from my nuclear family by settling eight hundred miles from where I grew up, but the problem of where I will go after I die, what I will be like, and who will be with me has not gone away It has only magnified.

Megarhyssa, the Latin name for the genus to which the giant ichneumon wasp belongs, translates to "large-tailed." The species that I saw was likely the most common of the eighteen species of this genus, *Megarhyssa macrurus,* which translates to "large-tailed, long-tailed." These genus and species names, then, provide no information that an observer couldn't pick up in a single, fleeting interaction with the insect itself. The tail is more precisely called an ovipositor, an appendage used by many female insects—and some

fish and other creatures—to place their eggs in a required location. That place might be soil, leaf, wood, or the body (inside or
out) of another species.

The ichneumon's process of depositing eggs with her long ovipositor goes from mystical to complicated to bizarre. First, she lo-
cates her host by sensing vibrations made from its chewing beneath
the wood. Her antennae stretch out before her like dowsing rods,
occasionally tapping the bark, and she divines the presence of the
horntail, catches it snacking like a child beneath the bed sheets
who has made a midnight trip to the kitchen. She "listens" for the
subsurface mastication of an individual caterpillar encapsulated in
old wood.

Now the pregnant female begins the increasingly complex actions that will transport the eggs from her body through as much
as two inches of woody tissue to the horntail's empty channel.
Keeping her head and thorax parallel to the wood, which she grips
with her legs, she first curves her abdomen under, into a circle,
touching its tip to her thin waist. Her ovipositor, as if its outrageous
length were not surreal enough, now performs a magician's feat: it
separates into three long threads. The center one is the true ovipositor; the other two are protective sheaths that will help steady
the insect's abdomen and guide the ovipositor as it enters the
wood. (When she is finished laying and flies off, you will sometimes
see these three threads trailing separately behind her.) The two
sheaths, one on each side, fold back and follow the curve of her
abdomen, then come together again at the very tip of her thorax
and head straight for the wood, sandwiching her body in two
broadly looped capital Ps. The ovipositor extends directly into the
two sheaths where they join and disappears between them. In order to allow the ovipositor's acrobatics, the exoskeleton at the tip
of the abdomen splits somewhat and pulls back. At this stage in her
laying, with her ovipositors perpendicular to the tree, her wings
flat and still, and her legs spread-eagle, the ichneumon looks as if
she has pinned herself to the wood as an entomologist might pin
her to a cork for observation.

Before we hang up from our once-weekly phone call, my mother
says she has one more little story to tell. This one is about Kristen,
my niece, at age five my mother's youngest granddaughter.

The week before Easter, she and Kristen drove to the church

where my grandparents and my mother's little brother, who died when he was a baby, are buried. My mother wanted to put flowers on the headstones. Before they got out of the car, Kristen began talking about her own mother and her older sister, Katie.

"Mommy and Katie want the same," Kristen said, "but I want to be different."

"What do you mean?" my mother asked.

"I want to be buried," Kristen replied. "But Mommy and Katie want their bones . . ." She paused for a minute, thinking, then continued. "They want their bones burned." Kristen paused again, then concluded, "But, really, I don't want to die."

My mother said she had to stifle a laugh. And I laughed, too, when she related Kristen's words. Yet I can't help but think our laughter was cover for some deeply rooted disquiet. It's merely the brain's best method for dealing with this cruel yet basic fact of life —that it ends—stated here so rationally by a little person just in the process of recognizing it.

My mother, always prepared for the teachable moment, put forward to Kristen, "Well, Jesus is going to give you your body back, you know."

Kristen was not appeased. "That's weird," she replied.

The very intricacy—and weirdness—of the ichneumon's egg laying makes it difficult for most of us not to wonder who came up with the complex series of steps involved. Part of that is because humans seem to be, as professor of psychology Paul Bloom puts it, "natural-born creationists." His essay "In Science We Trust," from the May 2009 issue of *Natural History,* posits that where humans see order—anything that is not random—we immediately assume that an intelligent being has created that order. Bloom sums up the research beautifully: children aged three to six who were shown pictures of both neat and messy piles of toys, along with a picture of a teenage girl and a picture of an open window with curtains blowing, reported that both the sister *and* the wind could have caused the messy pile, but only the sister could have stacked the toys neatly; likewise, shown a cartoon of a neat pile of toys created by a rolling ball, babies as young as one year old stared longer than normal, which, according to developmental psychologists, indicates surprise.

I once found at the mouth of a sizable hole along a favorite trail a mashed garter snake, a flattened mole, and a deceased opossum. They were each uneaten and—I knew intuitively—could not possibly have all died there coincidentally. Rather, I soon found out, they were a stack of "toys," planted neatly by a mother and father fox at the den entrance, to occupy their kits in the dusk and dawn while the parents hunted and scavenged for food. (When the family moves to a new den, which they frequently do, the parents will actually move the toys as well.) A pile of sticks pointed on both ends, with the bark removed to reveal the white wood underneath, mortared together with mud and lined up across a stream, has never been the work of the wind in the entire history of the earth, but always the work of an intelligent being— *Castor canadensis,* the American beaver.

But being "created" does not inherently imply the existence of a creator, as evidenced in Darwin's work on Natural Selection. Bloom explains, "Darwin showed how a nonintelligent process driven by random variation and differential selection can create complex structure—design without a designer." So this instinctive assumption that complexity is the work of an intelligent being is true *most,* but not *all,* of the time.

Natural Selection, though, in itself, does not inherently negate the existence of a creator. It is possible to imagine that a creator put into motion several set laws—the laws of Newton, for instance, and the laws of Natural Selection—then, without interfering, let creation unspool itself.

But even this belief begs a question. I asked my mother this question once, when I was seven or eight. We were in the car, on the way home from my organ lesson. "What was there before God?" I asked. "Who created him?"

"There was nothing," my mother said, and her hands left the steering wheel for a moment. Her fingers spread, like the fingers of an illusionist, as if she were scattering something, everything in the known world, I guess. These religious discussions of ours were delicate and infrequent, almost, like discussions of sex in our family, too intimate to occur between parent and child. When we did have them, it felt as if we were too close to uncovering something —for her, something too hallowed to be near; for me, something possibly too tragic. "I know," she conceded, "it's hard to imagine."

But I did imagine it, using the only sequencing skill I had then: a

two-frame comic strip. In the right frame there was a profile of a
cartoon God, and in the left frame just blackness.

Megarhyssa macrurus is a mixture of mustard yellow and auburn,
with chestnut brown accents. From a distance, the wasp may look
just dark, but pinned, as the female is during egg laying, and patient, as the male is when waiting for the virgins to emerge, you can
easily get close enough to notice the mostly yellow legs, yellow-and-
auburn-striped abdomen, and brown antennae and wing veins.

When the female is well into her egg laying, and possibly at the
point of no return, she becomes even more colorful and, at the
same time, more bizarre. We left her with her three tails separate
and in position, and her abdomen curled in a downward circle.
Once it is time to deposit the eggs, she uncurls and raises her abdomen so that it is nearly perpendicular to the tree and her body.
Her tails remain in their same positions. But two of the segments
near the tip of her abdomen open wide, like the first cut in an im-
promptu self–cesarean section, revealing a thin yellow membrane.
The membrane, taut like the surface of a balloon, is about two centimeters in diameter. It pumps gently. It is as attention-getting as
a peacock's display, but wetter, more intimate. Within that membrane you can see what look like portions of the ichneumon's three
tails as they exist *inside* her body. Though the ovipositor appears to
begin at the tip of her abdomen, as an appendage — like an arm
or a leg or a tail — it must in fact be more tonguelike, and extend
into her inner recesses. It's as if you're witnessing an X-ray, but
even so, it's very difficult to figure out exactly what is going on.
There are too many parts, too many steps, too much intertwining.
Watching the ichneumon lay her eggs is like trying to decipher
one of those visual-spatial problems on an IQ test: if the following
object is rotated once to the left, and twice vertically, will it look
like option A, B, or C? Give me the 3.5 billion years that Natural
Selection has had — whether here or in the afterlife — and I just
might figure it out.

Belief in an afterlife, and the manner of behavior, prayers, rituals,
and burial practices necessary for navigating one's way to it, can be
considered a universal in human cultures. But belief in an afterlife
cannot be considered the essence of all religions. Certainly there

were cultures obsessed with it—the Egyptians, for instance, who took part in elaborate processes of mummification in order to preserve the dead and aid them in making the physical journey to heaven. But, hard as it may be for Christians, for whom a belief in resurrection and the afterlife takes center stage, to understand, other cultures and religions either simply didn't address the afterlife or had a less-than-attractive view of it. Those who originally penned the Hebrew Bible, for example, did not conceive of any type of survival after death; God harshly punished those who did not listen to his Word in this life with plagues, fevers, famine, and exile, and rewarded those who did with immortality only through their physical descendants. Were Natural Selection an option for the early Hebrews, I believe they would have been more accepting of the theory than today's Americans.

Other cultures did conceive of an afterlife, but not the type that came as a reward for moral behavior or religious faith or acceptance of a certain savior. For the Babylonians and the ancient Greeks, immortality was reserved for the gods alone. Death for mortals meant a sort of eternal, shadelike, underground existence, where food and water would be merely sufficient. Incidentally, the Babylonian "afterlife" was so unappealing that it actually became the paradigm for hell in Christianity.

The concept that an afterlife is a reward for, or at least related to, moral acts carried out in this life was made popular by Plato and later by Judaism, Hinduism, Buddhism, Christianity, and Islam. In Hinduism and Buddhism, one can achieve immortality only by breaking the cycle of rebirth, something I am not sure, were I Hindu or Buddhist, I would even want to do. (The only thing more comforting to me than a religion with an afterlife would be the ability to exist on earth forever; returning even as a dung beetle could be quite exhilarating for someone who'd already had thrills at observing eight-inch wasps in this life.)

An appendix to *How Different Religions View Death and Afterlife,* by Christopher Jay Johnson and Marsha G. McGee, contains my entire former worldview. In response to the question "Will we know friends and relatives after death?" the spokesman surveyed on behalf of the United Methodist Church says: "We will know friends and relatives in the afterlife and may know and love them more perfectly than on earth." This was a worldview I picked up during

sixteen years of thirty-minute weekly lessons in a tiny basement
Sunday school room at Patapsco United Methodist Church, where
my mother was the organist and my parents purchased their bewildering number of burial plots. The church was high on a hill above
a creek and across from a junkyard, whose collage of rusted colors
I viewed every week through the window during Sunday services:
old cars tethered to the earth by kudzu and honeysuckle, seemingly inert, but easily unfettered when a father or brother came in
search of a hubcap, a passenger's door. Belief in an afterlife has
been the grounding expectation of my existence, the hope I find
so hard to give up even after giving up the Father, the Son, and the
Holy Ghost.

Before and even throughout my adolescence I was a believer. I'd
always held a sort of patient expectation for the Second Coming or
some other miracle. As soon as I learned to write, I tried to speed
things up a bit. *Sign here if you exist,* I wrote to God on lined white
paper in a collage of yellow capital and lowercase letters. The color,
which made the note barely legible, was not chosen for its symbolic connotation—enlightenment—but, rather, never even considered, in the way that children, caught up in the greatness of an
act, overlook the details necessary to achieve it. I slid the note under my dresser, and checked it every day. One morning, I found
that God had answered.

There came a moment of astonishment, then almost assurance,
when I pulled the note from its hiding place. But too soon I recognized the blue Bic pen, the neat, curvy letters, the same arcs from
the thank-you notes Santa wrote for the cookies we'd left him. Perhaps I would have believed it was God who'd answered if my mother
had simply done what the letter requested. Instead she wrote me a
note about love and faith, probably a series of X's and O's, like she
put in our Valentine's cards. She was, and is, a believer; she would
not forge his name.

Tim Lewens, author of *Darwin,* a book that deals with the impact
Darwin's thinking has had on philosophy during the last 150 years,
has discussed the very same question I asked my mother in the car
as a child on the way home from my organ lesson. Although the
question applies to any type of creator, in Lewens's interview for

the Darwin Correspondence Project he specifically addresses the idea of the laissez-faire God who sets up the laws of physics and of Natural Selection and then lets them do their own work, the kind of God who might appeal to most scientists, the God that Darwin himself, Lewens says, likely believed in.

To deal with this question, Lewens draws from the rationale of the philosopher David Hume. If you subscribe to this type of God, you are still left with the question of who or what was responsible for God, and who or what was responsible for whoever or whatever was responsible for God, and so on down the line, endlessly. At some point, Lewens says, if you want to be a theist you have to stop asking the question of what came before God or created him and just accept his existence as—in Lewens's own words—a sort of "brute, inexplicable fact." And if you allow for the existence of brute, inexplicable facts, then you might as well just accept the brute, inexplicable laws of physics and Natural Selection. If the only purpose for a creator is to set in motion the laws of science, Lewens asks, then why on earth do you need one? According to Hume and Lewens, whether God exists or not doesn't solve anything.

When the question of a creator's existence became for me just a matter of semantics and personification, it was easy enough to put the idea of that creator to rest. Would the ichneumon become any less immanent if it were created not by some*one* called God but by some*thing* called Natural Selection? Both warrant capitalization as text, and both require faith of a sort. This part of the equation is easy, but I find it much harder to let go of the one thing God gave me that I coveted: an afterlife, and a clear path to it. I blame nature for this.

About a week after my second sighting of an ichneumon, I encounter another on the same path, on the same tree, in what I have come to call the "pinned" position. I wait for her to curve her abdomen up, split it open, and reveal its inner workings, but I see no movement. I poke at her gently with a twig. She barely stirs. (I have read that occasionally during the drilling process, which can take half an hour, a wasp's ovipositor will become stuck in the wood, and she will be left there, a snack for some predator that will pluck her body from her tail as if detaching a bean from its thin tendril.

The ovipositor is left protruding from the wood like a porcupine quill.) A week later she is gone, and yet another ichneumon is performing her ancient task, already flexing the yellow circle. She is close to the ground, and the leaf of a small plant is obstructing my view. When I attempt to move it, my thumb and forefinger coming at the wasp like a set of pincers, she bats at me with her front legs, then takes off, detaching herself fully from the wood. Her membrane looks like a tiny kite. Her ovipositor and its sheaths, as well as her body, still warped into laying position, are like distorted and cumbersome tails, yet she flies, unimpeded, up and up, as if to another world.

Once, when I was nine or ten, I opened the screen door to share with my dog, who was lying on the back porch, the remnants of a grilled-cheese sandwich that I couldn't finish. The family parakeet, Sweetie, was perched on my head. The scene must have looked very Garden of Eden–esque: a primate accompanied by a parrot feeding a canine. But I'd forgotten about the bird in my hair, and when I opened the door he escaped to a high branch on one of the oaks that grew behind the clothesline. A crow landed next to him, looking huge and superhero-ish. But nothing could coax him to return, not even his open cage, which for the next several days we stood next to in the yard, calling his name. Immediately after his escape, I had run into the forest in tears and was gone for the afternoon. Later, on the porch steps, I asked my mother if Sweetie would go to heaven.

Like many years before, she could not lie.

"The Bible tells us that animals don't have souls," she replied. Perhaps seeing my devastation, she added, "But it also says that God knows when even the smallest sparrow falls."

According to Alan F. Segal, in *Life After Death*, my position is not unique: more Americans believe in an afterlife than in God himself. Furthermore, the General Social Survey he cites, conducted by the National Opinion Research Center, shows that Jewish belief in an afterlife has jumped from 17 percent (as recorded by those born between 1900 and 1910) to 74 percent (as recorded by those born after 1970). Segal's hypothesis, which is that a culture's conception of the afterlife reveals what that culture most values as a

society, fits right in with these statistics. Americans define themselves as defenders of freedom and individual rights. We believe we
should be happy, wealthy, and healthy all the time. Why, then, even
after giving up God himself, or even when subscribing to a religion
that doesn't pay much attention to the afterlife, would we consent
to imprison ourselves with mortality? Why would we give up our
individual right to eternal life?

Paul Bloom would attribute my tendency to believe in life after
death simply to being human. Humans, Bloom maintains, seem to
be born already believing in an afterlife. In his essay "Is God an
Accident?" published in *The Atlantic* in December 2005, he argues
that humans are natural dualists. He does not mean that we are
born with a Zoroastrian belief in the opposing forces of good and
evil, but that we hold two operating systems in our minds—one
with expectations for physical objects (things fall down, not up)
and another with expectations for psychological and/or social beings (people make friends with people who help them, not people
who hurt them). These expectations are not learned but built-in
and can be observed in babies as young as six months old. These
two distinct, implicit systems cause us to conceptualize two possible
states of being in the universe: soulless bodies and, no less possible,
just the opposite—bodiless souls. The two systems are separate.
Therefore, when a human's body dies, humans are predisposed to
believe that the soul does not, necessarily, die with it. Bloom cites a
supporting study by the psychologists Jesse Bering and David Bjork-
lund, in which children who were told a story about a mouse that
was eaten by an alligator rightfully believed that the mouse's ears
no longer worked after death and that the mouse would never
need to use the bathroom again. But more than half of them believed the mouse would still feel hungry, think thoughts, and desire things. Children, more so even than adults, seem to perceive
psychological properties as existing in a realm outside the body
and therefore believe these properties exempt from death. According to Bloom, "The notion that life after death is possible is not
learned at all. It is a by-product of how we naturally think about the
world."

I want my mother to read Bloom's essay. I'm not sure why.
Throughout her childhood my mother walked of her own volition
up the hill from her house to go to the little church across from

the junkyard. Her mother and sister went; her father and much younger brother did not. When she was nine or ten she was chosen by Mrs. Myrtle, the church organist, to receive free piano lessons. My mother had an old upright piano in her family's unheated front room. Some of the keys did not play, so the ramshackle instrument was eventually chopped up for firewood. Then she had to practice at Mrs. Myrtle's, and was sometimes politely sent home when she played too long. My mother has been the volunteer organist at Patapsco United Methodist Church for more than thirty years.

She is sixty-two years old. Why, at this late stage, do I want to push Bloom's essay into the face of her contentment? Why does it feel slightly cruel, as though I am a bully, and why do I want to do it anyway?

I bring "Is God an Accident?" with me during a summer visit. The first page has been somehow lost on the plane and I have to make a special trip to the local library to download and print out another copy. I'm hesitant to give my mother the essay, but when I find her one afternoon on the couch preparing for her Bible study — they've reread the entire book of Genesis, she says — it feels like an open invitation. She puts the essay to the side to read later, and the discussion that ensues between us is as unfulfilling, as inconclusive, as this discussion is everywhere — in high school biology classrooms; in rural, local papers' letters to the editor. It ends only with her decreeing, exasperated, "Your faith is stronger than mine." She does not mean my faith in Natural Selection, but my faith in God. She believes, perhaps counterintuitively, that all of my research and questions indicate some sort of allegiance to the religion I was raised with, or, at the very least, an inability to simply cast it aside.

My mother could live for forty more years, another entire life, longer than my own life has been so far. Or she could be dead in just eight years, at seventy, before her youngest grandchild reaches high school, possibly before she meets any child of mine. The current life expectancy for American women lies somewhere in the middle of that. Of course I know I could die in the next fifteen minutes from a brain aneurysm or be murdered by the man tuning my piano, but those are exceptions, and each would be a shock. If I were certain of the afterlife, what should not be a shock, what should be normal, is this: *our mothers are going to die, and for a while we are going to have to live without them.* But if I abandon my belief in life after death, am I putting my mother to rest before I really have to?

My mother tells me about an e-mail she sent to another of my nieces, Julie, who is seventeen. Julie had just lost her guinea pig, Charlie, a faithful pet for seven years. Twice in the last month she had found him lying in his cage, with his neck askew, unable to move. The second time, he died the next morning.

My mother wrote to Julie that she would never forget her cutting up vegetables for Charlie, making his daily salad. *He had a good life,* she wrote, *and you just might see him again one day.*

I can't help but notice the difference between my mother's un-solicited response to the death of Charlie and her answer to my di-rect question years ago of whether my escaped and presumed-dead parakeet would eventually make it to heaven. Perhaps, with age, my mother's views have softened a bit. Honesty has become less im-portant than comfort. What she would like to believe has super-seded what she once took as fact. We come to see that death is less about losing the self than about losing what was built between selves when they were alive. When she dies, my mother will be in a casket on the hill outside Patapsco United Methodist Church. The question I would like to ask is: Will we find each other again? But that is not it at all. Rather, I must find a way to live now knowing that one day we will no longer be mother and daughter.

Now that I have discovered the ichneumon nursery standing in my woods, I will revisit it often. In spring I want to see the newly metamorphosed wasps born from the horntail's tunnels, these bur-rows that double as grave and womb. I'd like to see how many will emerge, how big they are, how long they rest on the dying tree be-fore their first, seraphic flight. I'd like to search for whether there is any evidence of the horntail larva that nourished them—an exo-skeleton, perhaps—or whether the larva is now present some-how only in the new body it has helped to form. All winter I will anticipate the decaying oak's promise: the giant ichneumon wasps' emergence.

There is life, it seems, after death—but it may be only here on earth. Nature provides too many metaphors for us to so easily give up on this idea. It pummels us with them season after season, and has done the same to others, I suppose, for thousands of years, playing on the human mind's ability to compare unlike things in its search for truth. I once shook a nuthatch from torpor on the side of a red pine, where it had stood unmoving, face-down—as only its kind can—all through my breakfast after a night of below-zero weather. I lifted a mourning dove, seemingly frozen into my cross-country ski trail during a surprise snow squall, in my own two palms, from which it flapped as if from Noah's hands. In my parents' woods, where my mother pitched a flower left from her mother's funeral, a patch of daffodils came up the next year—and has come up every year since—on its own, through a foot of dry leaf litter, shaded and unwatered. And there is the freezing and thawing of wood frogs in northern climes; the hibernation of chipmunks and groundhogs and jumping mice; the estivation of turtles in summer; metamorphoses of all kinds, but in particular the monarch butterfly (what Sunday school classroom has not used this metaphor at Easter?); the dormancy of winter trees; the phases of the moon; the seasons themselves; sleeping and waking; monthly bleeding; and, though it is too early to remember, probably birth even.

This is what it comes down to. My mother, in giving me life after birth, also engendered in me the idea of God, and the unstoppable desire for life after death. We live on an earth where it seems nearly impossible for humans to have ever avoided inventing heaven, an earth that throws things back at us so reliably it is hard not to imagine that one day we will be resurrected, too, and that we will live forever. *Sign here if you exist,* I once wrote, but the question of whether God exists is really a question of whether *we* do. Without God and the promise of resurrection, you become extremely short-lived. Or the other option: you live forever, but what you currently perceive as yourself is a mere phase, a single facet: once oak, now horntail, soon-to-be ichneumon.

What type of afterlife do I need to survive—not so much in the next life, I realize now, but to get me through this one? I got my elements from stars: mass from water, muscles from beans, thoughts from fish and olives. When Edward Abbey died, his body was buried in nothing more than an old sleeping bag in the southern Arizona desert. He said, "If my decomposing carcass helps nourish the roots of a juniper tree or the wings of a vulture—that is im-

mortality enough for me. And as much as anyone deserves." Abbey
is right. Your trillions of cells—only 10 percent of which, remember, were yours anyway—will become parts of trillions of things.
And even the 10 percent wasn't really yours to begin with. You were
only borrowed. We've had it backward all along: the body is immortal—it is the soul that dies.

OLIVER SACKS

Face-Blind

FROM *The New Yorker*

IT IS WITH OUR FACES that we face the world, from the moment of birth to the moment of death. Our age and our gender are printed on our faces. Our emotions, the open and instinctive emotions that Darwin wrote about as well as the hidden or repressed ones that Freud wrote about, are displayed on our faces, along with our thoughts and intentions. Though we may admire arms and legs, breasts and buttocks, it is the face, first and last, that is judged "beautiful" in an aesthetic sense, "fine" or "distinguished" in a moral or intellectual sense. And, crucially, it is by our faces that we can be recognized as individuals. Our faces bear the stamp of our experiences and our character; at forty, it is said, a man has the face he deserves.

At two and a half months, babies respond to smiling faces by smiling back. "As the child smiles," Everett Ellinwood writes, "it usually engages the adult human to interact with him—to smile, to talk, to hold—in other words, to initiate the processes of socialization . . . The reciprocal understanding mother-child relationship is possible only because of the continual dialogue between faces." The face, psychoanalysts consider, is the first object to acquire visual meaning and significance. But are faces in a special category as far as the nervous system is concerned?

I have had difficulty recognizing faces for as long as I can remember. I did not think too much about this as a child, but by the time I was a teenager, in a new school, it was often a cause of embarrassment. My frequent inability to recognize schoolmates would cause bewilderment and sometimes offense—it did not occur to them (why should it?) that I had a perceptual problem. I usually recognized close friends without much difficulty, especially my two best friends, Eric Korn and Jonathan Miller. But this was partly because I identified particular features: Eric had heavy eyebrows and thick spectacles, and Jonathan was tall and gangly, with a mop of red hair. Jonathan was a keen observer of postures, gestures, and facial expressions, and seemingly never forgot a face. A decade later, when we were looking at old school photographs, he still recognized literally hundreds of our schoolmates, while I could not identify a single one.

It was not just faces. When I went for a walk or a bicycle ride, I would have to follow exactly the same route, knowing that if I deviated even slightly I would be instantly and hopelessly lost. I wanted to be adventurous, to go to exotic places—but I could do this only if I bicycled with a friend.

At the age of seventy-seven, despite a lifetime of trying to compensate, I have no less trouble with faces and places. I am particularly thrown if I see people out of context, even if I have been with them five minutes before. This happened one morning just after an appointment with my psychiatrist. (I had been seeing him twice weekly for several years at this point.) A few minutes after I left his office, I encountered a soberly dressed man who greeted me in the lobby of the building. I was puzzled as to why this stranger seemed to know me, until the doorman greeted him by name—it was, of course, my analyst. (This failure to recognize him came up as a topic in our next session; I think that he did not entirely believe me when I maintained that it had a neurological basis rather than a psychiatric one.)

A few months later, my nephew Jonathan Sacks came for a visit. We went for a walk—I lived in Mount Vernon, New York, at the time—and it started raining. "We had better get back," Jonathan said, but I couldn't find my house or my street. After two hours of walking around, during which we both got thoroughly soaked, I heard a shout. It was my landlord; he said that he had seen me pass the house three or four times, apparently failing to recognize it.

In those years I had to take the Boston Post Road to get from Mount Vernon to the hospital where I worked, on Allerton Avenue in the Bronx. Though I took the same route twice a day for eight years, the road never became familiar to me, I never recognized the buildings on either side, and I often took the wrong turn, realizing it only when I came to one of two landmarks that were unmistakable, even for me: at one end, Allerton Avenue, which had a large sign, or the Bronx River Parkway, which loomed over the Boston Road.

I had been working with my assistant, Kate, for about six years when we arranged to rendezvous in a midtown office for a meeting with my publisher. I arrived and announced myself to the receptionist, but failed to note that Kate had already arrived and was sitting in the waiting area. That is, I saw a young woman there but did not realize that it was her. After about five minutes, smiling, she said, "Hello, Oliver. I was wondering how long it would take you to recognize me."

Parties, even my own birthday parties, are a challenge. (More than once, Kate has asked my guests to wear name tags.) I have been accused of "absent-mindedness," and no doubt this is true. But I think that a significant part of what is variously called my "shyness," my "reclusiveness," my "social ineptitude," my "eccentricity," even my "Asperger's syndrome," is a consequence and a misinterpretation of my difficulty recognizing faces.

My problem with recognizing faces extends not only to my nearest and dearest but also to myself. Thus, on several occasions I have apologized for almost bumping into a large bearded man, only to realize that the large bearded man was myself in a mirror. The opposite situation once occurred at a restaurant. Sitting at a sidewalk table, I turned toward the restaurant window and began grooming my beard, as I often do. I then realized that what I had taken to be my reflection was not grooming himself but looking at me oddly.

In 1988 I met Franco Magnani, the "memory artist," and during the next couple of years I spent weeks with him, talking about his paintings, his life, and even traveling to Italy with him to visit the village where he grew up. When I finally submitted an article about him to *The New Yorker*, Robert Gottlieb, who was then the magazine's editor-in-chief, read the piece and said, "Very nice, fascinating—but what does he *look* like? Can you add some description?" I parried this awkward (and, to me, unanswerable) question by saying, "Who cares what he looks like? The piece is about his work."

"Our readers will want to know," Bob said. "They need to picture him."

"I will have to ask Kate," I said. Bob gave me a peculiar look.

*

I assumed that I was just very bad at recognizing faces, as my friend Jonathan was very good at it—that this was within the limits of normal variation, and that he and I just stood at opposite ends of a spectrum. It was only when I went to Australia to visit my older brother Marcus, whom I had scarcely seen in thirty-five years, and discovered that he, too, had exactly the same difficulties recognizing faces and places, that it dawned on me that this was something beyond normal variation, that we both had a specific trait, a so-called prosopagnosia, probably with a distinctive genetic basis.

(Our other two brothers seemed to have normal powers of facial recognition. My father, a general practitioner, was immensely gregarious and seemed to know hundreds of people, not to mention the thousands of patients in his practice. My mother, in contrast, seemed almost pathologically shy. She had a small circle of intimates—family members and colleagues—and was very ill at ease in large gatherings. I cannot help wondering, in retrospect, if some of her "shyness" was due to a mild prosopagnosia.)

That there were others like me was brought home in various ways. The meeting of two people with prosopagnosia, in particular, can be very challenging. A few years ago, I wrote to one of my colleagues to tell him that I admired his new book. His assistant then phoned Kate to arrange a dinner, and they settled on a weekend meeting at a restaurant in my neighborhood.

"There may be a problem," Kate said. "Dr. Sacks cannot recognize anyone."

"It's the same with Dr. W.," his assistant replied.

"And another thing," Kate added. "Dr. Sacks cannot find restaurants or other places; he gets lost very easily—he can't even recognize his own building sometimes."

"Yes, it's the same with Dr. W.," his assistant said.

Somehow, we did manage to meet, and enjoyed dinner together. But I have no idea what Dr. W. looks like, and he probably wouldn't recognize me, either.

Although such examples may seem comical, they are sometimes quite devastating. A person with very severe prosopagnosia may be unable to recognize his spouse or to pick out his own child in a group of people.

Jane Goodall also has a certain degree of prosopagnosia. Her problems extend to recognizing chimpanzees as well as people—

thus, she says, she is often unable to distinguish individual chimps by their faces. Once she knows a particular chimp well, she ceases to have difficulties; similarly, she has no problem with family and friends. But, she says, "I have huge problems with people with 'average' faces . . . I have to search for a mole or something. I find it very embarrassing! I can be all day with someone and not know them the next day."

She added that she, too, has difficulty recognizing places: "I just don't know where I am until I am very familiar with the route. I have to turn and look at landmarks so I can find my way back. This was a problem in the forest, and I often got lost."

In 1985 I published a case history called "The Man Who Mistook His Wife for a Hat," about Dr. P., who had a very severe visual agnosia. He was not able to recognize faces or their expressions. Moreover, he could not identify, or even categorize, objects; thus, he was unable to recognize a glove, to distinguish it as an article of clothing, or to perceive that it resembled a hand.

After Dr. P.'s story was published, I began to get letters from correspondents who would compare their difficulties in recognizing places and faces with his. In 1991 Anne F. wrote to me, describing her experiences:

I believe that three people in my immediate family have visual agnosias: my father, a sister, and myself. We each have traits in common with your Dr. P., but, hopefully, not to the same degree. The most striking behavior we all share in common with Dr. P. is the prosopagnosia. My father, a man who has had a successful radio career here in Canada (his particular gift is an ability to mimic voices), was unable to recognize his wife . . . in a recent photograph. At a wedding reception, he asked a stranger to identify the man sitting next to his daughter (my husband of five years at the time).

I have walked past my husband, while staring directly at his face, on several occasions without recognizing him. I have no difficulty recognizing him, however, in situations or places where I am expecting to see him. I am also able to recognize people immediately when they begin to speak, even if I've heard their voice only once in the past.

Unlike Dr. P., I feel I can read people well on an emotional level . . . I don't have the degree of agnosia for common objects that Dr. P. had. [However], like Dr. P., I am totally incapable of establishing a topographical representation of space . . . I have no memory for where I put things unless I verbally encode the location. Once an object leaves my hands, it drops off the edge of the world into a void.

While Anne F. seems to have prosopagnosia and topographical agnosia on a genetic or familial basis, others may develop this (or any other form of agnosia) in consequence of a stroke, a tumor, an infection, or an injury—or, like Dr. P., a degenerative disease such as Alzheimer's—that has damaged a particular part of the brain. Joan C., another correspondent, had an unusual history in this regard: she had developed a brain tumor in the right occipital lobe, and this was removed when she was fourteen. It seems likely, though it is difficult to be certain, that her prosopagnosia was the result of the tumor or the surgery. Her inability to recognize faces has often been misinterpreted by others: "I've been told that I'm rude, or a space cadet, or (according to a psychiatrist) suffering from a psychiatric disorder."

As I continued to receive more and more letters from people with prosopagnosia or topographical agnosia, it became clear to me that "my" visual problem was not uncommon and must affect many people around the world.

Face recognition is crucially important for humans, and the vast majority of us are able to identify thousands of faces individually or to easily pick out familiar faces in a crowd. A special expertise is needed to make such distinctions, and this expertise is nearly universal, not only in humans but in other primates. How, then, do people with prosopagnosia manage?

In the past few decades, we have become very conscious of the brain's plasticity—how one part or system of the brain may take over the functions of a defective or damaged one. But this does not seem to occur with prosopagnosia or topographical agnosia—they are usually lifelong conditions that do not lessen as one grows older. People with prosopagnosia, therefore, need to be resourceful and inventive in finding strategies for circumventing their deficits: recognizing people by an unusual nose or beard, for example, or by their spectacles or a certain type of clothing.

Many prosopagnosics recognize people by voice, posture, or gait; and, of course, context and expectation are paramount—one expects to see one's students at school, one's colleagues at the office, and so on. Such strategies, both conscious and unconscious, become so automatic that people with moderate prosopagnosia can remain unaware of how poor their facial recognition actually is, and are startled if it is revealed to them by testing (for example, with photographs that omit ancillary clues like hair or eyeglasses).

The artist Chuck Close, who is famous for his gigantic portraits of faces, has severe, lifelong prosopagnosia. He believes it has played a crucial role in driving his unique artistic vision. "I don't know who anyone is and essentially have no memory at all for people in real space," he says. "But when I flatten them out in a photograph I can commit that image to memory."

Perhaps this "flattening" allows him to memorize certain features. Although I myself may be unable to recognize a particular face, I can recognize various things *about* a face: that there is a large nose, a pointed chin, tufted eyebrows, or protruding ears. Such features become the identifying markers by which I recognize people. (I think that for similar reasons I find it easier to recognize a caricature than a straightforward portrait or a photograph.) I am reasonably good at judging age and gender, though I have made a few embarrassing blunders. I am far better at recognizing people by the way they move, their "motor style." And even if I cannot recognize particular faces, I am sensitive to the beauty of faces, and to their expressions.

I avoid conferences, parties, and large gatherings as much as I can, knowing that they will lead to anxiety and embarrassing situations, since I not only fail to recognize people that I know well but tend also to greet strangers as old friends. (Like many prosopagnosics, I avoid greeting people by name, lest I use the wrong one, and I depend on others to save me from egregious social blunders.)

I am much better at recognizing my neighbors' dogs (they have characteristic shapes and colors) than my neighbors themselves. Thus, when I see a youngish woman with a Rhodesian ridgeback hound, I realize that she lives in the apartment next to mine. If I see an older lady with a friendly golden retriever, I know that this is someone from down the block. But if I should pass either woman on the street without her dog she might as well be a complete stranger.

*

The idea that "the mind," an immaterial, airy thing, could be em-
bodied in a lump of flesh—the brain—was intolerable to seven-
teenth-century religious thinking; hence the dualism of Descartes
and others. But physicians, observing the effects of strokes and
other brain injuries, had long had reason to suspect that the functions of the mind and the brain were linked. Toward the end of
the eighteenth century, the anatomist Franz Joseph Gall proposed
that all mental functions must arise from the brain—not from the
"soul," as many people imagined, or from the heart or the liver.
Instead, he envisioned within the brain a collection of twenty-seven
"organs," each responsible for a different moral or mental faculty.
Such faculties, for Gall, included what we now call perceptual functions, such as the sensation of color or sound; cognitive faculties,
like memory, mechanical aptitude, and speech and language; and
even "moral" traits, such as friendship, benevolence, and pride.
(For these heretical ideas, he was exiled from Vienna and would
eventually end up in revolutionary France, where he hoped a more
scientific approach might be embraced.)

The physiologist Marie Jean Pierre Flourens decided to investigate Gall's theory by removing slices of the brain in living animals,
chiefly pigeons. But he could not find any evidence to correlate
specific areas of the cortex with specific faculties (perhaps because
one needs very delicate and discrete ablations in order to do so,
especially in the tiny pigeon cortex). The cortex, he concluded,
was equipotential, as homogeneous and undifferentiated as the
liver. "The brain," it is said, "secretes thought as the liver secretes
bile."

Flourens's notion of an equipotential cortex dominated thought
until the studies of Paul Broca in the 1860s. Broca performed au-
topsies on many patients with expressive aphasia, all of whom, he
showed, had damage that was limited to the frontal lobes on the
left side. In 1865 he was able to say, famously, "We speak with the
left hemisphere," and the notion of a homogeneous and undifferentiated brain, it seemed, was laid to rest. Broca felt that he had
located a "motor center for words" in a particular part of the left
frontal lobe (an area that we now call Broca's area).

This seemed to promise a new sort of localization, a genuine correlation of neurological and cognitive functions with specific centers in the brain. Neurology moved confidently ahead, identifying

"centers" of every sort: Broca's motor center for words was followed by Wernicke's auditory center for words, and Déjerine's visual center for words—all in the left hemisphere, the language hemisphere —and a center for visual recognition in the right hemisphere.

But while visual agnosia of a general sort was recognized in the 1890s, there was little notion that there could be agnosia for particular visual categories, such as faces or places, until 1947, when Joachim Bodamer, a German neurologist, described three patients who were unable to recognize faces but had no other difficulties with recognition. It seemed to Bodamer that this highly selective form of agnosia needed a special name—it was he who coined the term "prosopagnosia"—and that such a specific loss must imply that there was a discrete area in the brain specialized for face recognition. This has been a matter of dispute ever since: is there a system dedicated only to face recognition, or is face recognition simply one function of a more general visual-recognition system? A 1955 paper by the English neurologist Christopher Pallis, with beautiful detail and documentation, brought the issue to the fore.

Pallis's patient, A. H., was a mining engineer at a Welsh colliery who had kept a journal and was able to give Pallis an articulate and insightful description of his experiences. One night in June 1953, A. H. apparently suffered a stroke. He "suddenly felt unwell after a couple of drinks at his club." He appeared to be confused and was taken home to bed, where he slept poorly. Getting up the following morning, he found his visual world completely transformed, as he reported to Pallis:

> I got out of my bed. My mind was clear but I could not recognize the bedroom. I went to the toilet. I had difficulty in finding my way and recognizing the place. Turning round to go back to bed I found I couldn't recognize the room, which was a strange place to me.
>
> I could not see colour, only being able to distinguish light objects from dark ones. Then I found out all faces were alike. I couldn't tell the difference between my wife and my daughters. Later I had to wait for my wife or mother to speak before recognizing them. My mother is 80 years old.
>
> I can see the eyes, nose, and mouth quite clearly but they just don't add up. They all seem chalked in, like on a blackboard.

*

Since Pallis's time, a number of patients with prosopagnosia have come to autopsy. Here the data are clear: virtually all patients with prosopagnosia, irrespective of the cause, have lesions in the right visual-association cortex, in particular on the underside of the occipitotemporal cortex. There is nearly always damage in a structure called the fusiform gyrus, and these autopsy results gained additional support in the 1980s, when it became possible to visualize the brain of living patients by CT scanning and MRI—here, too, prosopagnosic patients showed lesions in what came to be called the "fusiform face area."

In the 1990s, such lesion studies were complemented by functional imaging—visualizing the brains of people with fMRI as they looked at pictures of faces, places, and objects. These functional studies demonstrated that looking at faces activated the fusiform face area much more strongly than looking at other test images did.

That individual neurons in this area could show preferences was first demonstrated in 1969 by Charles Gross, using electrodes in the inferotemporal cortex of macaques. Gross found cells that responded dramatically to the sight of a monkey's paw—but also, less strongly, to a variety of other stimuli, including a human hand. Subsequently, he found cells that had a relative preference for faces.

Much that we now take for granted in neuroscience was very unclear when Gross began this work. Even in the late 1960s, it was widely believed that the visual cortex did not extend far beyond its main locus in the occipital lobes (as we now know it does). That the representation and recognition of specific categories of objects—faces, hands, and so on—might rely on individual neurons or clusters of neurons was considered improbable, even absurd; the idea was good-humoredly mocked by Jerome Lettvin in his famous comments about "grandmother cells." Very little attention, therefore, was paid to Gross's early findings, and it was not until the 1980s that they were confirmed and amplified by other researchers.

In humans, some ability to recognize faces is present at birth or soon after. By six months, as Olivier Pascalis and colleagues have shown in one study, babies are able to recognize a broad variety of individual faces, including those of another species (in this study, pictures of monkeys were used). By nine months, though, the babies had become less adept at recognizing the monkey faces unless they had received continuing exposure to them. As early as three months, infants are learning to narrow their model of "faces" to those they are frequently exposed to. The implications of this work for humans are profound. To a Chinese baby brought up in his own ethnic environment, Caucasian faces may all, relatively speaking, "look the same," and vice versa. One prosopagnosic acquaintance, born and raised in China, went to Oxford as a student and has lived in the United States for decades. Nonetheless, he tells me, "European faces are the most difficult—they all look the same to me." It seems that there is an innate and presumably genetically determined ability to recognize faces, and this capacity gets focused in the first year or two, so that we become especially good at recognizing the sorts of faces we are likely to encounter. Our "face cells," already present at birth, need experience in order to develop fully.

The fact that many (though not all) people with prosopagnosia also have difficulty recognizing places has suggested to some researchers that face and place recognition are mediated by distinct but adjacent areas. Others believe that both faculties are mediated by a single zone that is perhaps oriented more toward faces at one end and toward places at the other.

The neuropsychologist Elkhonon Goldberg, however, questions the whole notion of discrete, hardwired centers, or modules, with fixed functions in the cerebral cortex. He feels that at higher cortical levels there may be much more in the way of gradients, where areas whose function is developed by experience and training overlap or grade into one another. In his book *The New Executive Brain,* Goldberg speculates that a gradiential principle constitutes an evolutionary alternative to a modular one, permitting a degree of flexibility and plasticity that would be impossible for a brain that is organized in a purely modular fashion.

While modularity, he argues, may be characteristic of the thalamus—an assemblage of nuclei with fixed functions, fixed inputs and outputs—a gradiential organization is more characteristic of the cerebral cortex, and becomes more and more prominent as one ascends from primary sensory cortex to association cortex and, finally, to the highest level of all: the frontal cortex. Modularity and gradients may thus coexist and complement one another.

Some researchers have proposed that prosopagnosia is not purely a problem with face blindness but one aspect of a more general difficulty in distinguishing the individuals in any class, whether that class consists of faces, cars, birds, or anything else.

Isabel Gauthier and her colleagues tested a group of car experts and a group of expert birders, comparing them with a group of normal subjects. The fusiform face area, they found, was activated when all the groups looked at pictures of faces. But it was also activated in the car experts when they were asked to identify particular cars, and in the birders when they were asked to identify particular birds. The fusiform face area is tuned primarily for facial recognition, it seems, but some of it can be trained to distinguish individual items of other sorts. (If, then, an expert bird spotter or car buff is unlucky enough to acquire prosopagnosia, he will also, we might suspect, lose his ability to identify birds or cars.)

The fusiform face area does not work in isolation; it is a vital node in a cognitive network that stretches from the occipital cortex to the prefrontal area. Face blindness may occur even with an intact fusiform face area, if the lower occipital face areas have been damaged. And people with moderate prosopagnosia, like Jane Goodall or me, can, after repeated exposure, learn to identify those they know best. Perhaps this is because we are using slightly different pathways to do so, or perhaps, with training, we can make better use of our relatively weak fusiform face areas.

Above all, the recognition of faces depends not only on the ability to parse the visual aspects of a face—its particular features and their overall configuration—and compare them with others but also on the ability to summon the memories, experiences, and feelings associated with that face. The recognition of specific places or faces, as Pallis emphasized, goes with a particular feeling, a sense of association and meaning. While purely visual recognition of faces is mediated by the fusiform face area and its connections, emotional familiarity is probably mediated at a higher, multimodal level, where there are intimate connections with the hippocampus and the amygdala, areas that are involved in memory and emotion. Thus A. H., after his stroke, lost not only his ability to identify faces but also this sense of familiarity; every face and place appeared new to him and continued to do so even if seen again and again.

Recognition is based on knowledge, and familiarity is based on feeling, but neither entails the other. The two have different neural bases and can be dissociated; thus, although both are lost in tandem with prosopagnosia, one can have familiarity without recognition or recognition without familiarity in other conditions. The former occurs in instances of déjà vu and also in the "hyperfamiliarity" for faces described by Devinsky. Here a patient may find that everyone on the bus or in the street looks "familiar"—he may go up to them and address them as old friends, even while realizing that he cannot possibly know them all. My father was always very sociable and could recognize hundreds or even thousands of people, but his feeling of "knowing" people became exaggerated, and perhaps pathological, as he moved into his nineties. He often attended concerts at Wigmore Hall, in London, and there, during the intermission, he would accost everyone in sight, saying, "Don't I know you?"

The opposite occurs in people with Capgras syndrome, for whom faces, though recognized, no longer generate a sense of emotional familiarity. Since a husband or wife or child does not convey that special warm feeling of familiarity, the Capgras patient will argue, they cannot be the real thing—they *must* be clever impostors, counterfeits. People with prosopagnosia have insight; they realize that their problems with recognition come from their own brains. People with Capgras syndrome, in contrast, remain immovable in their conviction that they are perfectly normal and it is the other person who is profoundly, even uncannily, wrong.

Individuals with acquired prosopagnosia, like A. H. or Dr. P., are relatively rare—most neurologists are likely to encounter such a patient once or twice in their career, if at all. Congenital prosopagnosia (or, as it is sometimes called, "developmental" prosopagnosia), such as I have, is much more common, yet it remains completely unrecognized by most neurologists. This is not entirely surprising, for people with congenital prosopagnosia do not generally consult neurologists about their "problem." It is just the way they are.

Ken Nakayama, at Harvard, who investigates visual perception, has long suspected that prosopagnosia is relatively common but underreported. In 2001 he and his colleague Brad Duchaine, at

University College, London, began seeking subjects with face blindness through their website, and they received an impressive response. Nakayama and Duchaine are now investigating several thousand people with lifelong prosopagnosia, ranging from mild to cripplingly severe.

While congenital prosopagnosics do not have gross lesions in the brain, a recent study by Garrido and colleagues showed that they do have subtle but distinct changes in the brain's face-recognition areas. The condition also tends to be familial: Duchaine, Nakayama, and their colleagues have described one family in which ten members—both parents and seven of their eight children (the eighth could not be tested), as well as a maternal uncle, have it. Clearly, there are strong genetic determinants at work here.

Nakayama and Duchaine have explored the neural basis of face and place recognition, generating new knowledge and insights at every level from the genetic to the cortical. They have also studied the psychological effects and social consequences of developmental prosopagnosia and topographical agnosia—the special problems these conditions can create for an individual in a complex social and urban culture.

And the range seems to extend in a positive direction, too. Russell, Duchaine, and Nakayama have described "super-recognizers," people with extraordinarily good face-recognition abilities, including some who seem to have indelible memories of virtually every face they have ever seen. Alexandra L., one of my correspondents, described her own uncanny ability to recognize people:

It happened again yesterday. I was on my way down into the subway in Soho when I identified someone fifteen feet ahead of me (back turned, talking intimately with his friend) as a man I knew, or had seen before. In this case, it was Mac, who used to be a family friend's art dealer. I had last seen him (briefly) two years earlier, at an opening in midtown. I'm not sure I've ever spoken with him beyond an introduction a good ten years ago.

This is an integral part of my life—I catch a passing glimpse of someone and, with no real effort, *flash,* place the face—yes, that's the girl who served us wine at an East Village bar last year (again, in a totally different neighborhood, and at night not during the day). It is true that I'm a big fan of people, of humanity and diversity . . . but to my knowledge I make no effort to record the physical traits of ice cream servers, shoe salesmen and friends of friends of friends. Even a slim wedge of face, or the way someone walks two blocks away at dusk, can trigger my mind to zero in on a match.

The super-recognizers, Russell and colleagues write, "are about as good at face recognition and perception as [lifelong] prosopagnosics are bad"—that is, about two or three standard deviations above average, while the most severe prosopagnosics have face-recognizing abilities two or three standard deviations below average. Thus the difference between the best face recognizers and the worst among us is comparable to that between people with an IQ of 150 and those with an IQ of 50. As with any bell curve, the vast majority are somewhere in the middle.

Severe congenital prosopagnosia is estimated to affect 2 to 2.5 percent of the population—6 to 8 million people in the United States alone. (A much higher percentage, perhaps 10 percent, are markedly below average in face identification, but not cripplingly face-blind.) For these people, who have difficulty recognizing their husbands, wives, children, teachers, and colleagues, there is still no official recognition or public understanding.

This is in marked contrast to the situation with another neurological minority, the 10 percent or so of the population with dyslexia. Teachers and others are increasingly aware of the special difficulties, and often the special gifts, that dyslexic children may have and are beginning to provide educational strategies and resources for them.

But for now people with varying degrees of face blindness must rely on their own ingenuity, starting with educating others about their unusual, but not rare, condition. Increasingly, prosopagnosia is also the subject of books, websites, and support groups, where people with face blindness or topographical agnosia are able to share experiences and, no less important, strategies for recognizing faces and places when the usual "automatic" mechanisms have been compromised.

Ken Nakayama, who is doing so much to promote the scientific understanding of prosopagnosia, also has a personal acquaintance with the subject, and posts this notice in his office and on his website (faceblind.org):

> Recent eye problems and mild prosopagnosia have made it harder for me to recognize people I should know. Please help by giving your name if we meet. Many thanks.

EVAN I. SCHWARTZ

Waste MGMT

FROM *Wired*

ON CLEAR WINTER NIGHTS, when the trees are bare, Donald Kessler likes to set up a small telescope on the back deck of his house in Asheville, North Carolina, and zoom in on the stars shining over the Blue Ridge Mountains. It's not the most advanced home observatory, but the retired NASA scientist treasures his Celestron telescope, which was made in 1978. That also happens to be the year Kessler published the paper that made his reputation in aerospace circles. Assigned to the Environmental Effects Project Office at NASA's Johnson Space Center in Houston, the astrophysicist had gotten interested in the junk that humans were abandoning in the wild black yonder—everything from nuts and tools to defunct satellites and rocket stages the size of school buses.

In that seminal paper, "Collision Frequency of Artificial Satellites: The Creation of a Debris Belt," Kessler painted a nightmare scenario: spent satellites and other space trash would accumulate until crashes became inevitable. Colliding objects would shatter into countless equally dangerous fragments, setting off a chain reaction of additional crashes. "The result would be an exponential increase in the number of objects with time," he wrote, "creating a belt of debris around the Earth."

At age thirty-eight, Kessler had found his calling. Not that his bosses had encouraged him to look into the issue—"they didn't like what I was finding," he recalls. But after the paper came out, NASA set up the Orbital Debris Program Office to study the problem and put Kessler in charge. He spent the rest of his career tracking cosmic crap and forming alliances with counterparts in other nations in an effort to slow its proliferation. His description of a runaway cascade of collisions—which he predicted would happen in thirty to forty years—became known as the Kessler syndrome.

While the scenario was accepted in theory by NASA officials, nothing much was done about it. Capturing and disposing of space junk would be expensive and difficult, and the threat was too far in the future to trigger much worry. After Kessler retired in 1996, he grew a trim gray beard, peered through his telescope on those clear nights, and waited. "I knew something would happen eventually," he says.

Then, on February 10, 2009—just a little more than three decades after the publication of his paper—the Kessler syndrome made its stunning debut. Some 500 miles above the Siberian tundra, two satellites were cruising through space, each racing along at about five miles per second. Iridium 33 was flying north, relaying phone conversations. A long-retired Russian communication outpost called Cosmos 2251 was tumbling east in an uncontrolled orbit. Then they collided. The ferocious impact smashed the satellites into roughly 2,100 pieces. Repercussions on the ground were minimal—perhaps a few dropped calls—but up in the sky, the consequences were serious. The wreckage quickly expanded into a cloud of debris, each shard an orbiting cannonball capable of destroying yet another hunk of high-priced hardware.

As Kessler received reports of the collision from former colleagues at NASA, he realized that the situation had played out pretty much as he'd foreseen. After all, he had forecast that the first satellite collision would happen around this time between objects of roughly this mass. Like an opening shot in a war, the crash served as a signal that the syndrome had gone from theory to reality. "Some people weren't aware how fast these objects are going," he says. "At those speeds, even something quite small can create tremendous damage."

Almost immediately after the accident, a military unit called the Space Surveillance Network sprang into action. Run by the Joint Space Operations Center at California's Vandenberg Air Force Base, the network uses a system of radar installations and optical sensors to monitor space junk. Before the Iridium-Cosmos incident, it had been tracking 120 active satellites and worrying about an average of five potential collisions, or "conjunctions," per day.

The crash took everyone by surprise. "It wasn't even on their list of possibilities that day," an Iridium spokesperson says.

The operations center moved quickly to double its computer capacity. By early 2010 it was keeping a close eye on 1,000 active satellites, 3,700 inactive satellites and rocket pieces, and another 15,300 objects the size of a fist or larger—a level of awareness that revealed a much higher daily average of seventy-five possible collisions. And that's ignoring the danger posed by the estimated *half-million* smaller pieces of debris the size of a marble or larger. Too small to track from the ground, each of those tiny projectiles is capable of severely damaging a satellite.

Just a month after the Iridium accident, a stray motor chunk hurtled toward the International Space Station. Cruising at an altitude of 220 miles, astronauts aboard the $100 billion laboratory were going about their daily chores at around noon EDT when they received a warning—prepare for possible impact. The crew was directed to scramble into the station's equivalent of a lifeboat, an attached Russian-made Soyuz capsule. It would give them a chance to abandon ship, if necessary. After a few minutes, the motor zipped by, missing the ISS by just a few miles—in space terms, a close call.

Then on December 1, with almost no warning, a small chunk from a different Cosmos satellite hurtled toward the ISS, coming within a mile of a direct hit. Due to its speeding-bullet velocity, even this fragment could have had an impact equal to that of a truck bomb. "A ten-centimeter sphere of aluminum would be like seven kilograms of TNT," says Jack Bacon, a senior NASA scientist charged with keeping the ISS safe. "It would blow everything to smithereens."

Incidents like these served as clear signs from above that something must finally be done about space junk. Its proliferation threatens not only current and future space missions but also global communications—mobile phone networks, satellite television, radio broadcasts, weather tracking, and military surveillance, even the dashboard GPS devices that keep us from getting lost. The number of manufactured objects cluttering the sky is now expected to double every few years as large objects weaken and split apart and new collisions create more Kesslerian debris, leading to yet more collisions.

NASA's Bacon puts it bluntly: "The Kessler syndrome is in effect. We're in a runaway environment, and we won't be able to use space in the future if we don't start dealing with this now."

Since the dawn of the space age, NASA has operated under what it calls the big sky theory—the notion that, given the vastness of space, it's perfectly fine to discard mission waste or abandoned rocket stages up there. After all, these objects would likely fall out of orbit and burn up as they reentered Earth's atmosphere. The only question was when. Some would take just a few years, while those in higher orbits might not descend for decades. "We didn't think twice about it," says the former astronaut Bryan O'Connor, now one of NASA's top safety officers. Russia and other nations have also been launching missions under the same assumption— that even if a giant solar panel happened to fall off the back of a shuttle, you could simply wish it bon voyage.

In the midst of all this complacency, Kessler wasn't the only voice raising concerns about the big sky theory. Arthur C. Clarke, who in the 1940s conceived of communications satellites, wrote a 1979 novel, *The Fountains of Paradise,* in which all the space junk that had been accumulating "had to be located, and somehow disposed of." He imagined what he dubbed Operation Cleanup: space fortresses armed with high-powered lasers would sweep the skies, vaporizing the debris. If we didn't act, Clarke warned, Earth would be cut off from space and we'd lose the ability to communicate by satellite and explore the heavens. "We would sink back into a dark age," he wrote. "During the resultant chaos, disease and starvation would destroy much of the human race."

To stave off such a day, Kessler and his colleagues at the Orbital Debris Program Office developed some guidelines to slow the accumulation of space junk. The rules limited what could be abandoned, and they required satellite operators to help clean up the crowded geosynchronous belt 22,400 miles above Earth by maneuvering retired spacecraft into slightly higher "graveyard orbits" out of harm's way. By 2008 similar guidelines had been adopted by most of the major space agencies around the world.

The new rules did slow the growth of space debris. And thanks to gravity, stuff continued to fall from the sky. Roughly once a day, an object in the official U.S. catalog of debris drops out of orbit and turns to ash on reentry. Once a week or so, an object that's too big or dense will survive reentry and plunge to Earth, but it typically plops down unnoticed in an ocean or some unpopulated expanse. (There are exceptions: an upper rocket stage once landed in the desert of Saudi Arabia, to the surprise of local shepherds, and in 1997, a steel fuel tank slammed into a yard in Texas.)

Yet much of the progress from the international effort was undone in a single moment. On January 11, 2007, the Chinese government staged a demonstration of its military might by firing a projectile at one of its own retired satellites, the Fengyun-1C, in low Earth orbit. The "kinetic kill vehicle" scored a bull's-eye, blasting the satellite into 3,000 trackable pieces. (Even though its space program is relatively young, China already accounts for 31 percent of all debris traceable to specific launches—comparable to the shares of Russia and the United States.) It wasn't long before the Chinese test started causing trouble. In May 2009 a roughly four-inch fragment from the Fengyun explosion whizzed by the space shuttle *Atlantis,* which at the time was in the vicinity of the Hubble Space Telescope on a repair mission. Had it struck either one, it could have done catastrophic damage.

The Chinese debris, combined with the Iridium-Cosmos collision, finally revealed the bankruptcy of the prevailing philosophy governing space. "The big sky theory is no longer a viable concept for space operations," says Chris Moss, chief of strategy for the military's Joint Space Operations Center. Officials at NASA now acknowledge that orbital debris is the biggest threat to the International Space Station. And the call for action is global, says Heiner Klinkrad, the top debris expert at the European Space Agency: "Debris removal is the only cure to the Kessler syndrome."

Last December, at a Marriott not far from Darpa's headquarters in Arlington, Virginia, about 175 people filed into a meeting hall to take part in the first-ever International Conference on Orbital Debris Removal, sponsored by Darpa and NASA. The gathering attracted a group of idealistic innovators, members of a self-proclaimed "debris community" that no one knew existed in such numbers. "This reminds me of Boy Scouts, with their motto 'Leave no trace,'" said Patrick Moran, an engineer with the California-based Aerospace Corporation. "The same rigor must now be imposed on space." The three-day meeting featured an address by none other than Donald Kessler, the dean of space debris. The official purpose of the conference was an open "call for information," but in some respects it resembled an episode of *American Idol*, with rocket scientists as contestants—and with Kessler, his old NASA colleagues, and Darpa officials serving as judges. Darpa had had only scant involvement with the debris issue, but the agency's director of tactical technology, David Neyland, kicked off the conference by noting that Darpa was created "to look across different tech domains to prevent surprises, and orbital debris is a surprise" —one with national security consequences for U.S. spy satellites.

As in the early rounds of a talent contest, many of the acts didn't seem quite ready for prime time. A librarian from a university about ninety miles from Roswell, New Mexico, proposed launching giant sticky space balls that would adhere to objects and drag them out of orbit. A grad student outlined his plan to attach sails to the space trash, gently floating the pieces Earthward. One engineer from Colorado insisted that a giant inflatable doughnut would do the trick by bouncing rubbish down into the atmosphere.

Then there were the inevitable laser-based approaches. A former Los Alamos National Lab scientist presented a long-standing plan involving a giant laser station on a Hawaiian peak. A professor from Alabama proposed zapping debris with lasers attached to satellites, causing the junk to plummet into the atmosphere.

But three of the plans seemed to hold real promise. The first came from seventy-two-year-old Jerome Pearson, an engineer best known for imagining the space elevator concept in the 1970s. Pearson showed up with an animated demo that looked remarkably like the old *Space Invaders* video game. Now president of the South Carolina–based Star Technology and Research, he proposed a seven-year mission in which a dozen suitcase-sized spacecraft would piggyback on other launches. Every vehicle would hold one hundred lightweight nets, each as big as a house. Onboard video cameras would allow ground crews to drape the nets over hunks of junk by remote control. The craft would drag the debris out of orbit, then return to search for more. Pearson claimed his system could capture 2,500 or so objects in the most crowded and dangerous bands of low Earth orbit. The estimated price tag: $240 million.

The second plan came from NASA's Bacon, who outlined a

scheme to launch a ten-ton mother ship that would hold ten additional tons of mass-produced nanosatellites, stored like larvae in a honeycombed nest. Once released, the nanosats would approach objects, toss lightweight nets over them, and then tow the debris into the atmosphere. Bacon's plan is suitable mainly for pieces of junk lighter than two pounds in low Earth orbit. Heavier items—rocket stages, satellites—would remain. He estimates that each launch of a mother ship would cost $100 million, with as many as twelve missions needed.

The third concept was from Rob Hoyt, the CEO of Tethers Unlimited, a Seattle-area space contractor. For the past fifteen years, Hoyt has been developing a system called Rustler, to "round up space trash [for] low Earth orbit remediation." He envisions midsize space vehicles—four hundred pounds or so—piggybacking on satellite launches. Once in orbit, such a craft would sidle up to junk and attach an electrodynamic tether—a wire-mesh kite tail up to six miles long—then send an amp of current through the material. The current would interact with Earth's magnetic field, producing a drag effect and lowering the debris into the atmosphere. Hoyt's approach drew praise from Kessler and others for its simplicity and relatively low price. According to Hoyt, a test mission to take out a few tons of trash could cost just tens of millions of dollars.

Which brings up the problem of how to pay for deploying any of these technologies. One idea would be to assess a fee on commercial satellite companies—which would benefit, after all, from safer skies. Eventually, aerospace firms may equip their satellites with built-in disposal devices so that they can be pulled out of orbit when their missions are complete. Hoyt has just such a product, called Terminator Tape, a three-pound package containing a tether that can be unfurled automatically using the spacecraft's own electronics.

By the end of the conference, there was a sense that removing the junk is actually possible. "I've gone from being totally skeptical to thinking maybe something will work," Kessler says. "We can bring things down; it's just going to cost a lot."

Eric Christiansen knows just how damaging space junk can be. From his office at NASA's Johnson Space Center in Houston, he directs a team that studies what happens when orbiting objects get whacked, slammed, pierced, and pummeled. His lab has a wonderfully badass name: the Hypervelocity Impact Technology Facility. The impacting actually happens a few hundred miles away, at NASA's testing range near White Sands, New Mexico. There technicians operate a cannon that uses gunpowder and pressurized hydrogen to fire plastic slugs at shields and panels. Just like real space junk, the projectiles can approach speeds of five miles per second.

Christiansen and his colleagues study the results and use their findings to help develop stronger materials and designs for spacecraft. They're also working on shields that might provide protection from stray projectiles. "This is the lightest shield that will stop a two-centimeter projectile," he says, pointing to a multilayer Kevlar and ceramic panel for the ISS.

Some of the mangled hardware is on display at the space center. The most treasured of these objects is a piece of window glass from a 1983 *Challenger* mission. Dinged by debris, it entered the collection when Kessler worked at NASA. "Don gave this to me," Christiansen says, "and he said, 'You protect this and leave it to the next guy.'"

Christiansen collaborates with the Orbital Debris Program Office, which does much of its work in a corrugated metal building. The office—run by Kessler's successor, Nicholas L. Johnson—is improving its ability to forecast crashes and spot some of the smallest space bits using a huge telescope on a Chilean mountaintop, which will be joined in 2011 by a new scope on a Pacific atoll.

This happens to be the day after the Obama administration announced it was canceling Constellation, NASA's grand plan to return astronauts to the moon by 2020. Along with the final phaseout of the Space Shuttle Program and the move to privatize launches, it all seems to represent a massive resetting of NASA's priorities. Perhaps that makes it the opportune time to get serious about tidying up the skies. The main obstacle, of course, is funding. But included in NASA's proposed 2011 budget is a first-ever provision to award $400,000 research grants for orbital-debris removal and other projects. The allocation would be part of a new "mission directorate" called Space Technology, funded to the tune of almost $5 billion over the next five years. "We're hoping NASA will be able to take charge," Star Technology's Pearson says.

For Donald Kessler, now seventy, that would be welcome news. Standing on his back porch, he doesn't need his telescope to spot the man-made spacecraft whizzing overhead. "Anyone can see satellites with the naked eye," he says. "Several pass by every hour." Sooner or later, he says, word will arrive that another one has been smashed and shattered, and the fragments will escalate the danger to all space missions. The man with the syndrome named after him will keep watch and wait for the next call.

SANDRA STEINGRABER

The Whole Fracking Enchilada

FROM *Orion*

I HAVE COME to believe that extracting natural gas from shale using the newish technique called hydrofracking is *the* environmental issue of our time. And I think you should, too.

Saying so represents two points of departure for me. One: I primarily study toxic chemicals, not energy issues. I have heretofore ceded that topic to others, such as Bill McKibben, with whom I share this column space in *Orion*.

Two: I'm on record averring that I never tell people what to do. If you are a mother who wants to lead the charge against vinyl shower curtains, then you should. If the most important thing to you is organic golf courses, then they are. So said I.

But high-volume slick-water hydrofracturing of shale gas—fracking—is way bigger than PVC and synthetic fertilizer. In fact, it makes them both cheaply available. Fracking is linked to every part of the environmental crisis—from radiation exposure to habitat loss—and contravenes every principle of environmental thinking. It's the tornado on the horizon that is poised to wreck ongoing efforts to create green economies, local agriculture, investments in renewable energy, and the ability to ride your bike along country roads. It's worth setting down your fork, pen, cellular phone —whatever instrument you're holding—and looking out the window.

The environmental crisis can be viewed as a tree with two trunks. One trunk represents what we are doing to the planet through atmospheric accumulation of heat-trapping gasses. Follow this trunk along and you find droughts, floods, acidification of oceans, dissolving coral reefs, and species extinctions.

The other trunk represents what we are doing to ourselves and other animals through the chemical adulteration of the planet with inherently toxic synthetic pollutants. Follow this trunk along and you find asthma, infertility, cancer, and male fish in the Potomac River whose testicles have eggs inside them.

At the base of both these trunks is an economic dependency on fossil fuels, primarily coal (plant fossils) and petroleum (animal fossils). When we light them on fire, we threaten the global ecosystem. When we use them as feedstocks for making stuff, we create substances—pesticides, solvents, plastics—that can tinker with our subcellular machinery and the various signaling pathways that make it run.

Natural gas is the vaporous form of petroleum. It's the Dr. Jekyll and Mr. Hyde of fossil fuels: when burned, natural gas generates only half the greenhouse gases of coal, but when it escapes into the atmosphere as unburned methane, it's one of the most powerful greenhouse gases of them all—twenty times more powerful than carbon dioxide at trapping heat and with the stamina to persist for nine to fifteen years. You can also make petrochemicals from it. Natural gas is the starting point for anhydrous ammonia (synthetic fertilizer) and PVC plastic (those shower curtains).

Until a few years ago, much of the natural gas trapped underground was considered unrecoverable because it is scattered throughout vast sheets of shale, like a fizz of bubbles in a petrified spill of champagne. But that all changed with the rollout of a drilling technique (pioneered by Halliburton) that bores horizontally through the bedrock, blasts it with explosives, and forces into the cracks, under enormous pressure, millions of gallons of water laced with a proprietary mix of poisonous chemicals that further fracture the rock. Up the borehole flows the gas. In 2000 only 1 percent of natural gas was shale gas. Ten years later almost 20 percent is.

International investors began viewing shale gas as a paradigm-shifting innovation. Energy companies are now looking at shale plays in Poland and Turkey. Fracking is underway in Canada. But nowhere has the technology been as rapidly deployed as in the United States, where a gas rush is underway. Gas extraction now goes on in thirty-two states, with half a million new gas wells drilled in the last ten years alone. We are literally shattering the bedrock of our nation and pumping it full of carcinogens in order to bring methane out of the earth.

And nowhere in the United States is fracking proceeding more manically than Appalachia, which is underlain by the formation called the Marcellus Shale, otherwise referred to by the *Intelligent Investor Report* as "the Saudi Arabia of natural gas" and by the Toronto *Globe and Mail* as a "prolific monster" with the potential to "rearrange the continent's energy flow."

In the sense of "abnormal to the point of inspiring horror," *monster* is not an inappropriate term here. With every well drilled—and 32,000 wells per year are planned—a couple million gallons of fresh water are transformed into toxic fracking fluid. Some of that fluid will remain underground. Some will come flying back out of the hole, bringing with it other monsters: benzene, brine, radioactivity, and heavy metals that for the past 400 million years had been safely locked up a mile below us, estranged from the surface world of living creatures. No one knows what to do with this lethal flowback—a million or more gallons of it for every wellhead. Too caustic for reuse as is, it sloshes around in open pits and sometimes is hauled away in fleets of trucks to be forced under pressure down a disposal well. And it is sometimes clandestinely dumped.

By 2012, 100 billion gallons per year of fresh water will be turned into toxic fracking fluid. The technology to transform it back to drinkable water does not exist. And even if it did, where would we put all the noxious, radioactive substances we capture from it?

Here, then, are the environmental precepts violated by hydrofracking. 1. Environmental degradation of the commons should be factored into the price structure of the product (full-cost accounting), whose true carbon footprint—inclusive of all those diesel truck trips, blowouts, and methane leaks—requires calculation (life-cycle analysis). 2. Benefit of the doubt goes to public health, not the things that threaten it, especially in situations where catastrophic harm—aquifer contamination with carcinogens—is unremediable (the Precautionary Principle). 3. There is no away.

This year I've attended scientific conferences and community forums on fracking. I've heard a PhD geologist worry about the thousands of unmapped, abandoned wells scattered across New York from long-ago drilling operations. (What if pressurized fracking fluid, to be entombed in the shale beneath our aquifers, found an old borehole? Could it come squirting back up to the surface? Could it rise as vapor through hairline cracks?) I've heard a hazardous-materials specialist describe to a crowd of people living in fracked communities how many parts per million of benzene will raise risks for leukemia and sperm abnormalities linked to birth deformities. I've heard a woman who lives by a fracking operation in Pennsylvania—whose pond bubbles with methane and whose kids have nosebleeds at night—ask how she could keep her children safe. She was asking me. And I had no answer. Thirty-seven percent of the land in the township where I live with my own kids is already leased to the frackers. There is no away.

ABIGAIL TUCKER

The New King of the Sea

FROM *Smithsonian*

ON THE NIGHT of December 10, 1999, the Philippine island of Luzon, home to the capital, Manila, and some 40 million people, abruptly lost power, sparking fears that a long-rumored military coup d'état was underway. Malls full of Christmas shoppers plunged into darkness. Holiday parties ground to a halt. President Joseph Estrada, meeting with senators at the time, endured a tense ten minutes before a generator restored the lights, while the public remained in the dark until the cause of the crisis was announced, and dealt with, the next day. Disgruntled generals had not engineered the blackout. It was wrought by jellyfish. Some fifty dump trucks' worth had been sucked into the cooling pipes of a coal-fired power plant, causing a cascading power failure. "Here we are at the dawn of a new millennium, in the age of cyberspace," fumed an editorial in the *Philippine Star,* "and we are at the mercy of jelly-fish."

A decade later the predicament seems only to have worsened. All around the world, jellyfish are behaving badly—reproducing in astonishing numbers and congregating where they've supposedly never been seen before. Jellyfish have halted seafloor diamond mining off the coast of Namibia by gumming up sediment-removal systems. Jellies scarf so much food in the Caspian Sea that they're contributing to the commercial extinction of beluga sturgeon—the source of fine caviar. In 2007 mauve stinger jellyfish stung and asphyxiated more than 100,000 farmed salmon off the coast of Ireland as aquaculturists on a boat watched in horror. The jelly swarm reportedly was thirty-five feet deep and covered ten square miles.

Nightmarish accounts of "Jellyfish Gone Wild," as a 2008 National Science Foundation report called the phenomenon, stretch from the fjords of Norway to the resorts of Thailand. By clogging cooling equipment, jellies have shut down nuclear power plants in several countries; they partially disabled the aircraft carrier USS *Ronald Reagan* four years ago. In 2005 jellies struck the Philippines again, this time incapacitating 127 police officers who had waded chest-deep in seawater during a counterterrorism exercise, apparently oblivious to the more imminent threat. (Dozens were hospitalized.) This past fall a ten-ton fishing trawler off the coast of Japan capsized and sank while hauling in a netful of 450-pound Nomura's jellies.

The sensation of getting stung ranges from a twinge to tingling to savage agony. Victims include Hudson River triathletes, Ironmen in Australia, and kite surfers in Costa Rica. In summertime so many jellies mob the waters of the Mediterranean Sea that it can appear to be blistering, and many bathers' bodies don't look much different: in 2006 the Spanish Red Cross treated 19,000 stung swimmers along the Costa Brava. Contact with the deadliest type, a box jellyfish native to northern Australian waters, can stop a person's heart in three minutes. Jellyfish kill between twenty and forty people a year in the Philippines alone.

The news media have tried out various names for this new plague: "the jellyfish typhoon," "the rise of slime," "the spineless menace." Nobody knows exactly what's behind it, but there's a queasy sense among scientists that jellyfish just might be avengers from the deep, repaying all the insults we've heaped on the world's oceans.

"Jellyfish" is a decidedly unscientific term—the creatures are not fish and are more rubbery than jamlike—but scientists use it all the same (though one I spoke with prefers his own coinage, "gelata"). The word "jellyfish" lumps together two groups of creatures that look similar but are unrelated. The largest group includes the bell-shaped beings that most people envision when they think of jellyfish: the so-called true jellies and their kin. The other group consists of comb jellies—ovoid, ghostly creatures that swim by beating their hairlike cilia and attack their prey with gluey appendages instead of stinging tentacles. (Many other gelatinous animals are often referred to as jellyfish, including the Portuguese man-of-war, a colony of stinging animals known as a siphonophore.) All told, there are some 1,500 jellyfish species: blue blubbers, bushy bottoms, fire jellies, jimbles. Cannonballs, sea walnuts. Pink meanies, aka stinging cauliflowers. Hair jellies, aka snotties. Purple people eaters.

The bell-shaped jellies—distantly related to corals and anemones—launched their lifestyle long, long ago. Exquisite jellyfish fossils found recently in Utah display reproductive organs, muscle structure, and intact tentacles; the jelly fossils, the oldest discovered, date back more than 500 million years, when Utah was a shallow sea. By contrast, fish evolved only about 370 million years ago.

The descendants of those ancient jellies haven't changed much. They are boneless and bloodless. In their domelike bells, guts are squished beside gonads. The mouth doubles as an anus. (Jellies are also brainless, "so they don't have to contemplate that," one jelly specialist says.) Jellies drift at the mercy of the currents, though many also propel themselves by contracting their bells, pushing water out, while others—such as the upside-down jellyfish and the flower hat, with its psychedelic lures—can recline on the seafloor. They absorb oxygen and store it in their jelly. They can sense light and certain chemicals. They can grow quickly when there's food around and shrink when there isn't. Their tentacles, which reach up to 100 feet long in some species, are covered with cells called nematocysts, which fire tiny poison harpoons, enabling the animals to immobilize krill, larval fish, and other prey without risking their mushy bodies in a struggle. Yet if a sea turtle bites off a hunk, the flesh regenerates.

A breeding jellyfish can spit out unfertilized eggs at a prodigious rate: one female sea nettle may spew as many as 45,000 per day. To maximize the chances of sperm meeting egg, millions of moon jellies of both sexes assemble in one place for a gamete swapping orgy.

Chad Widmer is one of the world's most accomplished jellyfish cultivators. At the Monterey Bay Aquarium in California, he is the lord of the Drifters exhibit, a slow-motion realm of soft edges, rippling flute music, and sapphire light. His left ankle is crowded with tattoos, including Neptune's trident and a crystal jellyfish. A senior aquarist, Widmer labors to figure out how jellyfish thrive in captivity—a job that involves untangling tentacles and plucking gonads until his arm is swollen with venom.

Widmer has bred dozens of jellyfish species, including moon jellies, which resemble animated shower caps. His signature jelly is the Northeast Pacific sea nettle, displayed by the score in a 2,250-gallon exhibit tank. They are orange and incandescent, like dollops of lava, and when they swim against the current they look like glowing meteors streaming to Earth.

The waters of Monterey Bay have not been spared the gelatinous woes said to be sweeping the oceans. "It used to be that everything had a season," Widmer says. Spring was the time for lobed comb jellies and crystal jellies to arrive. But for the past five years or so, those species seem to be materializing almost at random. The orange spotted comb jelly, which Widmer nicknamed the "Christmas jelly," no longer peaks in December; it haunts the shoreline practically year-round. Black sea nettles, once seen mostly in Mexican waters, have started appearing off Monterey. Last August millions of Northeast Pacific sea nettles bloomed in Monterey Bay and clogged the aquarium's seawater intake screen. The nettles typically retreat by early winter. "Well," Widmer informs me gravely on my February visit, "they're still out there."

It's hard to tell what may be causing jellyfish to proliferate. The fishing industry has depleted populations of big predators such as red tuna, swordfish, and sea turtles that feed on jellyfish. And when small, plankton-eating fish such as anchovies are overharvested, jellies flourish, gorging on plankton and reproducing to their hearts' content (if they had hearts, that is).

In 1982, when the Black Sea ecosystem was already weakened by anchovy overfishing, the warty comb jelly (*Mnemiopsis leidyi*) arrived; a species native to the East Coast of the United States, it was most likely carried across the Atlantic in a ship's ballast water. By 1990 there were some 900 million tons of them in the Black Sea.

Pollution, too, may be fueling the jelly frenzy. Jellyfish succeed in all sorts of fouled conditions, including "dead zones," where rivers have pumped fertilizer runoff and other materials into the ocean. The fertilizer fuels phytoplankton blooms; after the phytoplankton die, bacteria decompose them, hogging oxygen; the oxygen-depleted water then kills or forces out other marine creatures.

The number of coastal dead zones has doubled every decade since the 1960s; there are now roughly five hundred. (Oil can kill jellyfish, but no one knows how jellyfish populations in the Gulf of Mexico will fare in the long run after the BP oil spill.)

Carbon-based air pollution may be another factor. Since the Industrial Revolution, the amount of carbon in the atmosphere from burning fossil fuels and wood as well as from other enterprises has risen by some 36 percent. That contributes to global warming, which, some researchers speculate, may benefit jellyfish at the expense of other marine animals. Moreover, carbon dioxide dissolves in seawater to form carbonic acid—a major threat to marine life. As the seas become more acidic, scientists say, ocean water will begin to dissolve animal shells, stunt coral reefs, and disorient larval fish by skewing their sense of smell. Jellies, meanwhile, may not even be inconvenienced, according to recent studies by Jennifer Purcell of Western Washington University.

Purcell and a graduate student, Amanda Winans, decided to breed moon jellyfish in water with the staggering acid levels that some scientists say will prevail in the years 2100 and 2300. "We took it to very severe acid, using the worst predictions," Purcell says. The jellyfish reproduced with abandon. She has also conducted experiments that lead her to suspect that many jellies reproduce better in warmer water.

With the world's human population expected to increase 32 percent by 2050, to 9.1 billion, a number of environmental conditions that favor jellyfish are predicted to become more common. Jellyfish reproduce and move into new niches so rapidly that even within forty years, some experts predict "regime shifts" in which jellyfish assume dominance in one marine ecosystem after another. Such shifts may have already occurred, including off Namibia, where, after years of overharvesting, the once fecund waters of the Benguela Current now contain more jellyfish than fish.

Steven Haddock, a zooplankton scientist at the Monterey Bay Aquarium Research Institute (MBARI), is concerned that researchers and the news media may be overreacting to a few isolated jelly outbreaks. Not enough is known about historical jelly abundances to distinguish between natural fluctuation and long-term change, he says. Are there really more of the creatures, or are people simply more prone to notice and report them? Are the jellyfish changing, or is our perspective? A self-described "jelly hugger," Haddock worries that jellyfish are taking the blame for messing up the seas when we're the ones causing the damage. "I just wish that people had the perception that jellyfish are not the enemy here," Haddock says.

Purcell, who sports jellyfish earrings the day I meet her in Monterey, says she is disgusted by what she sees as humanity's efforts to exploit the ocean, filling it with fish farms and oil wells and fertilizer. Compared with fish, jellies are "better feeders, better growers, more tolerant of all kinds of things," she told me, adding of the marine environment: "I think it's entirely possible we've made things better for jellyfish." Part of her likes the idea of unruly jellies causing a commotion and foiling our plans. She's cheering for them, almost.

Widmer's lab at the Monterey Aquarium is dominated by bubbling lime-green columns of algae, which he feeds to brine shrimp, which he then feeds to jellyfish. The algae come in six other "flavors," but he says he prefers the green type for its mad-scientist aesthetic. The room is full of jellyfish tanks ranging in size from salad bowls to wading pools. The containers rotate slowly, creating a current. "Let's feed!" Widmer cries. He scrambles up and down stepladders, squirting a turkey baster of pink krill into this tank and that.

Toward the back of the lab, haggard orange sea nettles stumble along the bottom of their tank, their bells brownish and transparent, their tentacles torn. These, Widmer says, have been taken out of the public display and retired. "Retired" is Widmer's euphemism for "about to be snipped up with fabric scissors and fed to other jellies."

He calls his prize specimens "golden children." He speaks to them in cooing tones usually reserved for kittens. One tank holds the petite but striking purple-lipped cross jellies, which Widmer retrieved from Monterey Bay. The species has never been bred in captivity before. "Oh, aren't you cute!" he trills. The other golden child is a small brown smudge on a pane of glass. This, he explains, dabbing artistically at the smudge's edges with a paintbrush, is a colony of lion's mane jellyfish polyps.

When jellyfish sperm and egg meet, the fertilized egg forms a free-swimming larva, what Widmer describes as "a fuzzy ciliated Tic Tac." It whizzes around before landing on a sponge or other sea-floor fixture. There it morphs into a weedy little polyp, an intermediate form that can reproduce asexually. And then—well, sometimes nothing happens for a good long while. A jellyfish polyp can sit dormant for a decade or more, biding its time.

When ocean conditions become ideal, however, the polyp starts to "strobilate," or bud off new jellyfish, a process Widmer shows me under a microscope. One polyp looks as if it is balancing a stack of Frisbees on its head. The tower of tiny disks pulses slightly. Eventually, Widmer explains, the top one will fly off, like a clay pigeon at a shooting range, then the next one, and the next. Sometimes dozens of disks launch, each disk a baby jellyfish.

To test the impact of warming oceans on polyp productivity, Widmer assembled a series of incubators and seawater baths. If he heated each a few degrees warmer than the last, what would the jellyfish do? At 39 degrees Fahrenheit, the polyps generated, on average, about twenty teeny jellyfish. At 46 degrees, roughly forty. The polyps in 54-degree seawater birthed some fifty jellies each, and one made sixty-nine. "A new record," Widmer says, awed.

To be sure, Widmer has also found that some polyps can't produce young at all if placed in waters significantly warmer than their native range. But his experiments, which confirm research on other jellies done by Purcell, also lend some credence to anxieties that global warming may induce jelly extravaganzas.

Two events ultimately stalled the *Mnemiopsis* invasion in the Black Sea. One was the fall of the Soviet Union: in the ensuing chaos, some farmers ceased fertilizing their fields and water quality improved. The other was the accidental introduction of a second exotic jellyfish that happened to have a taste for *Mnemiopsis*.

In lieu of dismantling superpowers or importing invasive species, countries have adopted jelly-proofing strategies. South Korea recently released 280,000 native jelly-eating filefish along the coast of Busan. Spain dispatched indigenous loggerhead sea turtles off Cabo de Gata. Japanese fishermen hack at the giant Nomura's with barbed poles. Mediterranean beaches have organized jellyfish hot lines, spotter-boat armadas, and airplane flyovers; the slimy troublemakers are sometimes sucked up by garbage scows, carted off by backhoes, or used for fertilizer. Bathers in the worst areas are advised to wear full-body Lycra "stinger suits" or pantyhose or to smear themselves with petroleum jelly. Most sting-treatment products feature vinegar, the best remedy for jelly venom.

When, nearly two decades ago, Daniel Pauly, a fisheries biologist at the University of British Columbia, started warning of the dangers of overfishing, he liked to alarm people and say we would end up eating jellyfish. "It's not a metaphor anymore," he says today, pointing out that not only China and Japan but also the U.S. state of Georgia have commercial jellyfish operations, and there's talk of one starting in Newfoundland, among other places. Pauly himself has been known to nibble jellyfish sushi.

About a dozen jellyfish varieties with firm bells are considered desirable food. Stripped of tentacles and scraped of mucous membranes, jellyfish are typically soaked in brine for several days and then dried. In Japan, they are served in strips with soy sauce and (ironically) vinegar. The Chinese have eaten jellies for a thousand years (jellyfish salad is a wedding banquet favorite). Lately, in an apparent effort to make lemons into lemonade, the Japanese government has encouraged the development of haute jellyfish cuisine—jellyfish caramels, ice cream, and cocktails—and adventuresome European chefs are following suit. Some enthusiasts compare the taste of jellyfish to fresh squid. Pauly says he's reminded of cucumbers. Others think of salty rubber bands.

The main edible variety in U.S. waters, cannonball jellies, are found on the Atlantic Coast from North Carolina to Florida and in the Gulf of Mexico. They scored quite high on a "hedonic scale" of color and texture in a study led by Auburn University. Another scientific paper hailed jellyfish flesh—which is 95 percent water, a few grams of protein, the barest hint of sugar, and, once dried, only 18 calories per 100-gram serving—as "the ultimate modern diet food."

The research vessel *Point Lobos* heaves in the swells of Monterey Bay. After a two-hour ride from shore, the engine idles as a crane lowers the *Ventana,* an unmanned submarine stocked with a dozen glass collecting jars, into the water. As the submarine begins its descent into the canyon, its cameras feed footage to computer monitors in the boat's dark control room. Widmer and other scientists watch from a semicircle of armchairs. Widmer is allotted only a few trips on the MBARI submarine each year for his research; his eyes are shining with anticipation.

On the screens we see the bright green surface water darken by degrees to deep purple, then black. White flecks of detritus called marine snow rush past, like a star field in warp speed. The submarine drops 1,000, 1,500, 5,000 feet. We are en route to what Widmer has modestly named the Widmer Site, a jellyfish mecca on the lip of an undersea cliff.

Our spotlight illuminates a Gonatus squid, which clenches itself into an anxious red fist. Giant gray-green Humboldt squid sail by, like the ghosts of spent torpedoes. Glimmering beings appear. They seem to be constructed of spider webs, fishing line and silk, soap bubbles, glow sticks, strands of Christmas lights, and pearls. Some are siphonophores and gelatinous organisms I've never seen before. Others are tiny jellyfish.

Every now and then, Widmer squints at an iridescent speck, and —if it isn't too delicate and the gonads look ripe—asks the pilot of the remote-controlled sub to give chase. "I don't know what it is, but it looks promising," he says. We bear down on jellyfish the size of jingle bells and gumdrops, slurping them up with a suction device.

"Down the tube!" Widmer cries in triumph.

"In the bucket!" the pilot agrees.

The whole boat crew pauses to stare at the screen and marvel at a piece of kelp studded with fuzzy pink anemones. We snatch a jelly here, a jelly there, including a mysterious one with a strawberry-colored center, always keeping a sharp eye out for polyps.

The submersible sails over the wreck of a blue whale, a gigantic rockfish curled up like a cat beside the great skull. We pass a frilly albino sea cucumber and a Budweiser can. We see squat lobsters and spot prawns, bleached sea stars, black owl fish, bouncy coils of eggs, a pale pink orb with tarantula-like legs, lemon-yellow mermaid's purses, English sole, starry flounders, and the purple bullet shapes of sharks. The California sunshine seems dismal in comparison.

When the submarine surfaces, Widmer quickly packs his tiny captives into chilled Tupperware containers. The strawberry jelly begins to languish almost immediately, as the sunlight disintegrates the red porphyrin pigment in its bell; soon it will be floating upside down. A second unknown specimen with pinwheel-shaped gonads looks perky enough, but we have caught only one, so Widmer will not be able to breed it for public display. He hopes to retrieve more on the next trip.

He has, however, managed to corral half a dozen *Earleria corachloeae*, a species he recently discovered. He named it after his two young landlocked nieces in Wichita, Kansas—Cora and Chloe. Widmer produces a YouTube series for them called "Tidepooling With Uncle Chad," which introduces ocean wonders—sea turtle nests, bull kelp trumpets, snail racetracks—he wants them to know.

Two days later, the *E. corachloeae* produce a smattering of eggs like grains of fine beach sand. He will dote on his captives until they die or go on display. They are officially "golden children."

TIM ZIMMERMANN

The Killer in the Pool

FROM *Outside*

To work closely with a killer whale in a marine park requires experience, intuition, athleticism, and a whole lot of dramatic flair. Few people were better at it than top SeaWorld trainer Dawn Brancheau, who, at forty, was blond, vivacious, and literally the poster girl for the marine park in Orlando, Florida, appearing on billboards around the city. She decided she wanted to work with killer whales at the age of nine, during a family trip to SeaWorld, and she loved animals so much that as an adult she used to throw birthday parties for her two chocolate Labs.

On February 24, 2010, Brancheau was working the "Dine with Shamu" show, featuring SeaWorld's largest killer whale, a six-ton, twenty-two-foot male known as Tili (short for Tilikum). "Dine with Shamu" takes place in a faux-rock-lined, 1.6-million-gallon pool that has an open-air cafe wrapped around one side. The families snacking on the lunch buffet that Wednesday were getting an eye-ful. Brancheau bounced around on the deck of the pool, wearing a black-and-white wetsuit that echoed Tilikum's coloration, as she worked him through a few of the many "behaviors" he had learned during his nearly twenty-seven years as a marine-park denizen. The audience chuckled at the sight of one of the ocean's top predators performing like a circus animal.

The show ended around 1:30 P.M. As the audience started to file out, Brancheau fed Tilikum some herring (he eats up to two hundred pounds a day), doused him a few times with a bucket (killer whales love all sorts of stimulation), and moved over to a shallow ledge built into the side of the pool. There she lay down in a few inches of water, talking to him and stroking him, conducting what's known as a "relationship session." Tilikum floated inert in the pool alongside her, his nose almost touching her shoulder. Brancheau was smiling, her long ponytail flaring out behind her.

One level down, a group of families gathered before the huge glass windows of the underwater viewing area. A trainer shouted up that they were ready for Tilikum. That was Brancheau's signal to instruct the orca to dive down and swim directly up to the glass for a custom photo op. It's an awesome sight when six tons of Tili come gliding out of the blue. But that day, instead of waiting for his cue and behaving the way decades of daily training in captivity had conditioned him to, Tilikum did something unexpected. Jan Topoleski, thirty-two, a trainer who was acting as a safety spotter for Brancheau, told investigators that Tilikum took Brancheau's drifting hair into his mouth. Brancheau tried to pull it free, but Tilikum yanked her into the pool. In an instant, a classic tableau of a trainer bonding with a marine mammal became a life-threatening emergency.

Topoleski hit the pool's siren. A "Signal 500" was broadcast over the SeaWorld radio net, calling for a water rescue at G pool. Staff raced to the scene. "It was scary," a Dutch tourist, Susanne De Wit, thirty-three, told investigators. "He was very wild." SeaWorld staff slapped the water surface, signaling Tilikum to leave her. The whale ignored the command. Trainers hurried to drop a weighted net into the water to try to separate Tilikum from Brancheau or herd him through two adjoining pools and into a small medical pool that had a lifting floor. There he could be raised out of the water and controlled.

Eyewitness accounts and the sheriff's investigative report make it clear that Brancheau fought hard. She was a strong swimmer, a dedicated workout enthusiast who ran marathons. But she weighed just 123 pounds and was no match for a 12,000-pound killer whale. She managed to break free and swim toward the surface, but Tilikum slammed into her. She tried again. This time he grabbed her. Her water shoes came off and floated to the surface. "He started pushing her with his nose like she was a toy," said Paula Gillespie, one of the visitors at the underwater window. SeaWorld employees urgently ushered guests away. "Will she be okay?" one asked.

Tilikum kept dragging Brancheau through the water, shaking

her violently. Finally—now holding Brancheau by her arm—he was guided onto the medical lift. The floor was quickly raised. Even now, Tilikum refused to give her up. Trainers were forced to pry his jaws open. When they pulled Brancheau free, part of her arm came off in his mouth. Brancheau's colleagues carried her to the pool deck and cut her wetsuit away. She had no heartbeat. The paramedics went to work, attaching a defibrillator, but it was obviously she was gone. A sheet was pulled over her body. Tilikum, who'd been involved in two marine-park deaths in the past, had killed her.

"Every safety protocol that we have failed," SeaWorld's director of animal training, Kelly Flaherty Clark, told me a month after the incident, her voice still tight with emotion. "That's why we don't have our friend anymore, and that's why we are taking a step back."

Dawn Brancheau's death was a tragedy for her family and for SeaWorld, which had never lost a trainer before. Letters of sympathy poured in, many with pictures of Brancheau and the grinning kids she'd spent time with after shows. The incident was a shock to Americans accustomed to thinking of Shamu as a lovable national icon, with an extensive line of plush dolls and a relentlessly cheerful Twitter account. The news media went into full frenzy, chasing Brancheau's family and flying helicopters over Shamu Stadium. Congress piled on with a call for hearings on marine mammals at entertainment parks, and the Occupational Safety and Health Administration (OSHA) opened an investigation. It was the most intense national killer whale mania since 1996, when Keiko, the star of *Free Willy*, was rescued from a shabby marine park in Mexico City in an attempt to return him to the sea. Killer whales have never been known to attack a human in the wild, and everyone wanted to know one thing: Why did Dawn Brancheau die?

Killer whales have been starring at marine parks since 1965. There are 42 alive in parks around the world today—SeaWorld owns 26 of them—and over the years more than 130 have died in captivity. Until the 1960s, no one really thought about putting a killer whale in an aquarium, much less in a show. The public knew little about them beyond the fact that they sounded dangerous. (Killer whales, or orcas, are the largest members of the dolphin family.) Fishermen tended to blast them with rifle fire if they came near salmon and herring stocks.

But Ted Griffin helped change all that. A young impresario who owned the Seattle Marine Aquarium, Griffin had long been obsessed with the idea of swimming with a killer whale. In June 1965 he got word of a twenty-two-footer tangled in a fisherman's nets off Namu, British Columbia. Griffin bought the 8,000-pound animal for $8,000. He towed the orca, which he named Namu, 450 miles back to Seattle in a custom-made floating pen. Namu's family pod — twenty to twenty-five orcas — followed most of the way. Griffin was surprised by how gentle and intelligent Namu was. Before long he was riding on the orca's back, and by September tens of thousands of people had come to see the spectacle of the man and his orca buddy. The story of their "friendship" was eventually chronicled in the pages of *National Geographic* and in the 1966 movie *Namu, the Killer Whale*. The orca entertainment industry was born.

Namu was often heard calling to other orcas from his pen in the sea, and he died within a year from an intestinal infection, probably brought on by a nearby sewage outflow. Griffin was devastated. But his partner at the aquarium, Don Goldsberry, was a blunt, hard-driving man who could see that there was still a business in killer whales. He and Griffin had already turned their energies to capturing orcas in the Puget Sound area and selling them to marine parks. Goldsberry first built a harpoon gun, firing it by accident through his garage door and denting his car. Eventually, he and Griffin settled on the technique of locating orca pods from the air, driving them into coves with boats and seal bombs (underwater explosives used by fishermen to keep seals away from their catch), and throwing a wall of net across their escape path. Goldsberry and Griffin would then choose the orcas they wanted and let the remaining ones go. They preferred adolescents, particularly the smaller females, which were easier to handle and transport.

In October 1965, Goldsberry and Griffin trapped fifteen killer whales in Carr Inlet, near Tacoma. One died during the hunt. Another — a fourteen-foot female that weighed 2,000 pounds — was captured and named Shamu (for She-Namu). In December a fast-growing marine park in San Diego, called SeaWorld, acquired Shamu and flew her to California. Goldsberry says he and Griffin were paid $70,000. It was the start of a billion-dollar franchise.

Over the next decade, around 300 killer whales were netted off the Pacific Northwest coast, and 51 were sold to marine parks across the globe, in Japan, Australia, the Netherlands, France, and elsewhere. Goldsberry, who became SeaWorld's lead "collector" until he retired in the late 1980s, caught 252 of them, sold 29, and inadvertently killed 9 with his nets. In August 1970, concerned about backlash, Goldsberry weighted some dead orcas down with anchors and dumped them in deep water. When they were dragged up on a Whidbey Island beach by a trawling fisherman, the public started to understand the sometimes brutal reality of the "orca gold rush."

In 1972 the Marine Mammal Protection Act prohibited the taking of marine mammals in U.S. waters, but SeaWorld continued to receive killer-whale-capture permits under an educational-display exclusion. In March 1976 Goldsberry pushed his luck and the limits of public opinion. He sighted a group of killer whales in the waters just off Olympia, Washington's state capital. In full view of boaters—and just as the state legislature was meeting to consider creating a Puget Sound killer whale sanctuary—he used seal bombs and boats to chase six orcas into his nets at Budd Inlet. Ralph Munro, an aide to Governor Dan Evans, was out on a small sailboat that day and remembers the sight. "It was gruesome as they closed the net. You could hear the whales screaming," Munro recalls. "Goldsberry kept dropping explosives to drive the whales back into the net."

The state of Washington filed a lawsuit, contending that Goldsberry and SeaWorld had violated permits that required humane capture, and as the heat and publicity built, SeaWorld agreed to release the Budd Inlet killer whales and to stop taking orcas from Washington waters. With the Puget Sound hunting grounds closing, Goldsberry flew around the world looking for other good capture sites. He settled on Iceland, where killer whales were plentiful. By October 1976 SeaWorld's first Icelandic orca had been captured.

Over the next few years, Goldsberry spent freely to help create the infrastructure to net and transport whales out of Iceland. In November 1983, in the cold, rough waters off Berufjördur, an Icelander named Helgi Jonasson drew a large purse-seine net around a group of killer whales. Three young animals—two males and a female—were captured and transported to the Hafnarfjördur Marine Zoo, near Reykjavik.

There they were placed in a concrete holding tank. The smaller male, who was about two years old and just shy of 11.5 feet, would remain there for almost a year, awaiting transfer to a marine park. In the pool he could either cruise slowly in circles or lie still on the surface. He could hear no ocean sounds, only the mechanical rush of filtration. Finally, in late 1984, the young orca was shipped to Sealand of the Pacific, a marine park just outside Victoria, on British Columbia's Vancouver Island. He was given a name to go with his new life: Tilikum, which means "friend" in Chinook.

Sealand, situated at Oak Bay Marina, was a wholly alien world for a wild orca. Its performance pool—about one hundred feet by fifty feet, and thirty-five feet deep—was created by suspending mesh netting from the floating docks. The pool was open to the marina water, and thus to any bilge oil or sewage pumped into it by boaters. Marina traffic and motors created a cacophony of artificial underwater background noise, obscuring the natural sounds Tilikum had known in the wild. In the fourteen years before his arrival, seven orcas had died under Sealand's care. Their average survival time was just shy of three and a half years.

At Sealand, Tilikum joined two female killer whales, Haida and Nootka, who were sorting out the social pecking order. (Orca society is dominated by females.) That meant conflict and tooth raking for all three orcas, and even after Haida established herself as dominant, both females continued to push the young Tilikum around. The stress was worse at night. Sealand's owner, a local entrepreneur named Robert Wright, who'd captured his share of Pacific Northwest killer whales in the early 1970s, worried that someone might cut the net to free his orcas or that they might chew through it themselves. So at 5:30 P.M., after the shows were over, the orcas were moved into a small metal-sided pool that was twenty-six feet in diameter and less than twenty feet deep. The trainers referred to it as "the module," and the orcas were left in it for the next fourteen and a half hours.

According to Eric Walters, who was a trainer at Sealand from 1987 to 1989 while working toward a bachelor's degree in marine biology at the University of Victoria, the module was so tight that the orcas had difficulty avoiding conflict, and their skin would get scratches and cuts from rubbing against the sides. About once a week, Walters says, one or more of the orcas would simply refuse to swim into the module and would have to be left in the performance pool overnight.

The orca show was performed every hour on the hour, eight times a day, seven days a week. Both Nootka and Tilikum had stomach ulcers, which had to be treated with medication. Sometimes Nootka's ulcers were so bad she had blood in her stool.

Walters was interested in the science of training and was encouraged when Sealand brought in Bruce Stephens, a former SeaWorld head trainer, to make recommendations to improve Sealand's practices. Stephens gave each trainer a handbook, which warned, "If you fail to provide your animals with the excitement they need, you may be certain they will create the excitement themselves." He emphasized that killer whales needed constant change to keep them engaged and responsive, and made a series of recommendations for new learning sessions and playtime for Sealand's orcas. But within a month, Walters told me, Sealand was back to its usual routines. "They basically ran it like you would run McDonald's," he says. "It just can't be good for an animal that is so intelligent to do the same thing every day." (Wright still runs a marina at Oak Bay but declined to speak to *Outside*.)

As Stephens had warned, bored killer whales look to make their own fun. If any unusual object ended up in the water, Haida, Nootka, and Tilikum would race for it and play keep-away with the trainers. Once the orcas took something, they were determined to hang on to it. Walters worried about what might happen if one of the trainers—who worked in rubber boots on a painted fiberglass deck— fell into the pool. Many marine parks try to defuse the danger with desensitization training that teaches the killer whales to stay calm and ignore anyone who falls in. The training might start with just a foot in the water (the orca is conditioned to ignore it) but ultimately requires gradually easing an entire person into the pool. According to Steve Huxter, who was the head of animal training and care at the time, desensitization was a Catch-22. After thinking about it carefully, "Bob [Wright] was not willing to take that risk."

Each whale had a distinctive personality. Tilikum was youthful, energetic, and eager to learn. "Tilikum was our favorite," says Eric Walters. "He was the one we all really liked to work with."

Nootka, with her health issues, was the most unpredictable. According to Walters, Nootka pulled a trainer into the water. (Walters quickly yanked her out.) Twice she tried to bite down on Walters's hands. Not even the audience was safe. A blind woman was once brought onto the stage to pat Nootka's tongue. Nootka bit her, too.

Frustrated, Walters quit in May 1989. A year later he wrote a letter to the Canadian Federation of Humane Societies, to share with participants at a conference on whales in captivity. In it he detailed Sealand's treatment of its marine mammals and the safety concerns he had. In closing, he wrote, "I feel that sooner or later someone is going to get seriously hurt."

On February 20, 1991, Sealand had just wrapped up an afternoon killer whale show. Keltie Byrne, a twenty-year-old marine-biology student and part-time trainer, was starting to tidy up when she misstepped and fell halfway into the pool. As she struggled to get out, one of the killer whales grabbed her and pulled her into the water. A competitive swimmer, Byrne was no match for three orcas used to treating any unusual object as a toy. "They never had a plaything in the pool that was so interactive," says Huxter. "They just got incredibly excited and stimulated." Huxter and the other trainers issued recall commands and threw food in the water. They tried maneuvering a life ring close enough for Byrne to grab, but the orcas kept her away from it. In the chaos and dark water, it was hard to see which killer whale had her at any one time. Twice she surfaced and screamed. After about ten minutes, she popped up a third time for an instant but made no noise. She had drowned.

Bryne was the first trainer ever killed by orcas at a marine park. It took Sealand employees two hours to recover her body from Nootka, Haida, and Tilikum. They had stripped off all of her clothes save one boot, and she had bruises from bites across her skin. "It was just a tragic accident," Al Bolz, Sealand's manager, told reporters at the time. "I just can't explain it."

Paul Spong, seventy-one, the director of OrcaLab, in British Columbia—which studies orcas in the wild—did part-time research at Sealand before Tilikum arrived. He is not so befuddled. "If you pen killer whales in a small steel tank, you are imposing an extreme level of sensory deprivation on them," he says. "Humans who are subjected to those same conditions become mentally disturbed."

Byrne's death led to a coroner's inquest, which recommended a series of safety improvements at Sealand. The park responded, but according to Huxter, "the wind came out of [Wright's] sails for the business." In the fall of 1991, Sealand contacted SeaWorld to ask if it would like to buy Nootka, Haida, and Tilikum. Sealand closed in 1992.

If you want to try to get an inkling of what captivity means for a killer whale, you first have to understand what their lives are like in the wild. For that, there's no one better than the marine biologist Ken Balcomb, sixty-nine, who has spent thirty-four years tracking and observing killer whales off the coast of Washington State.

In early May I meet Balcomb in his cluttered yard on San Juan Island. He's trying to find the source of a leak on his Boston Whaler. His wood-framed house, which also serves as headquarters for his Center for Whale Research, sits perched atop the rocky shores of the Haro Strait, a popular orca hangout; Balcomb says he sees them about eighty days a year from his deck. Inside there's gear all over the place — spotting scopes, cameras, tool kits — from a recent expedition to California. In the middle of it all, on a table, sits an enormous killer whale skull that he picked up in Japan in 1975, when he was a flier and oceanographic specialist for the U.S. Navy.

Balcomb, of medium build, with a ruddy, sun-baked face and a salt-and-pepper beard, has been carefully photographing, cataloging, and observing the Puget Sound orcas — also known as the Southern Residents — since he was contracted by the National Marine Fisheries Service in 1976 to assess the impact of the marine-park captures. Many people assumed there were hundreds of orcas around Puget Sound. After identifying each individual killer whale by its markings, Balcomb found that there were just seventy left.

Since then he's become the Southern Residents' scientific godfather, noting every birth and death and plotting family connections. The population, he says, is now at eighty-five orcas, but he won't know for sure until they show up this summer. Talking on his sun porch, Balcomb stresses that one of the most important things to know about killer whales like Tilikum is that in the wild they live in complex and highly social family pods of twenty to fifty animals. The pods are organized around the females. The matriarch is usually the oldest female (some live to eighty or more), who has a wealth of experience and knowledge about where food can be found. Within the pod, mothers are at the center of smaller family groups. Males, who can live to fifty or sixty years, stay with their mothers their entire lives and often die not long after she does. According to Balcomb, separation is not a minor issue.

The Southern Resident population is made up of three distinct pods. Each pod might travel some seventy-five miles a day, following the salmon and vocalizing almost constantly to keep the entire group updated on who's where and whether there are fish around. Killer whales are highly intelligent. They coordinate in the hunt, share food freely, and will help an injured or ill member of the pod stay on the surface to breathe. Most striking is the sophistication of their dialect. Each family group within a pod uses the same vocalizations, or vocabulary, and there are also shared vocalizations between pods. Balcomb says he can usually tell which pod is about to turn up simply by the sounds he hears through a hydrophone.

The social and genetic connections that bind orcas in the wild are intense. There is breeding between the Puget Sound pods. Sometimes they'll all come together at once and go through a distinctive greeting ceremony before mixing. But they will have absolutely nothing to do with the genetically distinct, transient killer whales that sometimes pass through their waters. (Transients travel in much smaller groups over vast distances and mostly feed on marine mammals instead of fish.) "When you get born into the family, you are always in the family. You don't have a house or a home that is your location," says Balcomb. "The group is your home, and your whole identity is with your group." Aggression between members of a pod almost never occurs in the wild, he adds.

Puget Sound is small enough that Balcomb used to run into Goldsberry from time to time. Despite their differences, the two men would talk killer whales, drink Crown Royal, and trade stories. Today Goldsberry, seventy-six, lives about one hundred miles away in a small, ground-level condo near Sea-Tac Airport. His only water view is of a man-made lake, and when I go to see him he's busy drilling a walrus tusk that's been made into a cribbage board. Goldsberry has a square head, with close-cropped white hair. His health is fragile, and he has an oxygen tube clipped to his nose. But he still has the beefy arms of a waterman, and he appears unmoved by the controversy of his hunting days. "We showed the world that killer whales were good animals and all of a sudden people said, 'Hey, leave these animals alone,'" he says, sipping a mug of vodka and ice. "I had to make a living."

Goldsberry has mostly kept his mouth shut about his work for SeaWorld and doesn't much like talking to reporters. "I'm only speaking with you because those idiots out there, mainly the politicians, want to release all the killer whales," he growls. "You might as well put a gun to the whales' heads." He spends the next couple of hours telling me about his cowboy days in the orca business: how he helped build the global trade, how he kept one step ahead of Greenpeace and activists, and how he battled the media, dropping one TV newsman's camera into the water, asking, "I wonder if this floats?"

Goldsberry says he always got the resources he needed to keep the killer whales coming, and he developed relationships with other marine parks around the world, which would often hold killer whales for him, many of which would eventually end up at SeaWorld. (Balcomb calls it Goldsberry's "whale laundry.") "I would go into SeaWorld and say, 'I need a quarter of a million' or 'a half-million dollars,' and they put it in my suitcase," he says with a grin. "It was good, catching animals. It was exciting. I was the best in the world. There is no question about it."

Asked about Goldsberry's work for SeaWorld, Fred Jacobs, vice president of communications, denies that killer whales were laundered. "Any killer whale that entered our collection from another facility did so in full accordance with their export and our import laws," he says. "We have imported whales that were collected by other institutions, but they were not collected on our behalf and held for us."

Goldsberry's last great haul of wild orcas came in October 1978, when he caught six off Iceland. (Five ended up in SeaWorld parks.) He continued to collect all sorts of other animals for SeaWorld for the next decade. When Goldsberry and SeaWorld finally parted ways in the late 1980s, Goldsberry says he was offered $100,000 to keep quiet about his work for two years. He happily took it. SeaWorld's Jacobs explains that Goldsberry's relationship with Sea-

World occurred under prior ownership. "I have no way of knowing if this is true or not," he says.

Whatever his methods, Goldsberry had helped SeaWorld turn killer whales into killer profits. The company currently has parks in Orlando, San Diego, and San Antonio, which are visited by more than 12 million people annually. Most of those visitors, paying up to $78 each for an entrance ticket, come to see killer whales. Last year Anheuser-Busch InBev sold SeaWorld's marine parks—and seven amusement parks housed with SeaWorld under the Busch Entertainment umbrella—to a private-equity giant, the Blackstone Group. The purchase price was reported to be $2.7 billion.

One of the keys to SeaWorld's success was its ability to move away from controversial wild orca captures to captive births in its marine parks. The first captive birth that produced a surviving calf took place at SeaWorld Orlando in 1985. Since then SeaWorld has relied mostly on captive breeding to stock its parks with killer whales, even mastering the art of artificial insemination. "Early in the morning, the animal-care crew would take hot-water-filled artificial cow vaginas and masturbate the males in the back tanks," says John Hall, a former scientist at SeaWorld. "It was pretty interesting to walk by."

Tilikum's sudden availability in 1991 was a boon to the captive breeding program. While preparing to transfer Haida, Nootka, and Tilikum to Orlando, SeaWorld, one of only a few facilities with the expertise to care for them, discovered that Tilikum had already impregnated Haida and Nootka. A sexually mature male, even one involved in a dangerous incident, was a welcome addition. "It was not the only reason [SeaWorld] had interest but definitely a part of the decision," says Mark Simmons, who worked as a trainer at Sea-World from 1987 to 1996 and was part of a team sent to Sealand to manage Tilikum's transfer. Media reports at the time pegged Tilikum's price at $1 million.

If Sealand was like a McDonald's, SeaWorld Orlando was like a five-star restaurant, with 220 acres of custom marine habitats, thrill rides, eateries, and a 400-foot Sky Tower. There were seven different killer whale pools, including the enormous Shamu show pool, and 7 million gallons of continuously filtered salt water kept at an orca-friendly 52 to 55 degrees. There was regular world-class veterinary care. Even the food was a custom blend, made up of restaurant-quality herring, capelin, and salmon.

The big question for SeaWorld was whether to teach Tilikum to perform with trainers in the pool. Called "water work," it has long been the most thrilling element of the Shamu shows. In contrast to Sealand's repetitive food-for-work equation, SeaWorld's training strategy was finely honed and based on intense variation. Daily activities were constantly altered, and the orcas were given a variety of rewards—sometimes food, sometimes stimulation (back rubs, hose-downs, toys, or ice), and sometimes nothing. "Variability makes the animals more flexible about what the outcome is and keeps them interested," says Thad Lacinak, who was SeaWorld's vice president and corporate curator for animal training when Tilikum arrived and who left in 2008 to found Precision Behavior, a consulting firm for zoos and other animal facilities.

Lacinak believed that Keltie Byrne died because Sealand's killer whales had never been trained to accept humans in the water. So when she fell in, they treated her like any other surprise object. Lacinak had confidence that Tilikum could be trained for Shamu-show water work. But he and SeaWorld's top management also knew that when it comes to killer whales (or any wild animal), there are no guarantees. Normally SeaWorld begins training in-water interaction when its killer whales are 1,000 pounds or less, but Tilikum was by now a very large bull. Plus Tilikum had been involved in a death. "If something did happen, you would look like a fool," Lacinak says. "It was too risky, and from a liability standpoint it was decided not to do [water work]."

Some of the trainers at least wanted to desensitize Tilikum in case someone fell in. "There were several of us that pushed for water de-sense training. You don't run from the storm; you harness the wind," says Mark Simmons, who left SeaWorld in 1996 to earn a business degree and later cofounded Ocean Embassy, which consults on conservation and marine parks. "We wanted to make humans in the water so commonplace that it didn't elicit any response. And if that had been done, it would be very unlikely that we'd be having this conversation today."

But SeaWorld faced the same vexing Catch-22 that had given Sealand pause. SeaWorld's head trainer, Flaherty Clark, says that it's impossible to prove or disprove what might have happened if

Tilikum had been desensitized. "It's easy for former trainers to frame that as a hypothetical," she says, "but we viewed water work with him and all the conditioning that might have permitted it to be effected safely as simply too great a risk."

Instead, SeaWorld focused on creating roles for Tilikum that showcased his size and power when no trainers were in the water. The sight of him rocketing into the air awed the crowds. One of his specialties was inundating the front rows—the "splash zone"—with a tidal wave pushed up by his enormous flukes. "He's a crowd-pleasing, showstopping, wonderful, wonderful wild animal," says Flaherty Clark.

Keeping Tilikum from water work made sense for another reason: as long as SeaWorld had been putting trainers in the water with killer whales, trainers had been getting worked over by them. Since the 1960s there have been more than forty documented incidents at marine parks around the world. In 1971 the first Shamu went wild on a bikini-wearing secretary from SeaWorld, who was pulled screaming from the pool. For every incident the public was aware of (the ones that occurred in front of audiences or that put trainers in the hospital), there were many more behind the scenes. John Jett was a trainer at SeaWorld in the 1990s. He left to pursue a PhD in natural resource management in 1995, having grown disillusioned with the reality of keeping large, intelligent animals in captivity. He says that getting nicked, and sometimes hammered, was just part of the price of living the killer whale dream: "There were so many incidents. If you show fear or go home hurt, you might be put on the bench." Flaherty Clark says SeaWorld gives trainers wide latitude: "The safety of our trainers and animals is paramount. Our trainers are empowered to alter any show or session plan if they have even the slightest concern."

In 1987 alone, SeaWorld San Diego experienced three incidents that hospitalized trainers with everything from fractured vertebrae to a smashed pelvis. Jonathan Smith was one of them. In March, during a show, he was grabbed by two killer whales, who slammed him on the bottom of the thirty-two-foot-deep pool five times before he finally escaped. "One more dunk for me and I would have gone out," he says. "They let me go. If they didn't want to let me go, it would have been over." Smith was left with a ruptured kidney, a lacerated liver, and broken ribs. In response to these serious injuries, as well as other incidents, SeaWorld shook up its management team, pulled trainers from the water, and reassessed its safety protocols. After a number of changes (including making sure that only very experienced trainers worked with killer whales), trainers were allowed back in the pools.

Despite the modifications, in 2006 another serious incident took place at SeaWorld San Diego, when head trainer Kenneth Peters was attacked by a killer whale called Kasatka. Kasatka grabbed Peters and repeatedly held him below the surface of the pool for up to a minute. He came close to drowning, and Kasatka joined Tilikum and a couple of other unruly SeaWorld orcas on the "no water work" blacklist.

Following the Peters incident, OSHA opened an investigation. After digging into the inner workings of SeaWorld's killer whale shows, OSHA issued a report in 2007 that warned, "The contributing factors to the accident, in the simplest of terms, is that swimming with captive orcas is inherently dangerous and if someone hasn't been killed already, it is only a matter of time before it does happen." SeaWorld challenged the report as filled with errors, and OSHA agreed to withdraw it.

In late March 2010, a month after Brancheau's death, I visit Orlando's SeaWorld park for the first time. I pause for an instant to take in the sheer size of the place, with its hundreds of diversions, but there is just one thing I really want to see: a killer whale show. I thread my way through families and packs of ecstatic kids. Shamu Stadium, SeaWorld's colossal amphitheater, looms before me.

The current Shamu show is called "Believe," and Dawn Brancheau was one of the stars. Music, video, and killer whales are wrapped around the story of a kid who paddles out to bond with a wild orca and is inspired to become a trainer. Every element is intimately choreographed, with whales exploding into the air and on-screen in perfect synchronicity. Even though Brancheau's death has prompted SeaWorld to temporarily reinvent "Believe" without trainers in the water, it is still absolutely mesmerizing. The show builds to a climactic finale with a pack of orcas lining up and using their flukes to sweep a tidal wave of water onto the shrieking and willing inhabitants of the splash zone.

After the show I sit down with Brad Andrews in front of the underwater viewing area of G pool. Two killer whales are amusing a crowd of people who probably have no idea of the scene the same windows revealed a month earlier. Andrews is SeaWorld's chief zoological officer, and he's been with the park since 1986. He explains that while part of the goal is entertainment, SeaWorld's aim is to use the shows to educate and inspire visitors, as a way to help conserve the environment and support wildlife.

There's a lot of criticism that flies back and forth between Sea-World and the hard marine-science community, but there's no question that SeaWorld's close contact with killer whales over the course of decades has contributed to the world's knowledge of them. "The gestation of killer whales was never known to researchers in the wild. It was always assumed it was like a dolphin, twelve months," Andrews says. "Then we found out it's seventeen to eighteen months. We supplied an answer to a part of their puzzle."

The advances SeaWorld has made in veterinary care have also paid off when it comes to rescuing stranded or sick marine animals, and SeaWorld's state-of-the-art breeding techniques could be useful in trying to preserve marine mammal populations on the brink of extinction, such as the vaquita porpoises in the Sea of Cortez. SeaWorld also nurtures multiple partnerships with leading conservation nonprofits, from the World Wildlife Fund to the Nature Conservancy. "Every year we spend $3 million to $4 million on research and conservation programs outside our park and another $1.5 million on rescuing stranded animals," Andrews says.

Head trainer Kelly Flaherty Clark still has faith in the benefits of SeaWorld's mission in the wake of Brancheau's death. One of her mantras, known around the park as "Kellyisms," is "Do the right thing." As we sit together in the stands of Shamu Stadium, "Believe" looks like pure family fun. But for the trainers, the shows are the product of countless hours of hard work and practice. They know there are risks. "These are not dogs," Flaherty Clark says. "Every day you walk into your job, you are walking into a potentially dangerous situation. You never forget that. You can't afford to forget that."

SeaWorld doesn't forget, and conducts safety and rescue training once a month. Among other things, trainers are taught to go limp if they are grabbed, so the whales will lose interest. The killer whales are taught to keep their mouths closed while swimming and are desensitized so they stay calm and circle the perimeter of the pool if someone accidentally falls in. They learn emergency re-call signals—transmitted via a tone box and hand slaps—and are trained to swim to a pool exit gate if a net is dropped in. Scuba gear is always nearby. SeaWorld's intensive regime helped its train-ers interact with killer whales more than two million times without a death. But when a killer whale breaks from its training, all bets are off.

It's hard to know exactly what triggers an incident. It could be boredom, a desire to play, the pent-up frustration of confinement, a rough night in the tank with the other orcas, the pain of an ulcer, or maybe even hormonal cycling. Whatever the motivation, some trainers believe that killer whales are acutely aware of what they're doing. "I've seen animals put trainers in their mouths and know exactly what the breaking point of a rib cage is. And how long to hold a trainer on the bottom," says Jeffrey Ventre, who was a trainer at SeaWorld Orlando from 1987 until 1995, when he was let go for giving a killer whale a birthday kiss, in which he stuck his head into an orca's mouth.

If you're a killer whale in a marine park, there's probably no bet-ter place than SeaWorld. Yet no matter how nice the facility, there's stress associated with being a big mammal in a relatively small pool. Starting at Sealand, Tilikum had developed the habit of grinding his teeth against metal pool gates. Many of his teeth were so worn and broken that SeaWorld vets decided to drill some of them so they could be regularly irrigated with antiseptic solution. And once again he had to deal with the stress of hostile females, particularly a dominant orca called Katina. "Tili was a good guy that got beat down by the women," says Ventre, now a doctor in New Orleans. "So there are a lot of reasons he might be unhappy."

John Jett, who was a team leader for Tilikum, says he sometimes would suffer a beatdown bad enough to rake up his skin and bloody him and would have to be held out of shows until he healed. Jett had a term for the blood left streaming in the water: "sky writing." After a good thrashing from the other orcas, Jett says, Tilikum might be "off" for days, "splitting" from his trainer to swim at high speed around the pool, acting agitated around the females, or opening his eyes wide and emitting distress vocals if asked to get into a vulnerable position (like rolling over on his back). "It's extremely sad if you think about being in Tili's situation," says Jett. "The poor guy just has no place to run."

SeaWorld's Fred Jacobs denies that Tilikum was ever held out of shows due to injuries from other orcas. "Injuries as part of the expression of social dominance are rare and almost never serious," he says. "We manage Tilikum's social interaction on a daily basis."

In 1999 Tilikum reminded the world that, at least when it came to humans, he could be a very dangerous animal. Early on the morning of July 6, Michael Dougherty, a physical trainer at Sea-World, arrived at his office near the underwater viewing area of G pool. He glanced through the viewing glass and saw Tilikum staring back, with what appeared to be two human feet hanging down his side. There was a nude body draped across Tilikum's back. It wasn't moving. As in the Brancheau incident, Tilikum was herded onto the medical lift in order for SeaWorld staff to retrieve the body. Rigor mortis had already set in. It was a young male, and again the coroner's and sheriff's reports are telling. He had puncture wounds and multiple abrasions on his face.

The victim was Daniel Dukes, a twenty-seven-year-old with a reddish blond ponytail, a scraggly beard and mustache, and a big red "D" tattooed above his left nipple. Four days earlier he'd been released from the Indian River County Jail after being booked for retail theft. On July 5 he apparently hid at SeaWorld past closing or sneaked in after hours. At some point during the night, he stripped down to his swim trunks, placed his clothes in a neat pile, and jumped into the pool. Perhaps he was simply crazy or suicidal. Perhaps he believed in the myth of a friendly Shamu.

The coroner determined the cause of death to be drowning. There were no cameras or witnesses, so it's not known if Tilikum held him under or hypothermia did him in. But it's clear Tilikum worked Dukes over. The coroner found abrasions and contusions —both premortem and postmortem—all over his head and body and puncture wounds on his left leg. His testicles had been ripped open. Divers had to go to the bottom of the pool to retrieve little pieces of his body. SeaWorld ramped up its security, posting a twenty-four-hour watch at Shamu Stadium. Keltie Byrne had not been an aberration.

If anyone was going to take care around Tilikum, it was Dawn Brancheau. She was one of SeaWorld's best and completely dedicated to the animals and her job. (She even met her husband, Scott, in the SeaWorld cafeteria.) She had worked at SeaWorld Orlando since 1994, spending two years working with otters and sea lions before graduating to work with the killer whales. She was fun and selfless, volunteering at a local animal shelter and often keeping everything from stray ducks and chickens to rabbits and small birds at her home.

Over time, Brancheau had become one of SeaWorld's most trusted trainers, one of the dozen or so authorized to work with Tilikum. "Dawn showed prowess from the minute she set foot here. There's not one of us who wouldn't say that she was one of the best," says Flaherty Clark. Brancheau knew the risks and accepted them: "You can't put yourself in the water unless you trust them and they trust you," she once told a reporter.

Perhaps she trusted Tilikum too much. Thad Lacinak, the former vice president of animal training at SeaWorld, thinks so. He says Brancheau was an exemplary trainer, one of the best he'd ever seen in the water. Still, Lacinak thinks Brancheau made a mistake lying down so close to Tilikum's mouth and letting her hair drift in the water alongside him. "She never should have put herself in that vulnerable a position," he says. "One of the things we always talked about at SeaWorld was you never want to get totally comfortable with any animal."

Former trainer Mark Simmons has been involved in deconstructing previous SeaWorld incidents between trainers and killer whales and was a friend of Brancheau's. He also thinks Brancheau's vulnerable position and hair (which he says she was growing long so she could give it to cancer patients for wigs) were the key factors that led to her being pulled into the pool. "Tilikum has never had an aggressive disposition," he says. "This was not the first time Dawn had lain down next to Tili in that position, but it was the first time her hair was that long and contacted Tili." Simmons believes Tilikum reacted to this "novel stimulus" by taking it into his mouth. When Brancheau tried to tug it free, as spotter Jan Topoleski described, Tilikum suddenly had a tempting game of tug of war, which he was bound to win. (After Brancheau's death, SeaWorld's long-standing policy that long hair be kept in a ponytail was revised to mandate that it be kept in a bun.)

At least two witnesses, however, told investigators that they saw Tilikum grab Brancheau by the arm or shoulder, which would sug-

gest a more intentional act. Asked how certain he was that Tilikum
pulled Brancheau in by her hair, SeaWorld's Fred Jacobs responds,
"Witness accounts support that conclusion, and we have no reason
to doubt it."

The second critical question is: Why did Tilikum get so violent
once Brancheau was in the water? The coroner cataloged a fractured neck, a broken jaw, and a dislocated elbow and knee. A
chunk of skin and hair was ripped from her scalp and recovered
from a pool. "When a 12,000-pound animal gets its hands on 'the
cookie jar' and responds with the excited burst of energy common
in such situations, it can have tragic consequences," Simmons says.
"Once the alarm was sounded and emergency net procedures were
initiated, Tilikum's behavior became agitated. This is what appears
to the untrained observer to [constitute] an 'attack.'"

Whether the emergency response increased Tilikum's agitation
or not, once Brancheau was in the water, her fate was up to a killer
whale that hadn't become accustomed to humans in the pool. "He
got her down and that was it—she wasn't getting out," says former
trainer Jonathan Smith. "I truly believe that they are smart enough
to detect and know what they are doing. He's going to know she is
trying to get to the surface." Former trainer Ventre agrees. "If they
let you out, it's because they decide to," he says. "We don't know
for sure what motivated Tilikum. But there's no doubt that he knew
exactly what he was doing. He killed her."

SeaWorld says it is conducting the most exhaustive review in its
history. At press time, the review was not complete, and OSHA's
report is not expected until late summer.* For the moment, Sea-
World is not taking any chances. No trainers are performing in the
water with orcas, and all direct human contact with Tilikum has
ceased. "We used to interact very closely with Tilikum but now
maintain a safe distance," Flaherty Clark wrote on the SeaWorld
blog in March. Where Tilikum once got regular rubdowns and

* OSHA completed its investigation into the death of Dawn Brancheau in August
2010 and cited SeaWorld for failing to protect trainers adequately in their work with
killer whales "from recognized hazards that were causing or likely to cause death or
serious physical harm." In addition to fining SeaWorld $75,000, OSHA stipulated
that the only way it could create an acceptably safe work environment would be to
end close work with killer whales or adopt a number of safety measures, such as
keeping physical barriers between trainers and killer whales, that would fundamentally alter SeaWorld's shows. SeaWorld flatly rejected OSHA's conclusions and is ap-
pealing its citations.

close contact during cleanings and other husbandry, now he's hosed down instead of hand-massaged, and his teeth are cleaned with an extension pole. His isolation has only increased, opening a wider debate about the future of killer whale entertainment.

After Brancheau's death, Jean-Michel Cousteau, president of the Ocean Futures Society, made a videotaped statement in which he said, "Maybe we as a species have outgrown the need to keep such wild, enormous, complex, intelligent, and free-ranging animals in captivity, where their behavior is not only unnatural; it can become pathological," he said. "Maybe we have learned all we can from keeping them captive."

Cousteau raises a profound point. But regardless of how this incident affects orca captivity, Tilikum's fate is likely sealed, despite calls for his release back into the wild. *Free Willy*'s Keiko underwent extensive retraining before being released into the seas off Iceland and appears to have foraged for food on his own. But he never reintegrated with a pod. A little over a year later, after swimming to Norway, he died, likely from pneumonia. Ken Balcomb still believes that most marine-park orcas can be taught what they need to know to be returned to the wild. (No real effort was made to find Keiko's family, Balcomb says, which is a key to success.) But even he rules Tilikum out. "Tilikum is basically psychotic," he told me as we looked out over Haro Strait in May. "He has been maintained in a situation where I think he is psychologically unrecoverable in terms of being a wild whale."

There is one other option. "We have proposed to Blackstone Group a sea-pen retirement," says Naomi Rose, a marine mammal scientist at the Humane Society International. "Tilikum needs more space, more stimulation to distract him. Living as he is, with minimum human contact in a small concrete tank, is untenable."

SeaWorld's Fred Jacobs dismisses the idea. In addition to citing worries about the impact of taking him out of the social environment he is now accustomed to and about potential threats to his health from pollution and disease, Jacobs says, "All the animals at SeaWorld allow people a really rare privilege to come into contact with these extraordinary animals and learn something about them and maybe when they leave SeaWorld carry that respect forward into their lives. Tilikum is a really important part of that."

Whether or not Tilikum ever performs again, he's still SeaWorld's most prolific breeder. He's sired thirteen viable calves, with two more on the way this summer. Most likely he will finish his life as he's mostly lived it, in a marine park. He's nearly thirty, and only one male in captivity, who is still alive, is known to have lived past that age.

Three thousand miles away, Balcomb often sees a pod of killer whales easing their way through the wilderness of water that is his Haro Strait backyard. They swim with purpose and coordination, huffing spumes of mist into the salty, spruce-scented air. The group is known as L Pod, and one, a big male designated L78, was born just a few years after Tilikum. Balcomb has been tracking L78 for more than two decades. He knows that his mother—born around 1960—and his brother are always close by. He knows that L78 ranges as far south as California with his pod in search of salmon.

L78's dorsal fin stands proud and straight as a knife, with none of Tilikum's marine-park flop. He hunts when he's hungry, mates with the females who offer themselves, and whistles to the extended family that is always nearby. He cares nothing for humans and is all but oblivious to their presence when they paddle out in kayaks to marvel as he swims. He knows nothing of the life of Tilikum or the artificial world humans have manufactured for him. But Tilikum, before spending twenty-six years in marine parks, once knew L78's life, once knew what it was like to swim the ocean alongside his mother and family. And perhaps, just perhaps, that also helps explain why Dawn Brancheau died.

Contributors' Notes

Other Notable Science and
Nature Writing of 2010

Contributors' Notes

Yudhijit Bhattacharjee has been a staff writer at *Science* since 2003. His 2008 story on brain trauma caused by explosions on the battlefield won an award for best research writing from the National Mental Health Association. His work has appeared in the *New York Times*, *The Atlantic*, the *New York Times Magazine*, *Time*, *Wired*, and *Discover*.

Burkhard Bilger has been a staff writer at *The New Yorker* since 2000. His articles have appeared in *The Atlantic*, *Harper's Magazine*, the *New York Times*, and other publications, and his book, *Noodling for Flatheads*, was a finalist for a PEN-Faulkner Award. A former senior editor of *Discover* and deputy editor of *The Sciences*, Bilger lives in Brooklyn with his wife, Jennifer Nelson, and his children, Hans, Ruby, and Evangeline.

Deborah Blum is a Pulitzer Prize–winning science writer and the author of five books, most recently the *New York Times* bestseller *The Poisoner's Handbook: Murder and the Birth of Forensic Medicine in Jazz Age New York*. A former president of the National Association of Science Writers, she writes for a wide variety of publications, including *Slate*, *Time*, *Scientific American*, the *New York Times*, the *Wall Street Journal*, and the *Los Angeles Times*. She blogs about chemistry and culture for the Public Library of Science at Speakeasy Science (blogs.plos.org/speakeasyscience). A professor of journalism at the University of Wisconsin, she lives in Madison, Wisconsin, with her husband, two sons, a Labrador, two ferrets, and a mouse.

Jon Cohen has been a correspondent with *Science* since 1990. He has written for *The New Yorker*, *The Atlantic*, the *New York Times Magazine*, *Smith-*

sonian, Outside, Slate, Technology Review, and many other publications. His books include *Shots in the Dark, Coming to Term,* and *Almost Chimpanzee.* Cohen has done mini-documentaries for *Science* and *Slate V* and contributed to "Ending AIDS," a PBS documentary based on *Shots in the Dark.* He earned a BA in science writing from the University of California, San Diego, in 1981. He lives in Cardiff-by-the-Sea, California.

Luke Dittrich is a contributing editor at *Esquire* and is currently working on a book about Henry Molaison and the history of memory science. He and his five-year-old daughter, Anwyn, live in Whitehorse, the capital of Canada's Yukon Territory.

Jonathan Franzen is the author of four novels *(The Twenty-Seventh City, Strong Motion, The Corrections,* and *Freedom),* a collection of essays *(How To Be Alone),* and a personal history *(The Discomfort Zone).* He has contributed journalism and other nonfiction to *The New Yorker* since 1994, and more recently he has become a serious bird watcher.

Ian Frazier is the author of *Great Plains, The Fish's Eye, On the Rez,* and *Family,* as well as *Coyote v. Acme, Lamentations of the Father,* and, most recently, *Travels in Siberia.* A frequent contributor to *The New Yorker,* he lives in Montclair, New Jersey.

David H. Freedman writes on health, science, and technology for a number of publications. He is the author of *Wrong: Why Experts Keep Failing Us,* and four other books.

Atul Gawande is a general and endocrine surgeon at Brigham and Women's Hospital in Boston and a staff writer at *The New Yorker.* He is an associate professor at Harvard Medical School and the Harvard School of Public Health. He also leads Lifebox, a not-for-profit organization, and the World Health Organization's Safe Surgery Saves Lives program. He is the author of three books: *Complications, Better,* and *The Checklist Manifesto.*

Malcolm Gladwell is the author of *What the Dog Saw, Outliers: The Story of Success, Blink: The Power of Thinking Without Thinking,* and *The Tipping Point: How Little Things Can Make a Big Difference,* all *New York Times* bestsellers. Gladwell has been a staff writer with *The New Yorker* since 1996. In 2007 he received the American Sociological Association's first Award for Excellence in the Reporting of Social Issues, and in 2005 he was named one of *Time's* 100 Most Influential People. Gladwell was born in England, grew up in rural Ontario, and now lives in New York City.

Andrew Grant is a reporter at *Discover* who especially enjoys writing about physics and astronomy. He is a recent graduate of the Science, Health and Environmental Reporting Program at New York University and lives in New York City.

Stephen Hawking has worked on the basic laws that govern the universe. With Roger Penrose, he showed that Einstein's General Theory of Relativity implied that space and time would have a beginning in the big bang and an end in black holes. These results indicated that it was necessary to unify general relativity with quantum theory. One consequence of such a unification, he discovered, was that black holes should not be completely black but rather should emit radiation and eventually disappear. Another conjecture is that the universe has no edge or boundary in imaginary time. He has written three popular books: his bestseller *A Brief History of Time, Black Holes and Baby Universes and Other Essays, The Universe in a Nutshell,* and, most recently, *The Grand Design.* Professor Hawking has twelve honorary degrees. He was awarded the CBE in 1982 and was made a Companion of Honour in 1989. The recipient of many awards, medals, and prizes, he is a Fellow of the Royal Society and a member of the U.S. National Academy of Sciences. He continues to combine family life (he has three children and three grandchildren) and research into theoretical physics with an extensive program of travel and public lectures.

Amy Irvine lives in southwestern Colorado. Her second book, *Trespass: Living at the Edge of the Promised Land,* won the 2009 Orion Book Award and the 2009 Colorado Book Award.

Rowan Jacobsen writes about place and how it shapes ecosystems, cultures, cuisines, and us. His quest to capture the spirit of place and people has led him from the volcanic mountains of Mexico to the misty marshes of Alaska's Yukon Delta, from the bayous of Louisiana to the rivers of Amazonia. He has written for the *New York Times, Harper's Magazine, Newsweek, Outside, Sierra,* and others. He is the author of *A Geography of Oysters,* which won a James Beard Award; *Fruitless Fall,* which received the 2009 Green Prize for Sustainable Literature; *The Living Shore;* and *American Terroir,* which was named one of the Ten Best Books of the Year by *Library Journal.* His newest book is *Shadows on the Gulf: A Journey Through Our Last Great Wetland,* which began life as "The Spill Seekers."

Christopher Ketcham has written for *Harper's Magazine, Earth Island Journal, Orion, Vanity Fair, GQ, CounterPunch,* and many other magazines and websites. Find more of his work at www.christopherketcham.com or contact him at cketcham99@mindspring.com.

Dan Koeppel is the author of *Banana: The Fate of the Fruit That Changed the World* and a contributor to *Popular Mechanics, Wired,* and *National Geographic.* He lives in Los Angeles, where he mostly travels by bike and bus. His website is www.dankoeppel.com.

Jaron Lanier, a partner architect at Microsoft Research and an innovator in residence at the Annenberg School of the University of Southern California, is the author most recently of *You Are Not a Gadget.*

Leonard Mlodinow received his PhD in theoretical physics from the University of California at Berkeley, was an Alexander von Humboldt fellow at the Max Planck Institute for Physics and Astrophysics in Munich, and was on the physics faculty at the California Institute of Technology. He is the author of seven books, which have appeared in twenty-five languages. His book *The Drunkard's Walk: How Randomness Rules Our Lives* was a *New York Times* bestseller and was short listed for the Royal Society Book Award. His most recent book is *The Grand Design,* coauthored with Stephen Hawking. He has also written for television, including *MacGyver; Star Trek, the Next Generation;* and the comedy *Night Court.* He lives in South Pasadena, California.

Jon Mooallem has been a contributing writer to the *New York Times Magazine* since 2006 and is currently working on a book, to be published by Penguin Press, about people and wild animals in America. He lives in San Francisco with his wife and daughter.

George Musser is a staff editor and writer for *Scientific American,* where he focuses mostly (though not exclusively) on space science and fundamental physics. He was one of the lead editors for the magazine's single-topic issue "A Matter of Time" (September 2002), which won a National Magazine Award for editorial excellence, and he coordinated the single-topic issue on sustainable development "Crossroads for Planet Earth" (September 2005), which was an Ellie finalist. In 2010 he received the Jonathan Eberhart Planetary Sciences Journalism Award from the American Astronomical Society. His first book, *The Complete Idiot's Guide to String Theory,* was published in 2008.

Jill Sisson Quinn is the author of *Deranged: Finding a Sense of Place in the Landscape and in the Lifespan.* Her essays have appeared in *Fourth Genre, Crab Orchard Review, Ecotone,* and elsewhere. She won the Annie Dillard

Award in Creative Nonfiction in 2003 and the John Burroughs Award for an Outstanding Published Nature Essay in 2011. Her essays have twice been nominated for a Pushcart Prize. She lives in Scandinavia, Wisconsin. Quinn's website is www.jillsissonquinn.com.

Oliver Sacks is a physician and the author of eleven books, including *The Man Who Mistook His Wife for a Hat* and *Musicophilia,* in which he describes patients struggling to adapt to various neurological conditions. His book *Awakenings* inspired the Oscar-nominated film of the same name and the play *A Kind of Alaska* by Harold Pinter. He practices neurology in New York City, where he is also a Columbia University Artist.

Evan I. Schwartz lives in Boston and writes about discovery and innovation. He is a former editor at *BusinessWeek* and *Technology Review* and is the author of five books, including *The Last Lone Inventor: A Tale of Genius, Deceit & the Birth of Television.* His first article for *Wired* appeared in 1994.

Sandra Steingraber, PhD, is the author of *Living Downstream: An Ecologist's Personal Investigation of Cancer and the Environment,* now the subject of an award-winning documentary. Her most recent book, *Raising Elijah: Protecting Children in an Age of Environmental Crisis,* explores the relationship between fossil fuel extraction, including fracking, and toxic chemical exposure and concludes that the ongoing environmental crisis is fundamentally a crisis of family life. A scholar in residence at Ithaca College, she lives with her husband and children in the Finger Lakes region of upstate New York, where 40 percent of the land is leased to gas drillers.

Abigail Tucker has been a *Smithsonian* staff writer since 2008. Previously she was a general assignment reporter for the *Baltimore Sun* and the *Post-Star,* in Glens Falls, New York. She lives in Washington, DC.

Tim Zimmermann is a correspondent for *Outside* and the founder and editor of the weekly *Adventure Fix* newsletter. He is a former senior editor and diplomatic correspondent for *U.S. News & World Report* and the author of *The Race: The First Nonstop, Round-the-World, No-Holds-Barred Sailing Competition.* He has been a National Magazine Award finalist (in the category of feature writing) and has received the Thomas Renner Award (for outstanding reporting on organized crime) from Investigative Reporters and Editors. Tim lives in the Washington, D.C., area with his wife and two children.

Other Notable Science and Nature Writing of 2010

SELECTED BY TIM FOLGER

ANIL ANANTHASWAMY
   Ice Fishing for Neutrinos. *Discover.* March.

MICHAEL BALTER
   Anthropologist Brings Worlds Together. *Science.* August 13.
BRUCE BARCOTT
   What's the Catch? *OnEarth.* Summer.
RICK BASS
   Into the Wild Again. *OnEarth.* Fall.
MATTHEW BATTLES
   A New Wrinkle in Time. *The Atlantic.* November.
JESSICA BENKO
   Mother of God, Child of Zeus. *VQR.* Fall.
TIM BERNERS-LEE
   Long Live the Web. *Scientific American.* December.
JENNIFER BOGO
   Digital Sight for the Blind. *Popular Mechanics.* November.
KENNETH BROWER
   The Danger of Cosmic Genius. *The Atlantic.* December.
D. GRAHAM BURNETT
   A Mind in the Water. *Orion.* May/June.

CRAIG CALLENDER
   Is Time an Illusion? *Scientific American.* June.
ADRIAN CHO
   Probing the Secrets of the Finest Fiddles. *Science.* June 18.
MARK COHEN
   My Illegal Heart. *Men's Health.* April.

DAVID QUAMMEN
  Great Migrations. *National Geographic.* November.

HUGH RAFFLES
  A Conjoined Fate. *Orion.* January/February.
HANNA ROSIN
  Earthbound. *The Atlantic.* September.
SHERMAN APT RUSSELL
  True Confessions of a Citizen Scientist. *OnEarth.* Spring.

OLIVER SACKS
  A Man of Letters. *The New Yorker.* June 28.
MARK SCHAPIRO
  Conning the Climate. *Harper's Magazine.* February.
JAMIE SHREEVE
  Evolutionary Road. *National Geographic.* July.
BRYAN SMITH
  Your Lethal Lawn. *Men's Health.* June.
RICHARD STONE
  Home, Home Outside the Range? *Science.* September 24.
GINGER STRAND
  The Economics of Estuary. *Orion.* September/October.
PATRICK SYMMES
  The Beautiful and the Dammed. *Outside.* June.

GARY TAUBES
  Live Long and Prosper. *Discover.* October.
JOHN TRAVIS
  In Search of Sitting Bull. *Science.* October 8.
ABIGAIL TUCKER
  The Truth About Lions. *Smithsonian.* January.

LOGAN WARD
  The Race to Save the Bats. *Popular Mechanics.* December.
CHARLES WOHLFORTH
  Conservation and Eugenics. *Orion.* July/August.